Innovations in Industrial and Engineering Chemistry

ACS SYMPOSIUM SERIES **1000**

Innovations in Industrial and Engineering Chemistry

A Century of Achievements and Prospects for the New Millennium

William H. Flank, Editor
Pace University

Martin A. Abraham, Editor
Youngstown State University

Michael A. Matthews, Editor
University of South Carolina

Sponsored by the
**ACS Division of Industrial and Engineering
Chemistry, Inc.**

American Chemical Society, Washington, DC

The paper used in this publication meets the minimum requirements of American National Standard for Information Sciences—Permanence of Paper for Printed Library Materials, ANSI Z39.48–1984.

ISBN: 978-0-8412-6963-7

Distributed by Oxford University Press

The logo on the cover is reproduced with permission from the ACS Division of Industrial and Engineering Chemistry.

PRINTED IN THE UNITED STATES OF AMERICA

Foreword

The ACS Symposium Series was first published in 1974 to provide a mechanism for publishing symposia quickly in book form. The purpose of the series is to publish timely, comprehensive books developed from ACS sponsored symposia based on current scientific research. Occasionally, books are developed from symposia sponsored by other organizations when the topic is of keen interest to the chemistry audience.

Before agreeing to publish a book, the proposed table of contents is reviewed for appropriate and comprehensive coverage and for interest to the audience. Some papers may be excluded to better focus the book; others may be added to provide comprehensiveness. When appropriate, overview or introductory chapters are added. Drafts of chapters are peer-reviewed prior to final acceptance or rejection, and manuscripts are prepared in camera-ready format.

As a rule, only original research papers and original review papers are included in the volumes. Verbatim reproductions of previously published papers are not accepted.

ACS Books Department

Contents

vii

Indexes

Preface

The American Chemical Society (ACS) Division of Industrial and Engineering Chemistry, Inc. (I & EC) celebrates the centennial of its establishment in 1908. I & EC was the first technical division established within ACS, and this volume is part of the commemorations marking this significant milestone at the 2008 New Orleans and Philadelphia National Meetings. The theme of the book is innovation and creativity in the chemical industry and related sectors, where industrial chemists and chemical engineers have made and are continuing to make major contributions.

Our vision for this volume was that it not only chronicles the creativity of our industry, but also reveals the nature of the discoveries and innovations behind the particular industrial example. For example, the work of Nobel Laureates in catalysis has led to global industries in ammonia synthesis, polyethylene, and industrial gas separations, and this volume chronicles the translation from basic science to practical applications.

We would like this book to become a true resource for professionals, with future editions adding to the content. Certain chapters are retrospective in nature, as is befitting of older and mature industries (for example, petroleum cracking). Other contributions reflect recent developments in newer fields, and are thus more prospective (for example, ionic liquids and green chemistry). Very generally speaking, the chapters are in three parts: a historical perspective of the process or product, a description of the current state of the enterprise, and a prospective on the breakthroughs anticipated or needed that will drive the field for the next 15 to 50 years. The amount of text dedicated to these three parts varies, depending on the field and its age.

The book covers traditional chemical and petrochemical industries, as befitting the Division of Industrial and Engineering Chemistry, Inc. But consistent with the Division, the book includes several chapters that relate to the subdivisions: the Separation Science and Technology Subdivision is represented by chapters on membrane separations and ionic liquids; the Green Chemistry and Engineering Subdivision is represented by the chapter on green chemistry, and the Novel Chemistry with Industrial Applications Subdivision is represented by a chapter on the production of the acetyls.

It is our hope that by recording the scientific and technological details behind key industries, we will provide readers with a certain perspective, essentially a process view, that will illustrate the directions and pathways that innovation can take. And by looking at the evolution of the chemical industry, and the I & EC Division, we can provide some indication of where the industry and the Division may be headed in the future.

The I & EC Division itself has been innovative in a number of ways, as described in the initial chapter, but none more pertinent to this book than the leadership shown in publishing the first ACS Symposium Series volume in advance of the symposium sessions themselves. Symposium Series 135 appeared in the early 1980s, and a similar effort, Symposium Series 368, was published in the late 1980s. Numerous other symposium series volumes have derived from the presentations sponsored by the I & EC Division. The present work, Symposium Series 1000, exemplifies the approach the I & EC Division takes to providing service to its members and to the profession.

We envision a second century of leadership as noteworthy as the first.

William H. Flank
Department of Chemistry and Physical Sciences
Pace University
861 Bedford Road
Pleasantville, NY 10570

Martin A. Abraham
College of Science, Technology, Engineering, and Mathematics
Youngstown State University
One University Plaza
Youngstown, OH 44555

Michael A. Matthews
Department of Chemical Engineering
University of South Carolina
Swearingen Engineering Center
Columbia, SC 29208

Chapter 1

Putting Science to Work: A History of I&EC Leadership and Innovation in Its First Hundred Years

William H. Flank

Department of Chemistry and Environmental Science, Pace University, 861 Bedford Road, Pleasantville, NY 10570

The first Division established by the American Chemical Society has a long and distinguished history. A number of sources were used to assemble a comprehensive picture of the genesis, growth and innovative leadership of the I&EC Division over the past century. Befitting its role as the first applied Division of the Society, diversification into emerging areas has been the hallmark of I&EC, and the Division is credited with many "firsts." In parallel with this aggressive approach, numerous publications have been launched as an outgrowth of I&EC activities, and these are examined in detail as well.

Introduction

Several histories of the Industrial and Engineering Chemistry Division have been compiled over the last century, including the one written in 1951 by Division Chairman William A. Pardee, which was published in Vol. 43, No. 2 of *Industrial and Engineering Chemistry* (1951). Additional information was found in *A History of the American Chemical Society: Seventy-Five Eventful Years*, by Charles Albert Browne (Historian of the American Chemical Society) and Mary Elvira Weeks, published by the Society in 1952. An update was included in *A Century of Chemistry: The Role of Chemists and the American Chemical Society*, edited by Kenneth M. Reese and H. Skolnik, and published by the Society in 1976. I&EC historian David E. Gushee published a Division

Page article in June, 1994 in *CHEMTECH* titled *I&EC Division: An institution in periodic transition.* Reese also edited *The American Chemical Society at 125: A Recent History, 1976-2001,* published by the Society in 2002. It contained a section on I&EC contributed by former Chairs David E. Gushee and Steven J. Cooke, which became the basis for membership promotion material put together by the latter. The present history comprises a synthesis, enlargement and enhancement of those earlier efforts, upon which it has drawn heavily, and includes some details helpfully dug out of Society records by several ACS staff members.

It is clear from the available records that I&EC, initially known as the Division of Industrial Chemists and Chemical Engineers, was the first Division organized in the Society and followed close on the heels of the founding of the American Institute of Chemical Engineers as a separate professional organization. Other Divisions organized shortly thereafter include Agricultural and Food Chemistry, Fertilizer and Soil Chemistry (now defunct), Organic Chemistry, and Physical Chemistry (originally Physical and Inorganic Chemistry Division, with the Inorganic Division splitting off in 1957).

There is also a parallel history of I&EC-related publications, beginning with the gestation and publication of the *Journal of Industrial and Engineering Chemistry* starting in 1909, which will be discussed in more detail below.

According to the Division's Bylaws, and in addition to those of the Society, "further Objects of the Division shall be the advancement of industrial and engineering chemistry, and in specific furtherance thereof:

A. To encourage the highest standards of excellence in developing and applying knowledge of chemistry and chemical engineering to the products and processes of industry.
B. To promote the development of chemical science and technology in both academic institutions and in industry.
C. To improve the qualifications and usefulness of chemists and chemical engineers through high standards of professional ethics, education and attainment.
D. To increase the diffusion of chemical knowledge through its meetings, professional contacts, reports, papers, discussions and publications, thereby fostering public welfare and education.
E. To promote the mutual recognition of need and interest for the complete spectrum of chemical interests, from fundamental research to pragmatic technology."

Division Formation and Early History

Prior to the formation of this first formal Division in the Society, an Industrial Section had been established at least as early as 1904, with the following chairmen:

1904	Edward Hart
1905	S. W. Parr
1906	J. D. Pennock
1907	Arthur D. Little and W. H. Ellis
1908	William D. Richardson

The 1907 President of ACS, Marston T. Bogert, had appointed a committee to consider the advisability of undertaking the publication of a journal of industrial and engineering chemistry and the formation of an industrial division. At the Christmas 1907 national ACS meeting in Chicago, the committee recommended, and the Council approved, the publication of the journal and the formation of the Division of Industrial Chemists and Chemical Engineers. Industrial chemists were at that time, and still remain, the largest membership group in the Society, and this move was an attempt to avoid further fragmentation into specialized groups like the then-recently organized American Institute of Chemical Engineers.

The organizational meeting of this first ACS Division was held at Yale University on June 30, 1908, at the 38th general meeting of the Society. The meeting was presided over by William L. Richardson, the soon-to-be-elected editor-in-chief of the new Journal of Industrial and Engineering Chemistry. The first officers of the new Division of Industrial Chemists and Chemical Engineers were Arthur D. Little, chairman; A. H. Low, vice chairman; B. T. Babbit Hyde, secretary; and an executive committee of six members. Initially, several of the elected Chairs served for two consecutive years, but this has not been the practice since the end of World War II. Since that time, only Bill Flank has been elected to serve a second term, and that occurred 19 years after his initial term. The complete listing of Division Chairs to date is found in Table 1.

In the early days of the Division there were standing committees on:

Definition of Industrial Terms
Trade Customs
Standard Specifications and Methods of Analysis
   (a function later carried on by ASTM)
Research Problems
Descriptive Bibliographies
Publicity

In December, 1909 a special committee was appointed to confer with manufacturers for the purpose of determining at what prices various elements

Table 1. Chairs of the Division

Year	Chair		Year	Chair		Year	Chair
1908-1910	Arthur D. Little		1954	Charles J. Krister		1983	Kathleen C. Taylor
1911	George C. Stone		1955	Edward W. Comings		1984	Billy L. Crynes
1912-1913	George D. Rosengarten		1956	Charles M. Cooper		1985	William H. Flank
1914-1915	George P. Adamson		1957	Edmond L. d'Ouville		1986	John M. Storton
1916-1917	Harrison E. Howe		1958	DeWitt O. Myatt		1987	John L. Massingill
1918	William H. Walker		1959	James M. Church		1988	Wallace W. Schulz
1919	Harlan S. Miner		1960	Otto H. York		1989	Kathleen M. (Stelting) Schulz
1920-1921	Harry D. Batchelor		1961	Joseph E. Stewart		1990	Madan M. Bhasin
1922	Warren K. Lewis		1962	G. R. Seavy		1991	Melanie J. (Cravey) Lesko
1923-1924	D. R. Sperry		1963	Brage Golding		1992	David J. Pruett
1925-1926	W. A. Peters, Jr.		1964	Arthur R. Rescoria		1993	D. William Tedder
1927	William H. McAdams		1965	Arthur Rose		1994	Spiro D. Alexandratos
1928-1929	Robert J. McKay		1966	Robert B. Beckmann		1995	Lawrence A. Casper
1930-1931	Robert E. Wilson		1967	Robert Landis		1996	Ralph C. Gatrone
1932-1933	Donald B. Keyes		1968	Merrell R. Fenske		1997	Dale L. Perry
1934-1936	Walter G. Whitman		1969	Leo Friend		1998	Nancy B. Jackson
1937	Thomas A. Boyd, Jr.		1970	Robert N. Maddox		1999	Robin D. Rogers
1938-1939	Walter L. Badger		1971	James D. Idol		2000	Steven J. Cooke
1940-1941	Barnett F. Dodge		1972	William E. Hanford		2001	Dale Ensor
1942	Lawrence W. Bass		1973	Vernon A. Fauver		2002	Amy L. Manheim
1943-1944	R. Norris Shreve		1974	James R. Couper		2003	Martin A. Abraham
1945-1946	Thomas H. Chilton		1975	Peter K. Lashmet		2004	William H. Flank
1947	Francis J. Curtis		1976	David E. Gushee		2005	Dennis L. Hjeresen
1948	Henry F. Johnstone		1977	John Ehrenfeld		2006	Richard Sachleben
1949	Joseph C. Elgin		1978	James E. McEvoy		2007	Michael Matthews
1950	Lincoln T. Work		1979	Robert Squires		2008	Gregg Lumetta
1951	William A. Pardee		1980	Norman N. Li		2009	Joseph Zoeller (2008 Chair-elect)
1952	Melvin C. Molstad		1981	Charles M. Bartish			
1953	J. Henry Rushton		1982	Robert A. Stowe			

and compounds could be obtained if a large enough market was developed for them. In July, 1911 a committee was appointed to study the need for a professional code of ethics among chemists. These early committees, however, did not meet with much success. The requests of the committees on definitions and specifications were initially ignored when they asked for product data from the manufacturers. Information on such important materials of the time as bronze, inorganic chemicals grades, heavy chemicals, pharmaceutical products, Portland cement, petroleum products, iron and steel was not obtainable for some time. Subcommittees working on formulating specifications for soda ash, caustic soda, alum, mineral acids, solder and turpentine met with opposition from manufacturers.

It was only in later years that specifications were set for ACS Grade Analytical Reagents, when the need for information finally overcame the perceived need for trade secrecy. The task of setting standards for analytical methods and testing eventually fell to the American Society for Testing and Materials, later to become know as ASTM International. This group's activities, many of which involve industrial chemicals and chemistry, are predicated upon consensus procedures and a careful balancing of producer interests, user interests, and public or general interests.

The Division of Industrial Chemists and Chemical Engineers encouraged and sponsored numerous symposia and general papers covering subjects which then became the fields of specialization of new Divisions within ACS. Among these were Cellulose, Chemical Marketing and Economics, Fluorine, Fuel, Nuclear, Petroleum, and at least six others over the years. The first symposium was held in Boston in December, 1909 on the subject of paint. In September, 1918 a Symposium on the Chemistry of Dyestuffs was held under the chairmanship of R. N. Shreve, and interest in this program led to the formation of the Dye Section and later the Dye Division. Two symposia on cellulose, held in April and September of 1920 under the chairmanship of J. E. Crane and G. J. Esselen, respectively, played an important role in the formation of the Division of Cellulose Chemistry. A symposium on the chemistry of gases and fuels in 1921 under the chairmanship of C. H. Stone prompted the formation of a section and later a Division on Fuels.

Some of the other early symposia included Smelter Smoke (1910), Mineral Wastes (1911), Wood Wastes and Conservation (1915), Occupational Diseases (1916), Nitrogen Industry (1916), Metallurgy (1917), Potash (1918), and Refractories (1919). In addition, several less technically oriented symposia were held on Contributions of the Chemist to American Industries (1915), Industrial Chemists in Wartime (1917), Library Service in Industrial Laboratories (1919), Future of Certain American-Made Chemicals (1919), and Annual Patent Renewal Fees (1919).

The many symposia and papers relating to education and training of chemists and chemical engineers, fertilizers, petroleum, gases and fuels, sugar and rubber are indicative of the active interest of the I&EC Division in promoting chemistry and chemical engineering in these areas, eventually leading to the incorporation of such Divisions within ACS.

Division Maturation and Growth

In 1919 the ACS Council acted on a motion to change the name of the Division of Industrial Chemists and Chemical Engineers to Division of Industrial and Engineering Chemistry. It subsequently became informally known as the I&EC Division, but the abbreviation used in the National Meeting programs listed in Chemical and Engineering News was INDE. This persisted until 1985 when, at the request of Division Chair Bill Flank, the Meetings and Expositions Department agreed to change it to I&EC.

Many papers on unit operations and unit processes appeared in the early programs of the Industrial Section and then later in the Division. The first formal Symposium on Unit Operations appeared in the April 1921 program, on the subjects of drying and filtration. Following that, unit operations subjects were covered every year, until in 1934 a series of special symposia was inaugurated and held either separately from, or part of, National Meetings until the present time, and evolved into special topics of current interest having a chemical engineering focus. The first symposium on unit processes was presented in September, 1937 and became a regular feature after that.

While informal sections within the Divisional structure addressed special fields of interest, it was not until the 1940's that they operated on a formal basis. Around 1950, the Division's bylaws were amended to recognize these sections, which eventually became subdivisions. Among these earlier groups were sections on Chemical Marketing and on Fluorine Chemistry, later to become full-status Divisions.

In the mid- to late 1970's, energetic new leadership spearheaded by Jim McEvoy led to a number of initiatives and programs that attracted much attention and boosted membership. Interdisciplinary programming was emphasized, including a heavy emphasis on catalysis topics, and liaisons were established with other groups such as Corporation Associates, the Younger Chemists Committee and the Women Chemists Committee. McEvoy recruited Bill Flank to replicate the social programs he had developed in other professional societies, which strongly promoted interdisciplinary interaction, drew support from numerous chemical companies and drew overflow crowds. In recognition of the aging demographics of I&EC, successful efforts focused on attracting and encouraging students to become involved in poster sessions, professional interaction and networking, and other activities leading to involvement in the life of the Division.

An attractive newsletter edited by Jim McEvoy began to be published as well, and continues to be a prime source of information and communication with our membership, most of whom cannot regularly attend National Meetings. The

Division has been fortunate to have had a string of dedicated and talented editors for its newsletter, which is currently edited by Dustin James. A further innovation involved the establishment of a Public Affairs committee, which recognized the need for presenting the face of the chemical enterprise to the outside world.

The activities of the Division continued to drift away from an exclusive focus on traditional topics of industrial chemistry and chemical engineering, and embraced emerging areas of special interest. In the 1990's, Dan W. (Bill) Tedder organized a series of very successful free-standing symposia titled Emerging Technologies: Hazardous Waste Management; the publications from these symposia earned significant revenues for the Division. This was also true for the on-going series of symposia organized by Robin Rogers on Ionic Liquids in the first decade of the 21st century. The broad programming of special symposia that were held outside of ACS National Meetings to accommodate those interests was gradually brought back into the National Meeting format through the development of new subdivisions and the incorporation of new topics into the Division's programming cycle.

The Division experienced dramatic growth in the decade from 1975 to 1985, and increased its membership by over 50%, becoming the third-largest Division in the ACS. Some of the subdivisions grew large enough to become full Divisions in their own right, and over the years the I&EC Division has been the starting point for at least 11 other Divisions, including Cellulose (now Cellulose and Renewable Materials), Chemical Marketing and Economics (now Business Development and Management), Dye (now defunct), Environmental, Fluorine, Fuel, Nuclear, and Petroleum. Table 3 lists these, as best as can be reconstructed from the information extant.

Incorporation and Diversification of Activities

The Division was incorporated in 1983 due largely to the efforts of Kathleen Taylor, the first female Chair of the Division. Although female Chairs were unusual up to that time, there were some precedents in other Divisions, most notably in the History of Chemistry Division. As listed in the Articles of Incorporation, the incorporating I&EC Board of Directors comprised Lawrence A. Casper, Geoffrey K. Smith, Billy L. Crynes, Kathleen C. Taylor, William H. Flank, James D. Idol, Jr., James E. McEvoy and Norbert Platzer. The Division continues as an incorporated non-profit scientific organization, as defined in Section 501(c)(3) of the Internal Revenue Code.

In 1985, Division Chair Bill Flank appointed a liaison to AIChE and initiated a series of program co-sponsorships on topics of overlapping interest that were being presented at meetings of both societies and at free-standing conferences. This cooperative programming continues, and has more recently been spearheaded by Spiro Alexandratos and also Martin Abraham, who are members of both organizations. At the New Orleans National Meeting in 2008, which was held together with the AIChE meeting, several jointly organized symposia were presented which were attended by members of both organizations. Numerous benefits have resulted over the years from this close relationship, especially in the area of technical programming, and many chemical engineers hold membership in both organizations.

I&EC now has five sub-divisions, comprising a large and very active group in Separation Science and Technology, as well as Green Chemistry and Engineering, Advanced Materials and Nanotechnology, Industrial Bio-based Technology, and Novel Chemistry with Industrial Applications. Many of the officers and Executive Committee members of the Division are drawn from the subdivision ranks. In addition, the Division has participated in the programming and planning activities of the Catalysis and the Biotechnology Secretariats. Efforts are now being made to establish a Catalysis Division, and I&EC is supportive of this and has offered its assistance.

In the past several decades, I&EC has developed and presented several popular Short Courses, including a series on chromatography by Harold M. McNair and the long-running course on "Chemical Engineering for Chemists" by Richard G. Griskey. Many of the timely symposia on special topics such as molecular sieves, separations and "green" chemistry have resulted in published volumes which accelerated the growth in interest in these areas. Much of the early impetus in this period for book publication was due to the efforts of Jim McEvoy, who served the Division in a number of capacities over the years (as did several other former Chairs, including Dave Gushee, Bob Stowe, Bill Flank and Melanie Cravey Lesko).

One of I&EC's pioneering innovations in book publishing was the pre-publication of ACS Symposium Series #135 edited by William H. Flank in 1980, and ACS Symposium Series #368 edited by William H. Flank and Thaddeus Whyte, Jr. in 1988. These volumes covering molecular sieve science and technology were published in advance of the symposium presentations at the National Meetings and were available at the meeting sessions, a unique accomplishment in the annals of ACS Books.

In the 1990's, increased emphasis on student poster sessions by the Division, which boosted student participation, especially by local students, combined with the Division's innovative social activities, attracted attention and led to the initiation of the Society's popular SciMix sessions. Other innovative efforts, as a supporter and forum for the Vision 2020 program for developing practical solutions for the future, and collaboration with EPA in initiating green chemistry symposia, have led to broader attention and efforts directed at solving our nation's practical technological problems.

A more recent initiative has involved participation and financial support for programming at Regional Meetings, and the Division has budgeted funds and invited organizers to submit funding requests. A number of such grants have been made in the past several years. In another initiative, in 2007 the first two I&EC Division Fellow Awards were presented. This recognition for outstanding research effort by a chemist and by a chemical engineer was established through the efforts of Spiro Alexandratos.

In recognition of advances in technology, the Division has supported an informational website for a number of years, and publishes its semi-annual newsletter online. Robin Rogers was our first webmaster, and Dustin James is currently webmaster and newsletter editor. Close contact and follow-up regarding Divisional affairs are facilitated by the extensive use of an e-mail listserv and conference calls among Executive Committee members. With nine officers and Councilors, and thirteen additional Executive Committee members and committee chairs listed in the current Division bylaws plus several ad hoc committees, semi-annual meetings at the ACS National Meetings are simply insufficient to conduct the multiple activities of the Division.

Publication History and Activities

The *Industrial and Engineering Chemistry* journal was established in 1909, and in 1923 the *News Edition* of *Industrial and Engineering Chemistry* was started. An *Analytical Edition* was established in 1929, and in 1940 the *News Edition* was made an independent publication. The latter was renamed *Chemical and Engineering News* in 1942 and became a weekly in 1947. The *Analytical Edition* was made an independent publication in 1948 and renamed *Analytical Chemistry*. Table 2 shows the publications associated with the I&EC Division over the years.

In 1970, *Industrial and Engineering Chemistry* was temporarily interred – the Society euphemism was "publication suspension" – and editor Dave Gushee took the occasion to recite some history in marking the demise:

"*I&EC* and the Division whose name it shares were born in 1908 under rather hasty circumstances.

"Its mission was originally 'to serve the interest of the American technical chemist (industrially employed chemist) and chemical engineer engaged in manufacturing pursuits, more especially those having to do with processes and problems of a chemical character.' Over the years this mission has evolved as the needs of its audience have changed in character. As parts of the original audience developed into disciplines of their own, daughter publications were spun off. Today *[1970]* there are eight of these. The mother now passes to her reward.

"The father of this lusty brood - the technical chemist and chemical engineer - sired another such organization (AIChE) and publications family at the same time. Both families - AIChE and I&EC - have flourished for

Table II. Publications Associated With I&EC

Journal of Industrial and Engineering Chemistry launched	1909
Title shortened to *Industrial & Engineering Chemistry*	1923
I&EC News Edition started	1923
Analytical Edition of *Industrial and Engineering Chemistry* started	1929
News Edition became an independent publication	1940
News Edition renamed *Chemical and Engineering News*	1942
Chemical and Engineering News became a weekly publication	1947
Analytical Edition became *Analytical Chemistry*	1948
Journal of Agricultural and Food Chemistry spun off independently	1953
Chemical & Engineering Data series launched	1956
Journal of Chemical and Engineering Data founded	1958
I&EC Fundamentals Quarterly launched	1962
I&EC Process Design & Development Quarterly launched	1962
I&EC Product Research & Development Quarterly launched	1962
Accounts of Chemical Research launched	~1962
Environmental Science & Technology founded	1967
Industrial and Engineering Chemistry publication suspended	1970
Chemical Technology (CHEMTECH) launched	1971
I&EC Quarterlies collapsed into *I&EC Research*	1987
Web edition of *I&EC Research* started	1997
CHEMTECH metamorphosed into *Chemical Innovation*	2000
Biweekly publication instituted for *I&EC Research*	2001

Table III. Divisions Initially Associated With I&EC

	Founded
Environmental Chemistry	1915
Dye Chemistry (now defunct)	1920's
Organic Coatings and Plastics Chemistry	1920's
(now Polymeric Materials: Science and Technology)	1970's
Carbohydrate Chemistry	1921
Cellulose, Paper and Textile Chemistry	1922
(now Cellulose and Renewable Materials)	
Petroleum Chemistry	1922
Fuel Chemistry	1925
Analytical Chemistry	1938
Chemical Marketing and Economics	1954
(now Business Development and Management)	
Chemical Management	1950's/1960's
Nuclear Chemistry	1965
(now Nuclear Chemistry and Technology)	
Fluorine Chemistry	1965

considerably more than half a century. Technology has demanded, in essence, disciplinary bigamy - the father, two mothers (only one of which is being dealt with here), many sons (new disciplines and their practitioners), and many daughters (eight publications from the ACS lineage alone).

"The era immediately following World War II (the one with which we are personally familiar) was fertile with opportunity and demand. The process industries flourished so greatly that *Fortune* magazine suggested in the early fifties that this would be called 'the chemical century.' From 1946 - 1953, *I&EC* reflected the bonanza years with an explosive proliferation of contributed papers, staff-prepared or staff-assisted features, and journals, launched on material previously handled in I&EC (just as *I&EC* was grubstaked in 1908 with material previously carried in *JACS*). In 1942, the 'News Edition' became *C&EN*. In 1948, *Analytical Chemistry* became an entity. In 1953, the *Journal of Agricultural and Food Chemistry* spun off. In 1958, the *Journal of Chemical and Engineering Data* was founded, primarily from *I&EC*'s data content. In 1962 came the three *[I&EC]* quarterlies. And finally, in 1967, *ES&T* was born."

Gushee's final thought about his farewell editorial: "It is a reminder that every institution is a product of its time, and that as times change, so must institutions."

In other activities, the "State of the Art" Summer Symposium, held separately from the ACS national Meetings, was started about 1960 with one of its purposes being the generating of review articles for *Industrial and Engineering Chemistry*, the journal edited by David E. Gushee. That journal was phased out in 1970, as noted above, but the symposium series was continued from 1971 until 1980 under the leadership of Robert Shane.

The Society then launched a new journal, *Chemical Technology* (known more often as *CHEMTECH*), in 1971 under the editorship of Benjamin J. Luberoff, who presided over most of its 29-year history. An innovative feature was an advisory board comprised of initially 12 (and later as many as 22) Divisional representatives from the so-called applied Divisions of the Society. The I&EC liaison representative for many years was Bill Flank. The board was initially chaired by Leo Friend of I&EC. This group met regularly to provide feedback and monitoring, and to assist in securing relevant articles for the journal. After Friend's untimely death in an auto accident, the Leo Friend Award for the most significant article each year in *CHEMTECH* was established and funded by the I&EC Division.

In 2000, *CHEMTECH* metamorphosed into *Chemical Innovation* under the continued recent editorship of Michael J. Block, and continued (or started) publication as Volume 30. The farewell issue of *CHEMTECH* in December, 1999 included an extensive history by founding editor Ben Luberoff of the birth of the publication and much of its history and creative publishing innovations, of which there were many outstanding ones.

Recently, Donald R. Paul, Editor of today's *Industrial & Engineering Chemistry Research*, published an expressive anniversary editorial titled "100 years of *Industrial and Engineering Chemistry*" which he has given permission to present here:

"This journal has a long and rich history of providing quality technical content in the areas of chemical engineering and applied chemistry. The journal is approaching 100 years old, and this deserves a celebration.

"Here is a quick summary of the history of *Industrial and Engineering Chemistry Research* (abbreviated hereafter as *I&EC Research*). The first issue of *The Journal of Industrial & Engineering Chemistry* was published in January 1909 under the editorship of W. D. Richardson. It was the second journal started by the American Chemical Society; the first was the *Journal of the American Chemical Society* (or *JACS*), which was started in 1879. In 1923, the title was shortened to *Industrial & Engineering Chemistry*. In that same year, the *I&EC News* edition was started and then spun off in 1943 to form the current publication *Chemical & Engineering News*. In 1929, the analytical edition was started, and it also was spun off in 1949 to create the current journal *Analytical Chemistry*. In 1956, the *Chemistry & Engineering Data* series was initiated; and, in 1959, the current *Journal of Chemical & Engineering Data* was launched as a stand-alone publication. In 1962, the journal was split into the three I&EC quarterlies: *Fundamentals* (its first and only Editor was Robert L. Pigford), *Process Design and Development* (its first and only Editor was Hugh M. Hulburt), and *Product Research and Development* (its first Editor was Byron M. Vanderbilt, its second Editor was Howard L. Gerhart, and its third Editor was Jerome A. Seiner). In 1987, the three quarterlies were collapsed into a single monthly publication under the title *Industrial & Engineering Chemistry Research* (or *I&EC Research*). Its leadership comprised Editor Donald R. Paul; Senior Editor J. A. Seiner; and Associate Editors J. L. Anderson and J. D. Seader.

"In the early years of *I&EC Research* and its predecessors, manuscripts were sent to a Manuscript Office at ACS Headquarters in Washington, D.C., where staff handled the mechanics of manuscript processing, as was the case for other so-called 'applied' journals, including *Environmental Science and Technology*, *Journal of Chemical Engineering Data*, and *Journal of Agricultural and Food Chemistry*. This office was gradually phased out and eventually closed at the end of 1995, and its staff functions were transferred to the offices of the various editors. This increased the speed of publication and gave authors and reviewers more direct contact with the editors. In 2001, *I&EC Research* went to biweekly publication, because of tremendous growth in the number of manuscripts being published: In 1987, the total number of pages published in *I&EC Research* was 2579, whereas in 2006, that number was 10,072. This growth made it necessary to expand the number of Associate Editors; that list now includes Spiro Alexandratos, David Allen, John Anderson, Larry Biegler,

Mike Dudukivic, Benny Freeman, Massimo Morbidelli (the first outside the United States), and Tunde Ogunnaike. In 1997, the web edition of this journal was initiated, whereas in 2003, online submission of manuscripts was made possible; all of these milestones greatly shortened the time to publication. The American Chemical Society (ACS) scanned all the journal pages published prior to the online electronic version and made the ACS Journal Archives available in 2002. In 1996, we introduced a new category for papers on 'Applied Chemistry,' which has attracted many papers that previously would not have been submitted to *I&EC Research*.

"For many years, *I&EC Research* has had the practice of publishing collections of invited papers in a designated area, from selected symposia or to pay tribute to important people in the field on some significant occasion. A listing of these 'special issues' dating back to the year 2000 can be found at http://pubs.acs.org/journals/iecred/promo/special_issues/. We are taking several steps to celebrate and document this history and the development of the current version of *Industrial & Engineering Chemistry* over the past 100 years. We have launched a website that will be the focal point of the 100-year anniversary; the content of this website (http://pubs.acs.org/journals/iecred/promo/100th) will be preserved for posterity in some way. Among other things, we will highlight several papers published over the last 100 years that have had special impact on our field and there will be a link to these papers, so that all visitors to this website can have access to them. Second, this website will feature short essays by friends of the journal, reflecting on the past or looking forward to the future. This will be analogous to what was done at year 50; see *Ind. Eng. Chem.* **1958**, *50* (1), pp 2-4, to read some interesting commentaries and predictions (available at the website: http://pubs.acs.org/about.html, if your library has a subscription). Finally, in the lead-up to and during 2009, we will publish, in the journal, several special scholarly papers - generally, review papers - to commemorate our 100-year anniversary.

"It is significant to mention that, within the same time frame, the Division of Industrial and Engineering Chemistry of the American Chemical Society (see http://membership.acs.org/I/IEC/) and the American Institute of Chemical Engineers will also celebrate their 100-year anniversaries. We offer our congratulations to these organizations and hope to find ways to celebrate together this important time for chemical engineering and applied chemistry."

Conclusion

According to the 2002 historical update by Dave Gushee and Steve Cooke, "a stated objective of the I&EC Division, as defined in the Bylaws, is 'to support programs that promote the science, techniques, and technology of chemical process and product development.' Member surveys have shown that I&EC has been successful in doing this over the years, and has continued to 'act as an information resource, arrange meetings and symposia, provide information on new technologies, and provide newsletters and abstracts.' " As the Division enters its second century, it continues to evaluate its programs and strategically plan to meet the emerging needs of its varied membership. The recent development of a Division "Value Proposition," and the scheduling of a follow-up Strategic Planning Conference, are visible elements of that effort.

A lasting legacy for the parent American Chemical Society, deriving from the establishment of the Divisional structure whose centennial we now celebrate, is the format of the National Meetings held semi-annually in various cities around the country. Programming at these meetings is organized and presented by Divisions, and the strength of that programming is a tribute to the health and the value of the Divisional structure within the Society. As we mark 100 years of Division activity and I&EC's leadership role, we can note with pride that the success of National Meetings owes much to Divisions and their efforts.

Chapter 2

Ammonia Synthesis: Catalyst and Technologies

Svend Erik Nielsen

Haldor Topsoe A/SNymoelleves 55, DK–2800 Lyngby, Denmark

The following chapter gives a historical perspective of the ammonia synthesis technology including developments of the ammonia synthesis catalyst. Operating parameters in the ammonia synthesis are discussed and the various configurations with respect to selection of converter types, purge gas recovery systems, steam production, etc. are convered in detail for the technology currently in industrial use. Also, the most widely used commercial processes are described in this chapter, and finally prospectives and ideas for the future are mentioned.

Introduction

The development of the synthesis of ammonia was a landmark for the entire world and for the chemical industry in particular, since it not only solved the problem of securing food supply to the ever increasing world population, but it also had a significant impact on the industrial chemistry and laid the foundation for the theory and further developments in the industrial practice of heterogeneous catalysis. The ammonia industry has contributed significantly to improvements within chemistry and chemical engineering, and the very high R&D activity with ammonia catalysts had a major effect on other catalyst based chemical industries. The ammonia synthesis reaction is probably the best fundamentally understood and documented catalytic process.

Since the initial development of its synthesis, the industrial production of ammonia has been steadily increasing as can be seen from the curve (Figure 1), and in 2005 the world production has reached 147 million metric tonnes. Today 1.2% of the total world energy consumption is used for production of ammonia,

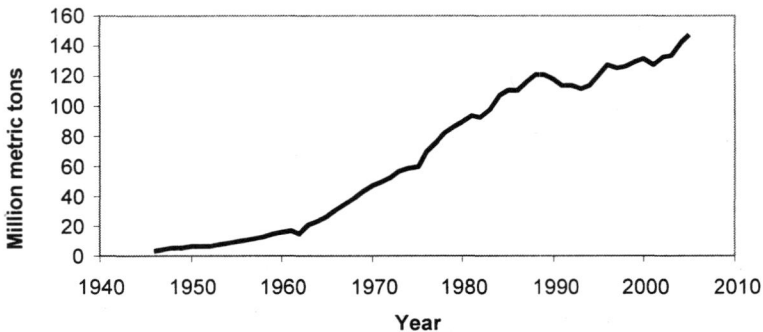

Figure 1. World Ammonia Production (Source Indexmundi, see (1))

and ammonia has become one of the largest chemical products since the production really started to increase rapidly in the beginning of the 1960's.

The present chapter describes ammonia synthesis technology. The first part gives a brief historical overview of the ammonia synthesis process including some early developments to better understand the current standing of the technology. The second part gives a description of the current state of ammonia synthesis technology, and the third part will deal with some ideas that may be addressed in the future in order to improve the technology even further. For more comprehensive details and surveys, please see *(2, 3)*.

Historical Perspective

At the end of the 19th century, it became evident that the known resources of nitrogen, which were primarily ammonium sulphate (as a by-product from coke and town gas production from coal) and natural deposits of saltpetre from Chile, were not sufficient to cover the demand for fertilizers required to increase the agricultural yield to satisfy the world food production. From here, the history of modern ammonia synthesis started in Germany just after year 1900, when Fritz Haber and his assistants developed and patented the process concept that forms the basis for all ammonia production today. Fritz Haber together with Carl Bosch from BASF jointly developed the Haber-Bosch process, and Alvin Mittasch discovered the promoted iron catalyst. They all played a significant role in the developments, and the splendid work of these gentlemen resulted in the first commercial plant at Oppau/Ludwigshafen near Mannheim in Germany. The plant was commissioned in 1913 and the capacity was 30 MTPD (Metric Tons Per Day). The capacity of this plant was rapidly increased to 250 MTPD in 1916. In 1917 a second plant was started at Leuna near Leipzig, which after further expansion produced 240,000 tonnes/year at the end of the First World War.

In parallel, independent development work took place in the USA. In 1918 a plant called "US Nitrate Plant No. 1" was constructed in Alabama. The plant was, however, not very successful, but the developments continued in the 1920's with the construction of several successful plants. In 1928 an American company (NEC) was commissioned to construct a plant in Europe, thereby entering worldwide competition.

During the 1930's and the Second World War the worldwide production capacity was increasing rapidly, and in 1945 about 125 ammonia plants existed with a total capacity of 4.5 million tonnes/year. The most important processes at that time were Haber-Bosch, Casale, Claude, Fauser, NEC and Mont Cenis.

Already at that time it was the impression that the ammonia technology was mature, and that no significant further developments could be expected. This was to a certain extent true, since most of the features that characterise modern synthesis technology – including catalyst type and most major converter designs – were already well proven in industrial applications. Since that time the developments in the ammonia technology have primarily been in the synthesis gas generation section and not the ammonia synthesis itself. Also the type of feedstock to the plants has changed, and natural gas is today the predominant feedstock for ammonia production.

In the early days the synthesis gas was produced at atmospheric pressure, and the synthesis gas was compressed in reciprocating compressors to pressures as high as 100 MPa in some cases. Capacities were limited to around 300 – 400 MTPD due to limitations in reciprocating compressors. However, with the development of steam reformer based front-ends and the introduction of centrifugal compressors, the ammonia plant capacities suddenly increased to 1000 MTPD with ammonia synthesis loop pressures typically around 15 MPa. Since the 1960's new developments have been in the ammonia converter designs, such as introduction of radial flow converters and introduction of converters with multiple catalyst beds to increase ammonia conversion.

Efforts to develop more active ammonia synthesis catalysts have been ongoing. As an example, a Ru-based ammonia synthesis catalyst has been introduced to the market. This catalyst is more active than the iron-based catalyst, but the cost is also much higher due to the high Ru price, and the use of the catalyst is very limited.

In spite of the fact that the same process scheme, with a few radical developments, has been used for decades, there has been a tremendous development in the scale of operation, and the largest plant today produces as much as 3,300 MTPD of ammonia. At the same time the energy consumption has been reduced from 50 – 63 GJ/MT NH_3 in the 1960'ies to below 29 GJ/MT NH_3 today, close to the theoretical minimum of 20 GJ/MT NH_3.

Ammonia Synthesis Catalyst

The formation of ammonia from hydrogen and nitrogen is a strongly exothermic reaction

$$N_2 + 3H_2 \rightleftharpoons 2NH_3 \quad \Delta H_{298} = -92.44 \text{ kJmol}^{-1} \quad (1)$$

The reaction is favoured by low temperatures and high pressures, in industrial practice ranging between 8-20 MPa and 350-480°C to yield a conversion per pass of reactants of 25-35% depending on reactor design and configuration. The reaction is reversible and under all practical conditions limited by equilibrium. Recycle of unconverted reactants is required to obtain high overall conversion.

Figure 2. Ammonia catalyst precursor manufacture

Traditional industrial ammonia catalysts are iron-based. They are prepared by fusion of iron ore (Figure 2) adjusted with various promoters, predominantly potassium, calcium and aluminium, which may be added to the melt as oxides, hydroxides, carbonates or nitrates. Upon solidification the product is crushed and sieved to the desired particle size, typically 1.5-3 mm irregular shaped particles (Figure 3).

A small particle size is essential to reduce diffusional limitation. The unreduced catalyst consists predominantly of magnetite, Fe_3O_4 containing minor amounts of wustite, FeO, or in some cases hematite, Fe_2O_3. In the boundaries between the iron domains are crystalline and amorphous phases containing mainly oxides of potassium and calcium, both of which are too large to enter the magnetite structure. The unreduced catalyst is virtually non-porous with a typical

Figure 3. Ammonia synthesis catalyst

surface area of less than 1 m^2g^{-1}. Wustite was reported *(4, 5)* as an alternative precursor leading to more active catalysts compared to those based on magnetite. The addition of modest amounts of cobalt, 3-6 wt%, increases the activity, and cobalt-containing catalysts have been commercialised *(6)*. Cobalt is incorporated into the crystal lattice and upon reduction forms an alloy with iron, from which partial segregation of Co takes place *(7)*.

The catalyst is supplied in either the as-prepared, unreduced form or in the prereduced, metallic state. In the prereduced form the catalyst is highly pyrophoric and must be passivated by controlled oxidation of the iron surface in order to enable safe handling upon exposure to air.

Metallic iron is the active and main phase in ammonia catalysts. In spite of this, pure iron exhibits only poor synthesis activity unless it is promoted by various amounts of oxides, primarily those of potassium, aluminium, calcium and magnesium, but numerous combinations involving additional elements, e.g., Si, Cr, Ti and Zr have been claimed. Thus, promoters are crucial to catalytic performance.

Promoters may be classified in two categories. Those which enhance the catalytic activity by inducing and preserving a high catalyst surface area are referred to as structural promoters, while those increasing the catalytic activity per unit surface area of iron are termed electronic promoters. To the first category belong the oxides of aluminium, calcium and magnesium, which during reduction segregate to form spacers that inhibit the growth of iron crystallites, and thereby assist in developing and maintaining a high surface area. In the absence of structural promoters rapid sintering of the pure iron phase under normal operating conditions would occur, whereas a small amount effectively inhibits crystal growth by forming a physical boundary between the iron crystallites. The second category, the electronic promoters, is represented by alkali compounds like potassium that strongly enhance the specific activity of the iron surface. Contrary to the structural promoters, potassium is highly dispersed in the reduced catalyst, and a uniform distribution over the iron surface is observed. The promoting effect of alkali is rationalised by electrostatic interactions and the effect applies to non-iron based catalysts as well.

Alternatives to the iron-based catalysts have been thoroughly researched. Although the high activity of Ru-based catalysts were early recognised (8), it was the work by Japanese researchers (9, 10) on alkali promoted, carbon supported ruthenium catalysts that triggered renewed interest and immense research efforts in identifying and researching alternative catalysts. Using Cs or Ba-promoted ruthenium catalysts, activities about an order of magnitude higher than seen for the iron catalyst may be achieved especially at high ammonia partial pressure, since the inhibition by ammonia is significantly less severe than for the traditional catalyst. Several other supports have been reported including MgO, $MgAl_2O_4$, boron nitride and zeolites. Although promoted ruthenium catalysts have been introduced in industry (11), it is still an open question whether the shorter lifetime and higher cost justify the use of Ru-based catalysts. Other alternatives, including non-iron, non-ruthenium catalysts, have been researched. Among these a novel class of high-active bimetallic nitrides have been discovered, e.g., Cs-promoted Co_3Mo_3N (12, 13) and Ba-promoted cobalt catalysts (14). However, common to any alternative to the classical catalyst is that they are competing against a class of traditional low cost iron-based catalysts characterised by high activity and with a durability that is unsurpassed in industrial catalysis.

The reduction of the magnetite catalyst precursor is endothermic:

$$Fe_3O_4 + 4H_2 \rightleftarrows 3Fe + 4H_2O \quad \Delta H_{298} = 149.9 \text{ kJmol}^{-1.} \quad (2)$$

It involves complete removal of oxygen from the magnetite lattice and a redistribution of iron atoms transforming the bulk non-porous magnetite into a porous metallic structure. Also, redistribution and segregation of promoters take place. The reduction does not cause any changes in overall particle dimensions. The reduction process is complex and numerous reduction studies have been published (15, 16, 17), and various models have been proposed. A more elaborate treatment of the reduction process is given by Schlögl (18).

Activation of the catalyst is carried out in a H_2/N_2 mixture at moderate pressure. The process must be carefully controlled in order to ensure the development of maximum surface area. The rate of reduction is controlled by adjusting the temperature; the lower the reduction rate, the better the development of the micro porous structure (r = 100 to 150 Å). Exposure to water/steam of the reduced part of the catalyst will cause sintering and loss of activity and counter-diffusion of water should be minimised by operating at highest possible gas velocities. Water is also reported to inhibit the magnetite reduction rate (15). It is removed by condensation before the exit gas is recirculated. As the catalyst becomes reduced ammonia formation begins, and recirculation rate to be increased. As the reduction accelerates, the loop pressure and recirculation rate may be gradually increased, providing a smooth transition from reduction to normal operation.

Different reduction procedures apply if the catalyst is prereduced or when a combination of prereduced and unreduced catalyst is used. Whereas reduction of the bulk magnetite catalyst goes on over days, the reduction of the superficial oxidic layer of the prereduced catalyst is facile and may be accomplished within approximately one day if solely prereduced catalyst is charged. Often the first bed is charged with prereduced catalyst to enable fast reduction and onset of the ammonia synthesis reaction, which thereby liberates heat to support the endothermic reduction in the remaining part of the bed.

Upon reduction, the crystalline density increases from about 5 to almost 8 g cm^{-3}. However, the reduction is not accompanied by any changes in particle dimension, hence settling of the catalyst bed does not occur thereby making radial flow converters the preferred choice in consideration of pressure drop issues.

Under normal operating conditions sintering is negligible and catalyst lifetimes exceeding ten years are far from unusual. However, the combination of high temperature and the presence of water and other oxygen-containing compounds may lead to accelerated crystal growth. Under mild conditions oxygen-containing gaseous components such as water and carbon oxides are classified as temporary poisons because the decrease in activity caused by oxygenates is regained when the oxygen species are no longer present in the feed [19]. Kinetic rate expressions accounting for the effect of partial poisoning by water have been established [20, 21]. Under more severe conditions or at prolonged exposure, high concentrations and excessive temperatures do lead to permanent loss in activity.

Sulphur compounds, halides, phosphorus and arsenic are permanent poisons to ammonia catalysts [22]. However, in most plants the upstream low-temperature shift catalyst and the Ni-based methanation catalyst both serve as efficient guards by irreversibly adsorbing traces of such compounds. Thus, permanent poisons are normally not a severe problem.

It was early recognised that the rate limiting step in the ammonia synthesis is the dissociative adsorption of nitrogen [23] and that hydrogenation proceeds at a much faster rate [24]. Temkin and Pyzhev [25] proposed a rate expression,

$$r = k_1 P_{N2} \{(P_{H2})^3/(P_{NH3})^2\}^\alpha - k_{-1} \{(P_{NH3})^2/(P_{H2})^3\}^{1-\alpha} \; ; \; (0.5 < \alpha < 0.8) \qquad (3)$$

which provided a good correlation with experimental data over a broad range of reaction conditions. The expression can be derived assuming that dissociative adsorption of nitrogen is rate-determining and that the surface coverage of dissociated nitrogen is high. The Temkin-Pyzhev rate equation – and elaborations thereof, e.g. by Brunauer *(26)*, Temkin *(27)*, Ozaki *(28)*, Temkin & Morozov *(29)* and Nielsen *(30, 19)* – have proved successful in fitting experimental and industrial data.

On the microscopic level the reaction may be represented by the elementary steps:

$$H_2(g) + 2* \leftrightarrow 2H*$$
$$N_2(g) + * \leftrightarrow N_2*$$
$$N_2* + * \leftrightarrow 2N*$$
$$N* + H* \leftrightarrow NH* + *$$
$$NH* + H* \leftrightarrow NH_2* + *$$
$$NH_2* + H* \leftrightarrow NH_3* + *$$
$$NH_3* \leftrightarrow NH_3(g) + *$$

where the dissociation of chemisorbed N_2, is the rate limiting step. This scheme was originally proposed by Emmett *(31)* and later on substantiated and detailed in studies by the groups of Ertl *(32, 33)*, Bowker *(34)* and Stolze *(35, 20)* applying statistical mechanics and surface science techniques.

Single crystal studies based on pure Fe show that the ammonia synthesis is a highly structure sensitive reaction *(36)*. Different crystal planes display pronounced variations in reaction rate, and by studies of the most densely packed surfaces the difference in reactivity was found to decrease by more than two orders of magnitude in the sequence $(111) > (100) > (110)$ *(37)*.

The role of alkali promoters has been subject to many studies, and it has been commonly accepted that the effect of alkali is to lower the activation barrier for the dissociation of chemisorbed N_2. Recently, the role of potassium has been accounted for by density functional theory calculations indicating that on iron catalysts destabilisation of NH_x species, creating more free sites for N_2 adsorption, may be the main reason for the rate enhancement observed, whereas on Ru and Co-based catalysts promotion becomes effective through weakening of the $N \equiv N$ bond of the adsorbed species *(38)*.

The Ammonia Synthesis Technology

Since the ammonia reaction is limited by equilibrium with a conversion per pass of reactants of $25 - 35\%$ as mentioned earlier, the synthesis of ammonia takes place in a recycle loop.

Basically two different layouts of the ammonia synthesis loop exist, depending upon the quality of the make-up synthesis gas feeding the loop. In most cases the make-up gas also contains (apart from H_2 and N_2 in proper ratios) some inerts (CH_4, Ar and minor traces of rare gases), which have to be purged from the loop to avoid build-up of inerts. (Inerts containing loop layout). This layout is primarily used in plants where the synthesis gas is generated from reformer based frond-ends with primary and secondary reforming.

In other cases, the make-up gas contains only H_2 and N_2 maybe with minor traces of inerts like CH_4, Ar, rare gases etc. (Inerts free loop layout). In these cases the synthesis gas is normally generated in front-ends based on coal gasification or with final clean-up of the synthesis gas based on cold box or PSA (Pressure Swing Adsorption) technologies.

The make-up gas addition point to the loop will vary depending upon the make-up gas quality. Figure 4 shows the layout used if inerts are present in the make-up gas. In case the loop is inert-free, the make-up gas is added directly to the loop recirculating gas, and no purge gas system is needed.

Generally speaking the ammonia synthesis loop will contain the following main elements:

− A reactor system comprising one or more catalytic synthesis converters with associated temperature control and heat recovery equipment
− Heat exchangers for cooling the product gas to recover waste heat, preheat converter feed gas and to condense the product ammonia.
− Vessels for the separation of product ammonia from unreacted gas and for the adjustment of product properties (degassing, adjustment of temperature and pressure).
− Equipment for compression and addition of make-up gas and for the recirculation of non-converted gas.
− In case the make-up gas contains inerts, equipment for removal of purge gas to prevent build-up of inerts in the loop is also needed.

The Ammonia Synthesis Converter(s)

The heart of the ammonia synthesis loop is the ammonia converter. The ammonia converter normally comprises a basket (internals) enclosed in a pressure shell. In the ammonia synthesis converter(s), due to the exothermal nature of the reaction taking place, it is normally not possible to go from the inlet to the outlet conditions in the reactor in one adiabatic step, if a reasonable conversion is desired. Therefore, some type of cooling is required. In principle, the cooling can be applied in three different ways:

24

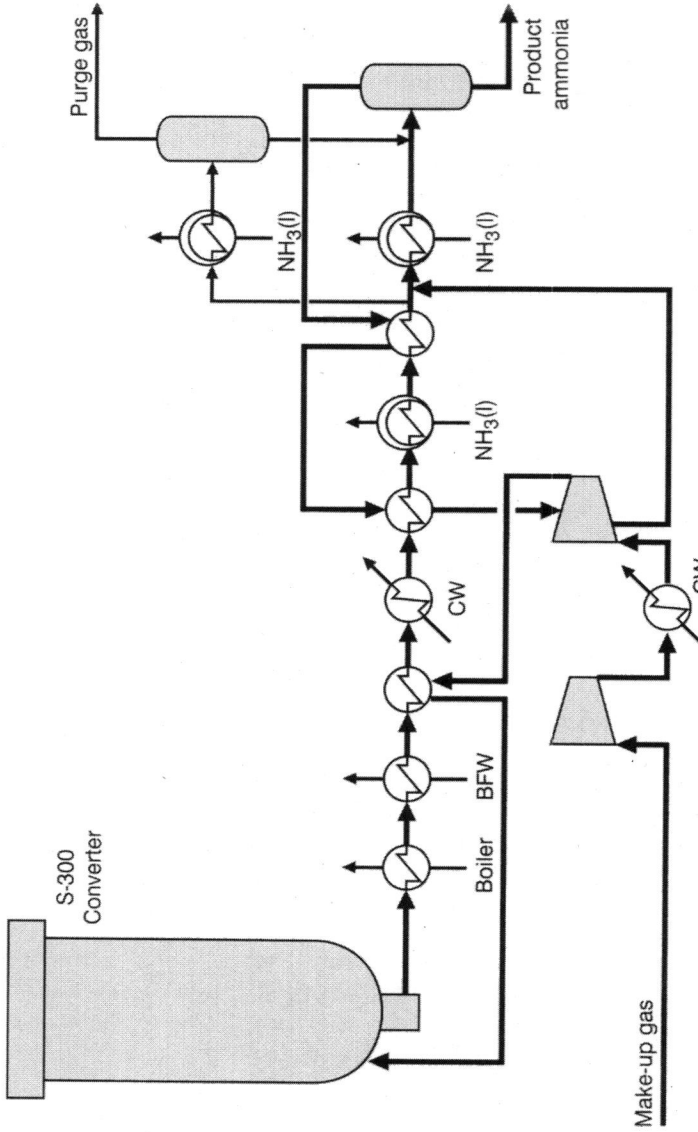

Figure 4. Typical layout of Topsøe ammonia synthesis loop with inerts containing make-up gas

– External cooling by heat exchange between catalyst beds. In most cases the heat exchangers are located inside the converter (interbed heat exchangers). The cooling medium can be either synthesis gas or some other medium, e.g. boiling water. Flow in the catalyst beds can be either axial or radial in vertical converters or downwards in horizontal converters.

– Internal cooling with cooling tubes or plates in the catalyst bed or with catalyst in tubes surrounded by a cooling medium. The internal cooling can be effected either with gas flow in the same direction in the catalyst bed and cooling channels (co-current flow), with gas flow in opposite directions in the catalyst bed and cooling channels (counter-current flow), or with gas flow in the catalyst bed perpendicular to the flow in cooling channels (cross flow). The cooling medium can be either synthesis gas or some other medium, e.g. boiling water.

– Quench cooling by injection of cold gas. The injection of quench gas can be either between adiabatic beds or into a catalyst bed at different locations. Flow in the catalyst beds can be either axial or radial in vertical converters or downwards in horizontal converters.

In other cases, when more than one converter is put in series, the last converter is normally a one bed converter without any cooling.

More detailed reviews of the various types of ammonia synthesis converters can be found in *(2, 3)*.

Today, almost all new ammonia plants use radial flow converters with indirect cooling by heat exchange between catalyst beds and/or between converters.

As an example of an ammonia synthesis converter, the Haldor Topsøe S-300 converter is illustrated in Figure 5. The ammonia synthesis converter consists of a pressure shell and a basket. The basket consists of three catalyst beds and two interbed heat exchangers placed in the centre of the first and second catalyst bed respectively.

The main part of the synthesis gas is introduced into the converter through the inlet at the bottom of the converter and passes upwards through the outer annulus between the basket and the pressure shell, keeping the latter cooled. It then passes to the bottom tube sheet of the first interbed heat exchanger through transfer pipes in the heat exchanger and passes the tubes in upward direction thereby cooling the exit gas from the first bed to the inlet temperature to second bed.

Another part of the synthesis gas is introduced through the bottom central inlet and flows upwards through the transfer pipe to the bottom tube sheet of the second interbed heat exchanger. It passes the tubes in upward direction thereby cooling the exit gas from the second bed to control the inlet temperature of the third bed.

26

A : MAIN GAS INLET
B : INLET FOR GAS TO
      LOWER IHE TUBE
C : COLD BY–PASS INLET
D : GAS OUTLET

1 : PRESSURE SHELL
2 : OUTER ANNULUS
3 : OUTER BASKET SHELL
4 : BASKET INSULATION
5 : BASKET COVER
6 : INTERBED HEAT EXCHANGER (IHE)
7 : TRANSFER PIPE
8 : SCREEN PANELS
9 : 1st CATALYST BED
10 : CENTRE SCREEN
11 : COVER PLATE
12 : CATALYST SUPPORT
13 : 2nd BED SUPPORT FLANGE
14 : 2nd CATALYST BED
15 : 3rd CATALYST BED

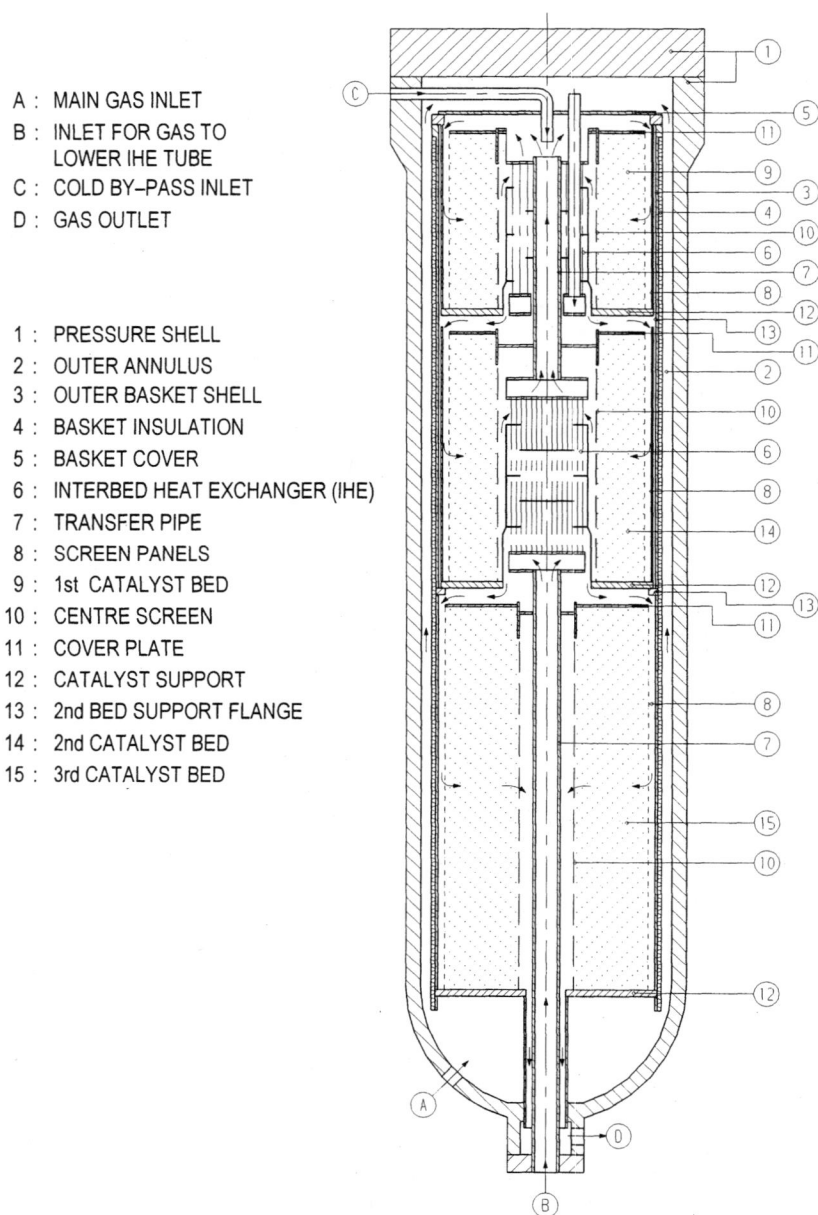

Figure 5. Topsøe S-300 Ammonia Synthesis Converter

The remaining part of the gas, the cold by-pass gas, is introduced at the top of the converter. In the top of the converter pipe it mixes with the gas leaving the tube side of the two interbed heat exchangers. The amount of cold by-pass gas controls the inlet temperature to the first bed.

After mixing, the gas flows through the space below the basket cover to the annuli of the panels around the first catalyst bed. From the panels it passes the first catalyst bed in inward direction and then flows to the annulus between the first catalyst bed and the first interbed heat exchanger. Even gas distribution in the catalyst bed is ensured by means of appropriate perforation in the panels. The effluent from the first catalyst bed passes the shell side of the first interbed heat exchanger for cooling to the proper inlet temperature to the second catalyst bed by heat exchange with gas introduced through the tube side of the first interbed heat exchanger as described above.

From the shell side of the first interbed heat exchanger, the gas is transferred to the second catalyst bed through the panels around the bed. The effluent from the second catalyst bed passes the shell side of the second interbed heat exchanger for cooling to the proper inlet temperature to the third catalyst bed by heat exchange with gas introduced to the tube side of the second interbed heat exchanger through the bottom inlet as described above.

Both second and third catalyst beds are passed in inward direction, the gas distribution being ensured by means of appropriate perforation at the walls of the bed.

The temperature inlet the third catalyst bed is controlled by means of the by-pass around the loop boiler feed water preheater, adjusting the gas temperature at converter inlet.

The gas leaving the third catalyst bed passes the perforated centre tube and flows to the converter outlet.

During start-up, hot gas from the start-up heater is introduced through the cold by-pass pipe at the top of the converter.

Selection of Construction Materials for Equipment in Ammonia Synthesis Loops

Synthesis loop converters and the loop waste heat boiler(s) are normally operated under high pressures, and are the most critical items in the ammonia loop. Especially the selection of construction materials is essential due to the high ammonia content and the high partial pressure of hydrogen in the reactor effluent.

The most common corrosion phenomena causing problems in ammonia synthesis loop equipment are hydrogen attack, nitriding, and stress corrosion.

Hydrogen attacks steel by the dissociation and diffusion of atomic hydrogen into the metal. The hydrogen atoms react with carbon in the steel causing the formation of methane. Methane cannot diffuse in steel, and very high pressure may build up in 'pockets' in the steel, ultimately causing destruction of the material. The reaction can be counteracted by alloying the steel with carbide-forming elements such as Cr and Mo. Hydrogen is to some extent soluble in all construction materials. Too rapid cooling may trap the dissolved hydrogen, whereby the material becomes brittle. Slow cooling will allow the hydrogen to escape, and the material stays ductile.

Nitriding is caused by ammonia reacting with the surface of the steel, forming a hard and brittle surface layer of nitride. It has been found by both experience and experiments that low alloy steel can be safely used in gas containing ammonia at temperatures below 380°C, whereas stainless steel, Inconel and Incoloy can be used at any temperature relevant for ammonia production.

Stress corrosion cracking may be caused by a number of reasons including presence of trace compounds, especially chlorine, poor heat treatment after welding, etc.

The above indicates that low alloy steel, e.g. 2½Cr ½Mo, can be used for most loop equipment including the converter pressure shell. Critical parts are the reactor internals (the 'basket'), which are most often made from stainless steel (SS 321), and the inlet part of the waste heat boiler which must either be made from or cladded by corrosion-resistant material, like Incoloy or similar.

Loop Operating Conditions

The operating conditions in the ammonia synthesis loop are described by a number of parameters, which in some cases may be independent variables, and in other cases a function of other parameters. The relationship between these parameters (and other parameters such as space velocity, inert level, concentrations and temperatures at various points in the synthesis loop, etc.) can be described in mathematical models that are used for design, simulation, and optimisation.

The most important parameters are loop pressure, separator temperature, recycle ratio, conversion per pass and inert level.

The loop pressure has an important influence on the performance of the ammonia synthesis loop because of its influence on the reaction equilibrium, reaction kinetics, and gas/liquid equilibrium in the product separation. Actual selection of loop pressure is in many cases a compromise between selecting a high pressure to favour the ammonia synthesis reaction, and on the other hand selecting a reasonable pressure to minimise the compression power of the synthesis gas compressor, which compresses the synthesis gas to the desired loop pressure. The loop pressure also has a significant impact on the ammonia refrigeration system, since a high loop pressure favours condensation of the ammonia product in the loop water cooler and saves compression power on the refrigeration compressor. On the other hand, a low loop pressure saves compression power on the synthesis gas compressor, but increases the compression power of the recycle compressor and the ammonia refrigeration compressor. A wide range of operating pressures has been used in practice, from less than 10 MPa to 100 MPa. The trend in modern plants has been to select operating pressure in the low to medium pressure range of 8 – 20 MPa. The recent trend for even higher capacities is demanding pressures around 20 MPa in order to avoid too large piping and equipment dimensions in the ammonia synthesis loop

The separator temperature, together with operating pressure and location of make-up gas addition point, determines the ammonia concentration at the converter inlet. A low temperature means a low ammonia concentration at the converter inlet, which again means either a low catalyst volume or high conversion. But low separator temperature can only be obtained by increased power consumption and cost of the refrigeration units. In the end the selection of loop pressure as well as the choice of separator temperature is a compromise between energy consumption and capital costs in various parts of the synthesis unit.

As previously mentioned, two types of ammonia synthesis loops exist. For the inert- containing loop, the inert level in the synthesis loop (most often measured at converter inlet) depends on the inert level in the make-up gas, the production of ammonia per unit make-up gas (the loop efficiency), and the purge rate. The inert level in the make-up gas is solely determined by the conditions in the synthesis gas preparation unit. The ammonia production is determined by conditions around the converter, the gas flow (which may be expressed by the recycle ratio), the inlet temperature and pressure, catalyst volume and activity, and converter configuration.

Since it is a loop, all of the parameters described above are interrelated. This means that for a given situation neither the purge rate nor the inert level can be changed without affecting operating pressure. It also means that if the operating pressure or the catalyst activity (by change of catalyst) is changed, then the other parameters will change until a stable operating point is obtained.

Steam Production in Ammonia Synthesis Loops

The heat of reaction in ammonia synthesis is 46.22 kJ/mol ammonia at standard temperature and pressure. It is of utmost importance to integrate the ammonia synthesis process with the steam generation system. The waste heat not used for preheating the converter feed gas and for preheating the inlet gas streams for each catalyst bed is most efficiently used for production of high pressure steam. Normally waste heat down to around 80-100°C can be utilised for these purposes.

In large ammonia plants, steam is most often produced at 10 – 12 MPa. The waste heat downstream the synthesis converter is typically available for steam production between 450°C and down to about 260°C with the exact value depending on the layout and the conversion within the converter system *(39)*. This means that the waste heat must be recovered partly in a boiler (down to about 340°C), and partly in a boiler feed water preheater. In low energy plants, the tendency is to increase the conversion by increasing the outlet concentration of ammonia from the last catalyst bed, resulting in a reduced converter outlet temperature (at constant pressure). At the same time, maximum high pressure steam production in the synthesis loop is desired in order to satisfy the overall energy balance in the plant.

In some cases, it may be necessary to recover part of the heat before the reactions in the converter system have been completed. This may be accomplished in a converter system with two or more converters, where cooling, not only after the last converter, but also between the converters, is done by recovering the waste heat for steam production. Such a converter configuration will allow for more duty available for boiling or superheating of steam, and might be a way to shift the waste heat available for boiling and boiler feed water preheating in the loop.

Ammonia Product Recovery

Ammonia is without exception recovered as a product from synthesis loops by cooling at synthesis pressure to condensation, followed by separation of the liquid product from the gas, which is recirculated to the converter. In high-pressure synthesis loops, the cooling or part of it can be done by water cooling and/or air cooling. In modern plants, this is in most cases supplemented by cooling from a refrigeration unit. The refrigeration is typically supplied from a mechanical ammonia refrigeration cycle with one, two or several refrigeration levels. Refrigeration down to -25°C has been used, which corresponds to cooling by evaporation of ammonia at almost atmospheric pressure.

The liquid ammonia from the separator contains a small amount of dissolved gases. These are partly released by pressure reduction in a 'let-down vessel' normally to about 2.5 MPa. After the let-down vessel the ammonia is further flashed to almost atmospheric pressure before being sent to the ammonia storage tank. In order to reduce the risk of stress corrosion cracking in the ammonia storage tank, the ammonia product should contain minimum 0.2% of water.

In order to increase the product yield, the ammonia vapour from the let-down gas is recovered by washing with water, and this ammonia water is distilled together with ammonia water obtained from washing of the purge gas. Remaining off-gas is normally used as fuel somewhere in the complex. The ammonia vapour from the last flash vessel is normally sent to the refrigeration compressor.

Purge Gas Recovery

In gasification based plants, or in plants using liquid nitrogen wash or PSA (Pressure Swing Adsorption) for final purification, the concentration of inerts in the make up gas to the synthesis loop may be so low – 'inerts free' loop with 100 vol ppm or less – that they are dissolved in the liquid ammonia leaving the separator. In such cases, no continuous purge stream is taken from the loop

In the inerts-containing loop, the synthesis gas will normally contain certain concentrations of 'inerts', i.e. compounds that are not consumed by the ammonia synthesis reaction and that do not interfere with the catalyst performance. The inerts are typically CH_4, Ar, and traces of other rare gases. In special cases, significant amounts of helium may be present originating from He-containing natural gas. In all inerts-containing loops, it is necessary to purge a gas stream from the synthesis loop in order to prevent excessive build-up of the inerts.

Along with the inerts in the purge gas stream also some hydrogen and nitrogen is present, which represents potential raw materials for ammonia production. And since energy has been invested in the production of especially the hydrogen and in bringing the gases to the synthesis pressure, it is obvious to do an effort to recover the 'lost' hydrogen again. It is therefore attractive to separate at least the hydrogen from the purge gas and return it back to the synthesis loop. Recovery of the nitrogen is less attractive, especially in plants based on steam reforming, because preferential recovery of hydrogen will shift the required hydrogen/nitrogen ratio in the synthesis gas to be generated in the front-end to lower values, from the normal value of 3.0 to, in some cases, as low as 2.8. This makes it possible to transfer some of the reforming duty from the primary to the secondary reformer, and the fuel firing in the front-end of the plant is thereby reduced.

Several ways exist to recover the hydrogen from the purge gas. One way to separate the inerts from the purge gas is to use a cryogenic unit. The purge gas is washed with water and then dried over molecular sieves to prevent freezing of water and ammonia on the heat exchange surfaces. The stream is thereafter cooled to partial condensation in a heat exchanger. Liquid and gas are separated, and the hydrogen-rich gas is reheated by heat exchange with the incoming feed gas. The liquid, which contains most of the inerts, is depressurised, and the resulting gas is also reheated by heat exchange with the incoming feed gas. This gas is further used for regeneration of the molecular sieves in the drying unit, before it is used as fuel. Refrigeration is supplied by the expansion of the off-gas and, and if required, by additional chilling from the ammonia refrigeration system.

Typical recoveries in a cryogenic unit (the fraction of a compound in the incoming gas which is found in the hydrogen-rich gas) are about 90 – 95% for hydrogen, 25% for nitrogen, 25% for argon, and 4% for methane. The recovered hydrogen is normally returned to the ammonia synthesis loop at relatively high pressure (7 – 8 MPa) to one of the inter stages of the synthesis gas compressor.

If energy is of lower importance, the hydrogen can be recovered from the purge gas in a membrane system, which has a lower investment cost than the cryogenic unit. However, in the membrane system, the hydrogen will be returned at a lower pressure than from the cryogenic unit requiring more compression energy in the synthesis gas compressor.

Some membranes are sensitive to ammonia, and the purge gas is therefore normally washed with water in the ammonia absorption unit to remove ammonia to a low level before being passed to the 'shell side' of the membrane assembly comprising a large number of hollow fibres. The fibres pass through a 'tube sheet' at one end while they are sealed at the other end. Hydrogen will – together with smaller amounts of other compounds – pass through the membrane and pass out on the 'tube side'. Hydrogen recovery and purity depend on the membrane material, on the pressure differential, and on space velocity. Typical recoveries in a membrane unit (the fraction of a compound in the incoming gas which is found in the hydrogen-rich gas) are about 85 – 95% for hydrogen, 5 -15% for nitrogen, 20% for argon, and 10% for methane. The recovered hydrogen is returned to the suction side of the synthesis gas compressor at 2.5 – 3 MPa. In special cases the recovered hydrogen can be returned to the synthesis gas compressor at two different pressures (2.5 – 3 MPa and 5 – 5.5 MPa) if energy consumption is of high importance.

Pressure Swing Adsorption (PSA) is another process for hydrogen recovery in ammonia plants. A recovery of hydrogen of about 90% can be obtained with 100% rejection of inerts, and the pressure loss through the PSA unit is on the off-gas, while the hydrogen is available approx. at feed gas pressure. However, the operating pressure of PSA units is with today's technology limited to about 6 MPa.

The off-gases from the hydrogen recovery units described above are normally used as fuel somewhere else in the plant (typically reformer or auxiliary boiler).

Commercial Processes

In the past, many different companies have offered technology for production of ammonia to the market. Comprehensive surveys may be found in *(2, 3)*. However, the competitive situation has been such that today (2007) three ammonia technology licensors dominate the market, namely:

- Haldor Topsøe A/S (Denmark)
- Kellogg Bown & Root (KBR) (USA)
- Uhde GmbH (Germany)

In the following, a brief description of these ammonia synthesis technologies will be given. For more detailed and currently updated descriptions, reference is made to the respective company websites *(40, 41, 42)*.

Haldor Topsøe A/S

The Topsøe Low Energy Ammonia Process is flexible and can be adjusted to specific project requirements. Most operating ammonia plants designed by

Topsøe use the so-called S-200 synthesis loop. However, other converter configurations are also available *(39)*.

Today the S-300 converter is the preferred choice, featuring a high conversion per pass. The S-300 converter is three bed radial flow converter with indirect cooling between the catalyst beds (see Figure 6). In some cases, a different configuration, the S-350 loop, is used featuring two converters, a three-bed (S-300) converter followed by a one-bed (S-50) converter. The S-350 synthesis loop is somewhat more energy efficient than the normal S-300 loop, and is also a converter system that can be used if the waste heat recovery should be maximised (generates more waste heat for steam boiling/superheating). The S-350 converter configuration is used mainly for mega size plants or in situations with high energy cost. Addition of a S-50 converter is also an alternative revamp option for existing, single converter loops in capacity revamps or for energy saving purposes *(43)*.

In addition to the converter system, the Topsøe loop comprises:

- Start-up heater
- Waste heat recovery by generation and/or superheating (especially in S-350 configuration) of high pressure steam and preheat of boiler feed water
- Feed effluent heat exchanger (hot exchanger) for preheat of the converter feed gas
- Water cooler in which part of the product ammonia is condensed
- Chillers at two temperature levels for further condensation of the product ammonia
- Gas-gas heat exchangers (cold exchangers) at the inlet to each chiller for recovery of the refrigeration energy
- Product ammonia separator

The waste heat recovery and the control of the temperature in the converter are integrated in a manner which ensures maximum plant efficiency under all operating conditions.

The make-up gas is added to the loop upstream of the last chiller, where most of the ammonia has already been condensed. Traces of carbon dioxide and water vapour are removed by co-condensation in the last chiller so that the risk of poisoning of the synthesis catalyst with these compounds is eliminated. In case of inert free make-up gas the make-up gas is mixed directly with the recirculating gas in the synthesis gas compressor.

If the make-up gas contains inerts, purge gas is drawn from the loop after the second cold exchanger just before addition of the make-up gas. At this point the gas has the maximum content of inerts. The purge gas is usually transferred to a purge gas recovery unit for recovery of hydrogen. Recovered hydrogen is recycled to the process and introduced at a relevant stage of the synthesis gas compressor.

The product ammonia is depressurised and taken to the let-down vessel where the main part of the dissolved gases is flashed off. The ammonia content of the flash gas is recovered either by absorption and distillation or by refrigeration.

From the let-down vessel the product ammonia is transferred either to further processing or to storage.

Recent literature about the Topsøe ammonia technology: *(44, 45, 46, 47, 48).*

KBR

KBR offer various schemes for production of ammonia. The most widely used converter is the KBR horizontal inter cooled ammonia converter using iron based ammonia synthesis catalyst. The ammonia loop is of the inert free type, because the synthesis gas is treated in a molecular sieve unit followed by a cold box (purifier unit). The make-up gas is mixed directly with the recirculating gas in the synthesis gas compressor. The ammonia chillers for condensing the ammonia product are combined in one unit (unitised chiller)

Alternatively, KBR offer the KAAP (Kellogg Advanced Ammonia Process) synthesis concept, where the synthesis gas is converted to ammonia in a low pressure synthesis loop (8 – 9 MPa) featuring a four bed synthesis converter loaded with conventional iron based catalyst in the first bed and Ru-based catalyst in the lower beds. This technology has so far only had limited use due to the high cost of Ru.

Recent literature about KBR's ammonia technology: *(49, 50, 51, 52).*

Uhde GmbH

The Uhde ammonia process features the same process steps and the same flexibility as described under the Topsøe process.

Apart from this, Uhde have recently commercialised a special ammonia synthesis concept, the Dual Pressure Process, especially for very large capacity plants. In this concept, compression of make-up gas is carried out in two steps, first in a two-stage inter-cooled compressor. This is the low pressure (LP) casing of the synthesis gas compressor. A three bed, inter-cooled, once-through make-up gas converter in this location, operating at about 10 MPa, can produce a third of the total ammonia. The effluent from this converter is cooled and most of the ammonia produced is separated from the gas.

In the second step, the remaining synthesis gas is compressed to the operating pressure of the synthesis loop (about 20 MPa) in the high pressure casing of the synthesis gas compressor. The high loop pressure is achieved by a combination of the chilled second casing of the synthesis gas compressor and a relatively high make-up gas pressure from the front-end.

Recent literature about the Uhde ammonia technology: *(53, 54)*.

Other Technologies

As mentioned above, Haldor Topsøe A/S, KBR, and Uhde GmbH share the major part of the market for new ammonia synthesis technology. For plant retrofits also others are active and especially for ammonia converter revamps, Ammonia Casale offer their ammonia synthesis technology. The main feature of the Casale technology is the so-called axial-radial flow synthesis converter *(55, 56)*, and *(57)*.

Prospectives

The ammonia synthesis is a rather mature technology. The process scheme has in principle been unchanged for many years. However, this does not mean that the technology has not been developed over the years. Just over the last 25 – 30 years the specific energy consumption for the overall ammonia plant has been reduced by around 30%, and still potentially approx. 30% can be gained before the theoretical minimum energy consumption is reached. The major part of this gain will, however, most probably be achieved in the synthesis gas generation part of the ammonia plant, and not in the ammonia synthesis itself, which accounts for "only" 14 % of the theoretical energy loss.

Generally, the ammonia synthesis is energy consuming because the synthesis reaction is favoured by high pressure. Ideally, it would be most beneficial if the ammonia synthesis could take place at the same pressure at which the synthesis gas is generated in the front-end. This would require a synthesis catalyst with an activity several orders of magnitude higher than known types *(58)*, in order to maintain a reasonable size of the ammonia synthesis converter. Even the Ru-based catalyst does not fulfil this requirement. With otherwise current technologies such a low pressure scheme would save the compression energy in the synthesis gas compressor. Most of this saved energy would, however, be lost again since the ammonia product is condensed by ammonia refrigeration, and the load on the refrigeration compressor would go up accordingly. Lowering of the loop pressure would also require larger equipment and pipe sizes, which in today's market with a tendency to significantly increase plant capacities, would be a direct drawback.

Since the ammonia synthesis reaction is limited by equilibrium, another obvious idea is to separate ammonia directly from its reactants – hydrogen and nitrogen for example in a membrane system. Several polymeric materials have been proposed that can separate ammonia from hydrogen and nitrogen *(59)*, however, such polymers are characterised by low selectivity and low permeability for ammonia, and they have poor thermal stability as well.

Researchers at the University of Bath are also working on finding suitable membrane configurations of operating conditions for effective recovery of ammonia from the synthesis gas loop. Tubular silica membranes as well as zeolite membranes have been tested with high ammonia permeance and selectivity *(60)*. The challenge is to find a 'cheap' membrane that can withstand the operating conditions in the synthesis loop where the operating temperature and pressure are high as in todays operating plants. It is as always a compromise between cost and selectivity.

However, as long as the permeate is not pure but contains both ammonia and hydrogen, the ammonia still need to be condensed and separated. This means use of energy for condensation, and energy for recycling the hydrogen back to the process. Hopefully future developments might result in better performance and thereby a potential to improve the current ammonia technology. In theory, if a membrane can be found with sufficiently high permeability and high selectivity that would allow all hydrogen and nitrogen to react and all the formed ammonia to be removed, a significant reduction in energy can be achieved in the ammonia synthesis loop (3.3 - 4.2 GJ/MT NH_3), if the membrane system can be operated at around 4 – 5 MPa with a highly active ammonia synthesis.

Alternatively, the separation of ammonia from hydrogen and nitrogen can be done using sorbents for ammonia removal. A liquid absorbent stream has to be pumped around, but the associated power requirement for pumping is lower than the power needed in the normal system based on recycling gas. In such a system chilling with refrigerated ammonia is not needed as usually, but instead some energy is required for desorbing the ammonia. The gained energy will be available as waste heat for steam production. Actual expected saving will be at most in the order of 0.8 – 1.3 GJ/MT NH_3 *(61)*.

Both the membrane system as well as the absorption system is only being studied in laboratory scale, and there is still a long way to commercialisation *(62)*.

Literature Cited

1. http://www.indexmundi.com/en/commodities/minerals/nitrogen/nitrogen_t12.html.
2. Dybkjær, I. In Ammonia - *Catalysis and Manufacture*, Nielsen, A. Ed., Springer-Verlag, New York , 1995, pp. 231-327.
3. Appl, M. In Ammonia – *Principles and Industrial Practice*. Wiley-Veh, Weinheim, Chichester, Brisbane. Singapore, Toronto 1999.
4. Liu, H.-Z.; Li, X.-N.; Hu, Z.-N. *Appl.Catal. A* **1996**, *142*, 209.
5. Pernicone, N.; Ferrero, F.; Rosetti, I.; Forni, L.; Canton P.; Riello, P.; Fagherazzi G.; Signoretto M.; Pinna F. *Appl. Catal. A* **2003**, *251*, 121.

6. Pinto, A.; Moss, J.M.S.; Hicks, T.C. *34th AIChE Ammonia Safety Symp.*, San Fransisco, 1989..
7. Kalenczuk, R.J. Int.J.*Inorg.Mater.* **2000**, *2*, 233.
8. Mittasch, A. *Adv.Catal.* **1950**, *2*, 81.
9. Ozaki, A.; Aiki, K.; Hori, H. Bull.*Chem.Soc.Jap.* **1971**, *44*, 3216.
10. Aiki, K.; Hori, H.; Ozaki, A. *J.Catal.* **1972**, *27*, 424.
11. Strait, R. *Nitrogen and Methanol* **1999**, *238*, 37.
12. Christensen, C.J.H.; Dahl, S.; Hansen, P.L.; Törnqvist, E.; Jensen, L.; Topsøe, H.; Prip, D.V.; Møenshaug, P.B.; Chorkendorff, I. J. *Mol.Catal. A* **2000**, *163*, 19.
13. Kojima, R.; Aiki, K. *Chem.Lett.* **2000**, 514.
14. Raróg-Pilecka, W.; Miskiewicz, E.; Kepinski, L.; Kaszku, Z.; Kielar, K.; Kowalczyk, Z. *J. Catal.* **2007**, *249*, 24.
15. Baranski, A.; Reizer, A.; Kotarba, A.; Pyrczak, E. *Appl.Catal. A* **1988**, *40*, 67.
16. Baranski, A.; Kotarba, A.; Lagan, J.M.; Pattek-Janczyk, A.; Pyrczak, E.; Reizer, A. *Appl.Catal. A* **1994**, *112*, 13.
17. Herzog, B.; Herein, D.; Schlögl, R. *Appl.Catal. A* **1996**, *141*, 71.
18. Schlögl, R. In *Catalytic Ammonia Synthesis, Fundamentals and Practice*, J.R. Jennings ed. New York, 1991, Plenum Press, 19.
19. Nielsen, A. *An Investigation of Promoted Iron Catalysts for the Synthesis of Ammonia*, 3rd edn. Copenhagen, 1968, Jul Gjellerups Forlag.
20. Stoltze, P. *Phys. Scr.* **1987**, *36*, 824.
21. Stoltze, P.; Nørskov, J.K. *J.Vac.Sci.Technol.* A5, **1987**, 581.
22. Højlund Nielsen, P.E. In *Ammonia, Catalysis and Manufacture* A. Nielsen, ed., Berlin Heidelberg, Springer-Verlag, **1995**, 191.
23. Emmet, P.H.; Brunauer, S. *J.Am.Chem.Soc.* **1933**, *55*, 1738.
24. Kozhenova, K.T.; Kagan, M.Ya. *J.Phys.Chem USSR*, **1940**, *14*, 1250.
25. Temkin, M.I.; Phyzev, V. *Acta Physicochim. USSR* **1940**, *12*, 327.
26. Brunauer, S.; Love, K.S.; Keenan, R.G. *J.Am.Chem.Soc.* **1942**, *64*, 751
27. Temkin, M.I. *Zhur.Fiz.Khim.* **1950**, *24*, 1312.
28. Ozaki, A.; Taylor, H.S.; Boudart, M. *Proc.Roy.Soc.(London)* **1960**, A258, 47.
29. Temkin, M.I.; Morozov, N.M.; Shapatina, E.N. *Kinetika I Kataliz* **1963**, *4*, 260.
30. Nielsen, A. *J.Catal.* **1964**, *3*, 68.
31. Emmet, P.H. In *The Physical Basis for Heterogeneous Catalysis*, E. Drauglis and R.I. Jaffee, eds. New York, 1975, Plenum, 3.
32. Weissg M.; Ertlg G.; Nitschkeg F. *Appl.Surf.Sci.* **1979**, *2*, 614.
33. Grunze, M.J.; Bozso, F.; Ertl, G.; Weiss, M. *Appl. Surf.Sci.* **1987**, *1*, 241.
34. Bowker, M.; Parker, I.; Waugh, K. *Appl. Catal.* **1985**, *14*, 101.
35. Stoltze, P.; Nørskov, J.K. *Phys.Rev.Lett.* **1985**, *55*, 2502.
36. Boudart, M. *Adv. Catal.* **1969**, *20*, 153.

38

37. Spencer, N.D.; Schoonmacher, R.C.; Somorjai, G.A. *J.Catal.* **1982**, *74*, 129.
38. Dahl, S.; Logadottir, A.; Jacobsen, C.J.H., Nørskov, J.K. *Appl.Catal. A*, **2001**, *222*, 19.
39. Dybkjær, I.; Jarvan, J.E. Advances on Ammonia Converter Design and Catalyst Loading. Nitrogen 97 Conference, 9-11 Feb. Geneva, Switzerland.
40. http://www.topsoe.com/site.nsf/all/BBNN-5PFH9M?OpenDocument
41 http://www.kbr.com/industries/energy_and_chemicals/downstream/ ammonia_and_syngas/ammonia_process_technologies.aspx
42. http://www.uhde.biz/competence/technologies/fertilisers/techprofile. n.epl?profile=2&pagetype=1&pagenum=1.
43. Nielsen, S.E. *Ammonia Plant Saf.* **2004**, *45*, 168-176.
44. Nielsen, S.E.; Kato, H. The Largest Ammonia Plant in Asia – the 2000 MTPD Topsøe-designed Ammonia Plant for P.T. Kaltim Pasifik Amoniak. AsiaNitrogen 1998, February 22-24, Kuala Lumpur, Malaysia.
45. Nielsen, S.E. *Ammonia Plant Saf.* **2002**, *42*, 304-311.
46. Nielsen, S.E. (Latest Developments in Ammonia Production Technology, p. 1 – 14, FAI International Technical Conference on Fertiliser Technology, April 12 -13, 2007, New Delhi.
47. Okuzumi, N.; Jones, B.D.; Nielsen, S.E. Start Up of the World's Largest Ammonia Plant. *Nitrogen 2001*, February 18-21, Tampa Bay, Florida, USA 49.
48. Okuzumi, N.; Kjær, H.G. *Ammonia Plant Saf.* **2002**, *43*, 271-279.
49. Abughazaleh, J.; Gosnell, J.; Strait, R. *Ammonia Plant Saf.* **2002**, *43*, 258-264.
50. Czuppon, T.A.; Knez, S.A.; Strait, R.B. *Ammonia Plant Saf.* **1997**, *37*, 34-49.
51. Malhotra, A.; Hackemesser, L. *Ammonia Plant Saf.* **2002**, *42*, 223-229.
52. Jovanovic, W.; Malhotra A., *Ammonia Plant Saf.* **2006**, *47*, 267-276.
53. Larsen, J.; Lippmann, D.; Hooper, C.W. *Nitrogen & Methanol* **2001**, *253*, 41-46.
54. Larsen, J.S.; Lippmann, D. *Ammonia Plant Saf.* **2002**, *43*, 80-89.
55. Davey, W.L.E.; Wurzel, T.; Filippi, E. *Nitrogen & Methanol* **2003**, *262*, 41-47.
56. *Nitrogen* (1994) A combination of proven technologies. 208, 44-49.
57. Iob, S. Ammonia Casale Technologies and Case Histories, p. 15 – 29, FAI, International Technical Conference on Fertiliser Technology, April 12 -13, 2007, New Delhi.
58. Dybkjær, I.; Gam, E.A. *Ammonia Plant Saf.* **1985**, *25*, 15-18.
59. Laciak, D.V.; Quimm, R.; Pez, G.P.; Appleby, J.B.; Puri, P.S. *Separation Science and Technology* 1990, Vol. 25, Issue 13-15, 1295-1305.
60. Camus, O.; Perera, S.; Crittenden, B.; van Delft, Y.C.; Meyer, D.F.; Pex, P.P.A.C.; Kumakiri, I.; Miachon, S.; Dalmon, J.-A.; Tennison, S.;

Chamanol, P.; Groensmit, E.; Nobel, W., *AIChE Journal*, 2006, Vol. 52, Issue 6, 2055-2065.

61. Westerterp, K.R.; Bodernes, T.N.; Vrigland, M.S.A.; Kurzynski, M., "Two New Methanol Converters", Hydrocarbon Processing, 1988, p. 673.

62. Jeroen de Beer, Potential for Industrial Energy - Efficiency Improvement in the Long Term (1998).

Chapter 3

Innovation and Enterprise: The Industrial Gases Industry in the United States

John Royal

Director, Cryogenic Equipment Technology, Praxair, Inc., 178 East Park Drive, Tonawanda, NY 14150

The centennial of the ACS's Industrial and Engineering Chemistry Division marks a century of accomplishment during which one of humanity's grand technical achievements, the American chemical industry, grew to maturity. Driven by society's needs for goods and services that enhanced the quality-of-life, chemists and chemical engineers produced ever-more innovative products to economically meet these demands.

The rise of the American industrial gases industry is part of this story of innovation and enterprise. Beginning from nothing at the beginning of the 20th century, the industry grew to become a critical sector of an advanced economy. See Figure 1.

By the end of the 20th century, the U.S. industrial gas industry's production of nitrogen and oxygen, taken together by mass, had become the largest commodity chemical produced in the U.S. economy (1).

Roots

Like the chemical industry, the industrial gases industry arose from a social need -- oxygen. Oxygen had been recognized as an element in the mid-18th century (2). By the mid-19th century, oxygen was being produced commercially. Applications were mainly theatrical lighting, which used an oxy-hydrogen flame. The flame was produced by heating quicklime to generate hydrogen (the origins of the term "limelight"). Modest amounts of oxygen were also used for medical applications.

Oxygen for these purposes was produced chemically. Thermal decomposition of potassium chlorate, sodium nitrate, or manganese dioxide was typical (4).

42

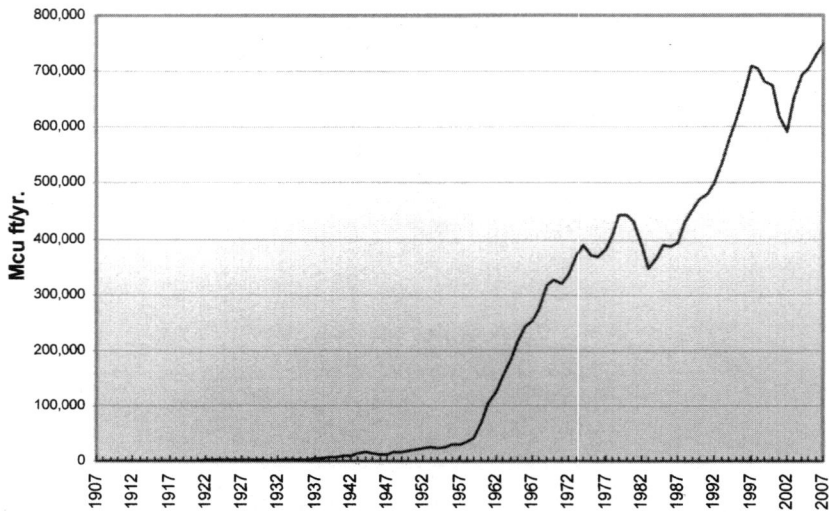

Figure 1. 100 Years of U.S. Oxygen Production (2, 3)

By the end of the 19th century, another grand technical achievement, the ferrous metals industry, had created the "Iron Age." The industry was producing products so efficiently that iron and steel were in use throughout society. However, the properties of these products created problems. Once the steel had left the mill, it was very difficult to shape and attach. Cutting was done mechanically, and fastening was primarily by riveting; both of these processes were laborious and limited in application.

In 1895, Henri Le Chatelier (1850-1936) discovered that an acetylene-fueled oxygen flame produced the highest-temperature flame then known, 3200°C (5). This discovery was recognized as having the potential to solve the metals industry's problems with shaping and fastening steel. However, the lack of adequate supplies of oxygen created difficulties.

Technical innovators addressed this opportunity. They recognized that a significant new market was available; all that was needed was production technology appropriate to the scale and costs required.

Technical Convergence

By the late 19th century, oxygen was available commercially, but production methods were expensive and production was limited to small volumes. First, thermal dissociation and electrolysis were used; later, a reversible reaction of barium oxide/barium dioxide was used to separate oxygen from air (6).

By the end of the century, technology to reach the temperatures required to liquefy air was available. Further, scientists discovered that liquid air could be distilled to produce oxygen.

Low-temperature refrigeration technology had emerged. The leading physicists of the day each tried to out-do the others in reaching lower temperatures. This scientific competition spurred the development of more refrigeration techniques, making possible lower temperatures and more refrigeration power.

By 1895, air was liquefied by compressing it to high pressures (100 to 200 bar), pre-cooling it to the lowest possible temperature, and expanding it through a throttle valve (7). In 1895, Carl von Linde (1842-1934) of Germany and William Hampson of Great Britain introduced the notion of regeneratively cooling the high-pressure air against the expanded gas (Figure 2) (8, 9). Soon, Linde's Gesellschaft fur Linde's Eismaschinen was selling air liquefying equipment.

In the United States, an entrepreneur, inventor, and perpetual motion enthusiast, Charles E. Tripler (1849-1906) (10), independently developed a similar technique. Soon, Tripler was producing liquid air in multi-liter quantities and shipping it in insulated containers across the United States (11).

An effective publicist (12), he raised $10 million and started the General Liquid Air and Refrigeration Company in 1899 (13). But his bankruptcy in 1902 and his association with perpetual motion soured the investment climate in the U.S. on further interest in industrial gases (14).

In France in 1902, Georges Claude (1870-1960) (15) and his colleagues developed a practical piston expansion machine capable of reliable operation at the temperatures required to liquefy air (16) (Figure 3). Recognizing that the efficiency improvements offered by this technology for the production of lower temperature refrigeration could provide a commercial advantage, Claude and his partner, Paul Delorme, formed the French industrial gas company L'Air Liquid that same year (17).

With both Air Liquide and Gesellschaft fur Linde's Eismaschinen's air liquefiers well established, the European cryogenic industry was launched. Further, their competition stimulated intense technical development that rapidly improved the state of cryogenic technology. However, attempts to produce high-purity oxygen took longer to reach fruition.

Air liquefaction made it possible to distill air. Attempts to do so using a single distillation column produced a gas product consisting of 40 percent to 50 percent oxygen. Cleverly exploiting his patent for this process, Linde marketed this oxygen-enriched gas as "Linde Air" (19).

While important to some chemical processes, Linde Air was insufficiently oxygen-pure for welding and cutting. Spurred by the demand, effective but inefficient processes for producing pure oxygen were developed by industry entrepreneurs.

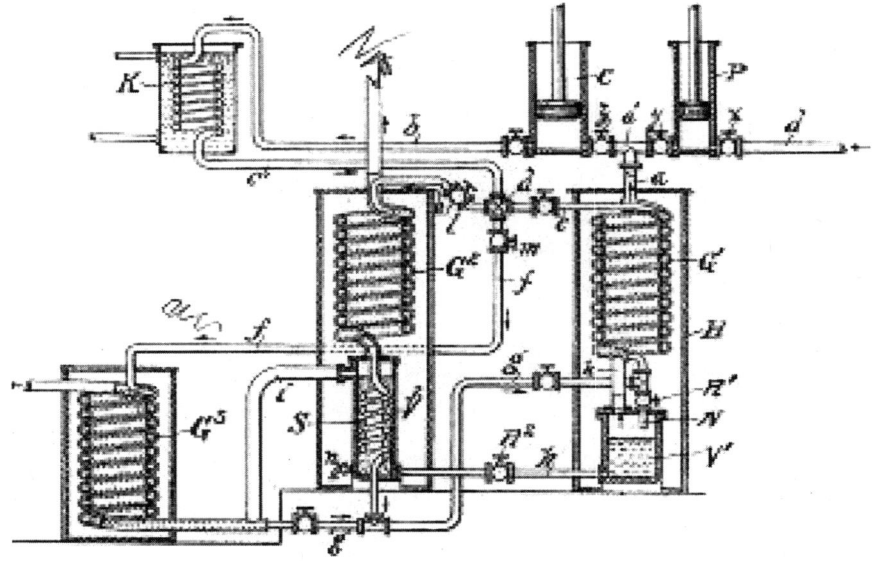

Figure 2. Linde's U.S. Liquefier Patent (8)

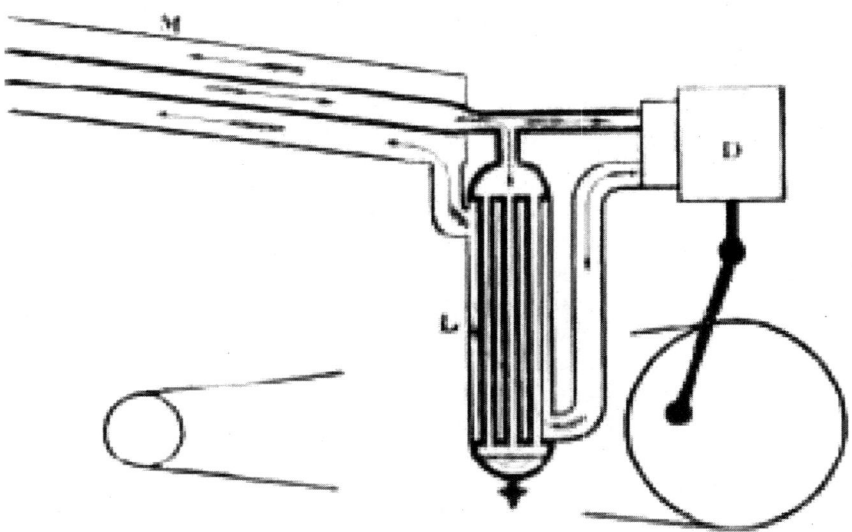

Figure 3. Claude's Liquefier (Reproduced from reference 18. Published 1919 Van Nostrand Company.).

By liquefying high-pressure air, reboiling the bottoms of a single column, and throttling the liquid air to reflux the column, Linde was able to produce nearly pure oxygen (*20*). Unfortunately, the process recovered so little of the oxygen contained in the feed air that the oxygen was expensive to produce.

Claude soon answered with an improvement (*21*) (Figure 4). By replacing the reboiler with a reflux condenser, producing an oxygen-enriched stream and a nitrogen-enriched stream, throttling these, and delivering the streams to a single distillation column as reflux at appropriate points, more of the oxygen contained in the feed was recovered.

Still, large-scale use required more efficient means of production. The search for economical, commercially significant quantities of pure oxygen continued.

In 1910, Linde found the answer: the double distillation column (Figure 5). The double-column system improved the production of high-purity oxygen by dramatically increasing the fraction of oxygen produced from the feed air (*22*). If one invention can be said to have created an industry, this one created the air separation industry.

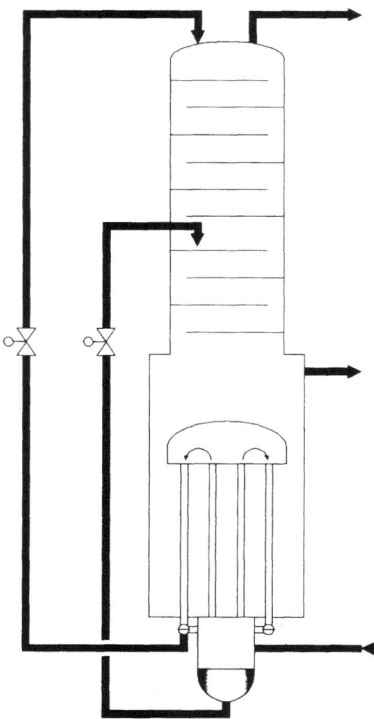

Figure 4. Claude's "retour en arrière"

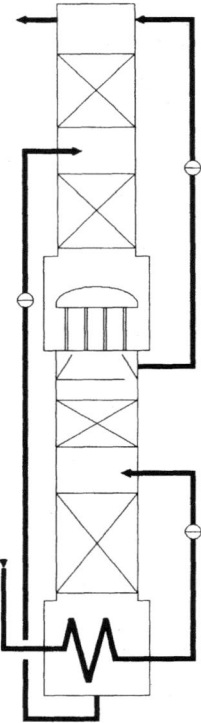

Figure 5. Linde's Double Column

The technologies were in place: adequate refrigeration to liquefy air in large quantities and the processes to distill it efficiently to produce pure-enough oxygen. The emerging welding industry could be served, but still served expensively.

Oxygen production by distillation is energy-intensive due to the low temperature requirements. Linde's system was particularly so because of the use of Joule-Thompson throttling to produce the refrigeration.

Combining Claude's expander technology with the double column brought air separation technology to a standard of effectiveness that made possible the growth of the modern air separation industry and the widespread applications for its products (Figure 6). The stage was set for the rapid growth of a new industry.

While oxygen was the primary product of interest, early practitioners soon learned to produce high-purity nitrogen and argon. Early single-column cryogenic distillation systems were capable of producing essentially pure nitrogen in large volumes. Rapidly growing chemical industries soon found uses for nitrogen as an inerting agent, a safe pressurizing fluid, and a safe pneumatic transport agent.

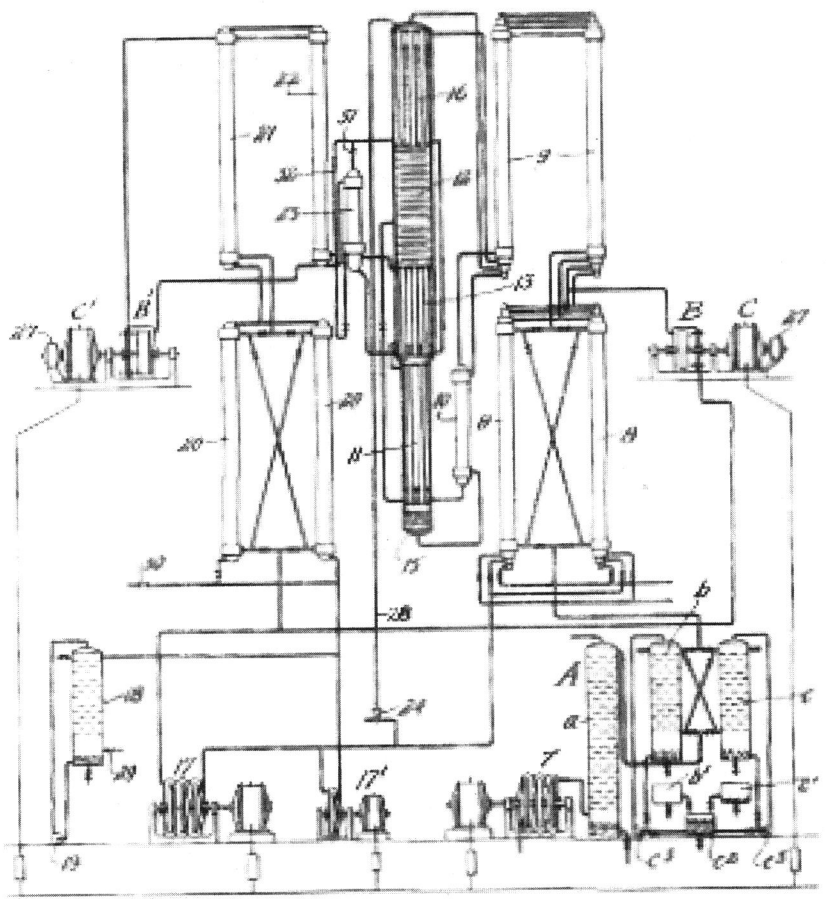

Figure 6. Classic Early ASU Process Schematic (23)

Soon, argon was also distilled from air, first as a scientific curiosity, then as an important product used in welding and metallurgical applications. This rare gas became and remains today a valuable byproduct of oxygen separation.

Roots of the Air Separation Industry in the United States

Among the industry's pioneers, Linde was one of the most visionary commercial and technical innovators. Early in his career, he recognized the industrial importance of his air-separation-related inventions and sought to commercialize them internationally through licensing.

Linde long had created a business based on technologies he pioneered. His creation of Gesellschaft fur Linde's Eismaschinen in 1879 allowed him to exploit Linde's refrigeration patents for use in the brewery industry and elsewhere. This was the beginning of today's Linde Group.

In 1906, Linde licensed his patents to Brin's Oxygen Company (BOC) in exchange for company shares and a seat on the board of directors. (The company later changed its name to the British Oxygen Company.) In 1907, BOC (as Linde's licensee) and Claude's Air Liquide were involved in a series of patent disputes that culminated in settlements that permitted their respective companies to thrive. Linde also enforced his patents against both German and Swiss competitors.

Building a global enterprise around his air separation patents led Linde to all the major industrial economies of the time. Of particular interest is his foray into the United States. In 1906, he sent Cecil Lightfoot (24) to the United States to evaluate the market, select a business location, and raise start-up capital.

Initial attempts were unsuccessful. The failure of Tripler's General Liquid Air and Refrigeration Company discouraged American investors' interest in cryogenics-based technologies. However, Lightfoot did confirm that a large market potential existed. He also located an ideal location for an American air separation works in Buffalo, N.Y. Located at an important transportation nexus and close to Niagara Falls (the largest source of low-cost electric energy in the United States), Buffalo seemed to be exactly what Linde was looking for.

In late 1906, Linde traveled to the United States. Determined to establish a U.S. air separation company, he inspected Lightfoot's site and purchased the land. Early in 1907, he persuaded a group of businessmen to invest in the Linde Air Products Company. With the company's incorporation in January 1907, the American air separation business was firmly rooted (25).

An oxygen plant soon was installed on the Buffalo site. Growth was rapid, and this first air separation unit was soon sold out. More German-sourced equipment was imported. Linde returned to the United States to supervise its installation.

Hands across the Sea

In 1910, Joe Fuzy, one of the first employees at the Linde Air Product's Company's new East Chicago plant, encountered Linde. Fuzy, an 18-year-old Midwesterner, worked with the professorial German international captain of industry installing new German-made air separation equipment on the plant floor.

Although Linde understood English, he spoke it with difficulty. Communication between Linde and Joe as they worked assembling the new complicated machinery was challenging. Eventually, they developed a system; according to Joe, it consisted mainly of the use of the phrase: "Ach so" (26).]

In 1912, the first domestically built air separation unit was produced in the Buffalo shops (27).

Other companies began entering the now booming business. Rockefeller interests joined with Air Liquide to form the Air Reduction Company in 1916 (28).

By the outbreak of World War I, there were approximately 20 air separation plants in the United State. (29). Growth had been dramatic; by any standard, air separation had become a growth industry. Figure 7 traces the early growth of oxygen production in the United States.

World War I had important consequences for the American industrial gases business. The war led to the expansion of the U.S. metalworking industry; initially to meet the needs of increased weapons production to support European and then, eventually, American involvement in the war. This expansion dramatically increased demand for oxygen for welding and other metalworking applications.

The war changed the industry's ownership. Partial ownership of the Linde Air Products Company by Linde ended when America and Germany went to war. In 1917, driven by a growing metalworking market, four companies joined to form the Union Carbide and Carbon Company (31). Union Carbide Company brought calcium carbide production to the combination. Prest-O-Lite brought acetylene cylinder and acetylene handling technology. National Carbon brought carbon, the precursor for the manufacture of calcium carbide. The Linde Air Products Company brought oxygen. The Union Carbide and Carbon Company thereby brought together under one organization the resources to serve this market. In 1992, the industrial gas division was spun off from Union Carbide and became today's Praxair.

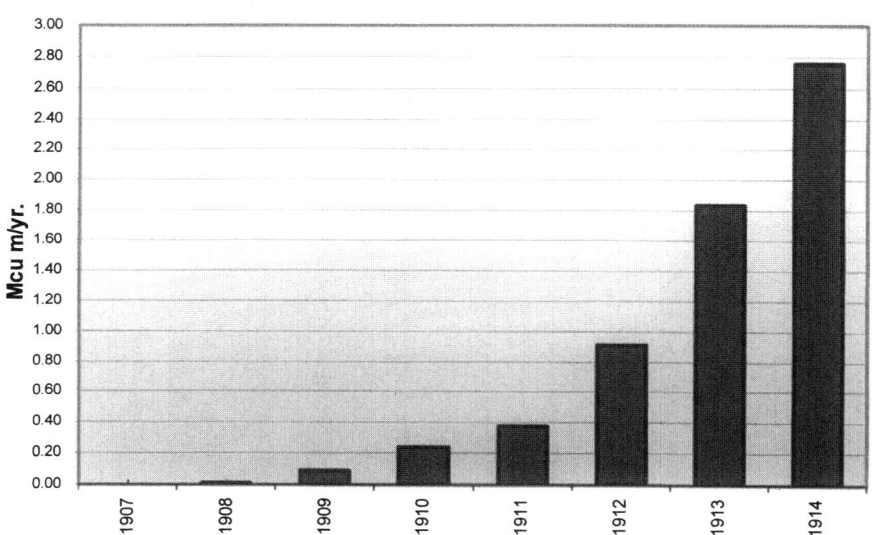

Figure 7. Early U.S. Oxygen Production (30)

Innovation Changes the Business Model

Market growth began to challenge the existing business model of oxygen delivery in pressurized cylinders. Borrowed from the pre-existing carbon dioxide business, gas delivery by pressurized cylinders was a well-established practice from the beginning of the industry.

As demand for oxygen mounted, air separation companies replaced their horse-drawn wagons with trucks and rail cars. Gas-cylinder technology was advanced, enabling increased gas pressure in the containers up to the metallurgical limits of the time. Manufacturing of such high-pressure gas cylinders was a noteworthy technology achievement of the age.

Still, demands for oxygen were constrained by the economics of supply; multiple truck or railcar deliveries of numerous cylinders could only meet so much demand. Another approach was necessary.

The industry's technical community understood that the answer lay in supplying liquid to their customers. Liquid supply would enable substantially larger quantities of gas to be delivered by truck or rail. Liquid provided approximately 7.5 times the gas volume per weight transported when compared to cylinders. For example, a state-of-the-art cylinder weighing 70 kg, filled to 150 bar, provided approximately 6.7 m^3 (9 kg) of oxygen at standard temperature and pressure (STP); a 70 kg mass of liquid oxygen provided almost 50 m^3 of gas. Even discounting for the necessary liquid containment, the economic and scale appeal of liquid supply was compelling.

But technical challenges remained. The efficiency of liquid production needed to be improved, safe and effective means of delivering cryogenic liquids were needed, and systems for containing, preserving, and evaporating the delivery liquids on demand were required. Paul Heylandt's (1884-1947) (*32*) pioneering work in Germany to create a viable liquids-based technology was instrumental in overcoming these challenges.

Liquid production required more energy-efficient technologies than gas production. While the expensive refrigeration required to separate air could be recovered in recuperative heat exchangers, the refrigeration required to produce cryogenic liquids was exported from the plant with the product. Efficient compression, efficient expanders, improved insulation, highly effective heat exchangers, and, above all, effective process arrangements, were required.

Heylandt's work addressed many of these areas (*33, 34*). By the 1920s, he had developed an effective technology to make a liquids business practical. Union Carbide licensed his technology and imported his equipment into the United States. Accompanying his equipment, Heylandt assisted with its installation and start-up.

Early Memories

"You will of course remember old 'Doc' Heylandt, his brother-in-law Schneider, and the Swicau Compressor

Company erector, Guenther, and the inability of these three to speak English, which resulted in some amusing incidents. I'm sure you will remember Guenther's shopping trip in which he attempted to buy a nice linen table cloth for his wife and the lack of success because he asked all the sales clerks for a bed spread for his table.

Do you still have a picture of Heylandt and his wife going shopping? Heylandt being 6'3" or 6'4," weighing about 230 or 240 pounds, followed one step behind by his wife who was short and carrying all the packages? (*35*)

To enable a successful liquids business, technologies for the safe transportation, stationary storage, and evaporation of cryogenic fluids also were required. These, in turn, required improved insulation and an engineering knowledge of materials properties at cryogenic temperatures. Union Carbide's Linde division hired Leo Dana (1896-1990) of the University of Leyden (*36*) to develop solutions to these problems.

Tank-based systems that would reliably evaporate liquids at pressure and deliver warm gas to the customer were instrumental in the growth of the liquids business. Dana developed equipment for vaporizing liquid using atmospheric temperatures to make possible simple, low-cost systems (Figure 8).

Managing heat leak and liquid withdrawal rates from cryogenic storage tanks while maintaining delivery pressure was a challenge. Clever engineering was required to create low-cost, durable systems that could be left exposed to the elements for many years. Figure 9 shows one of Dana's early patents addressing this challenge.

Advances in cryogenic insulation technology also enabled more efficient storage systems. Slivered glass vacuum dewars set the standard for insulating power but were too fragile to meet the demands of scale and ruggedness required for a vibrant liquids business. Metal systems of appropriate low-temperature-tolerant materials were required. Double-wall insulation systems using mineral wool, sawdust, and later, perlite, in the insulation space were found effective. Techniques were developed to desiccate the insulating materials to improve their long-term performance. Later, as manufacturing techniques improved, vacuum was applied to the annular space, further improving insulating power. Figure 10 illustrates an early vacuum-insulated system.

With the development of effective liquid production, a second business model was established. In the old business model, gas was sold by the cylinder with the customer paying for the gas and a modest, time-based rent for the cylinder itself. With the development of the liquids business, a similar model evolved: The customer purchased gas outright by volume and rented the

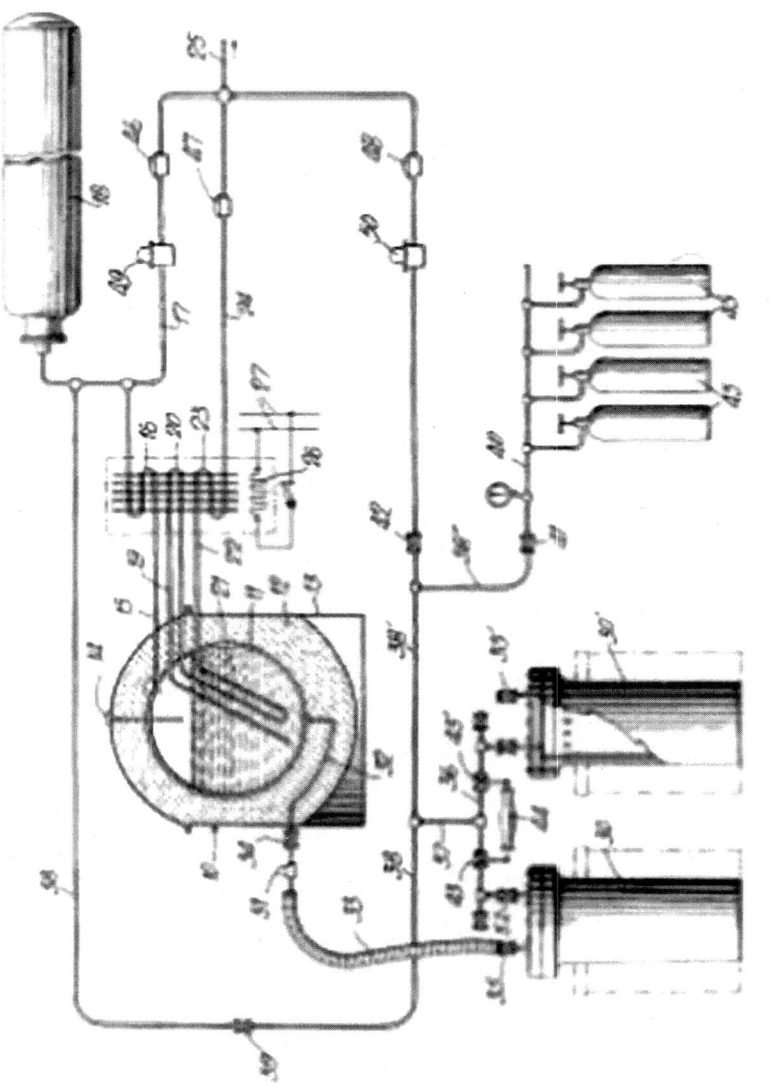

Figure 8. Apparatus for Dispensing Gas Material (37)

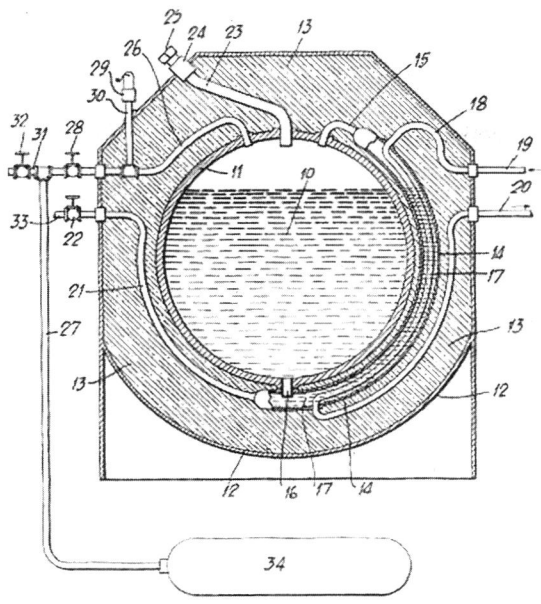

Figure 9. Method and Apparatus for Dispensing Gas Material (38)

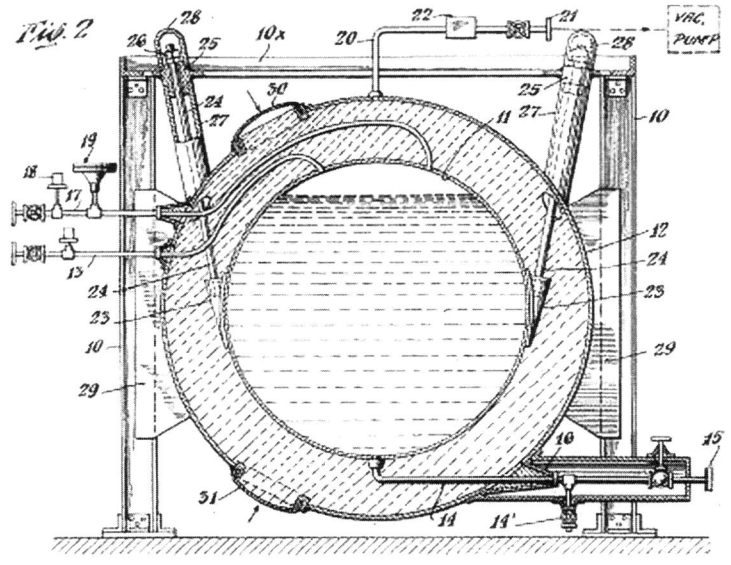

Figure 10. Insulated Container for Liquefied Gases and the Like (39)

necessary storage tanks and vaporizer hardware necessary for storage and use of the purchased product.

Because the storage tanks and evaporation equipment placed at a customer site represented a significant capital investment by the gas suppler, liquid supply contracts soon came to cover several years. This longer-term arrangement allowed the gas supplier to earn a competitive return on their investment in the equipment while offering the customer a competitive rental fee.

The liquids business, with its ability to economically supply unprecedented volumes of oxygen and other air gas constituents (nitrogen and argon), met the growing demands. Its large-volume capability and attractive product pricing led to expansion of the technology into new markets.

Adding to the still-growing metalworking and healthcare markets, chemical industry applications for oxygen and nitrogen for inerting purposes became economical. Oxygen and argon for metals production were also available economically. These demands accelerated industry growth and greatly increased production volumes. See Figure 11.

Innovation Alters Production Scale

Growing demand spurred innovations to increase production rates and reduce unit costs. As the industry shifted from an empirical basis to one grounded in science, understanding the physical and transport properties of the feedstock and products became more important. As this became apparent, some industry participants created physical property laboratories to determine these properties.

Accurate property data, which was carefully gathered and carefully protected, was an important competitive advantage for those who owned it. Such knowledge enabled more cost-effective designs, shorter development time, and, most importantly, more certainty of the actual performance of a proposed new plant.

Improvements in the removal of water and carbon dioxide from the feed air to an air separation plant also reduced costs. The feed air to a cryogenic air separation plant must be dehydrated and the carbon dioxide removed; otherwise, as the feed cools to cryogenic temperatures, these contaminants will condense, freeze, and plug heat exchanger passages, thereby shutting down the plant.

From the earliest days of industrial-scale air separation, this issue had been dealt with by chemical means. Typically, carbon dioxide was removed from the feed air using lime or caustic soda. Water was removed using caustic potash. The bed used for these purposes was replaced when expended; the expended beds were either disposed of or regenerated (40). Later, activated alumina and/or silica gel beds were used to remove water from the feed air.

As air separation units grew in size, these means of air feed clean-up were increasingly cumbersome. Other technologies were required. In the early 1930s,

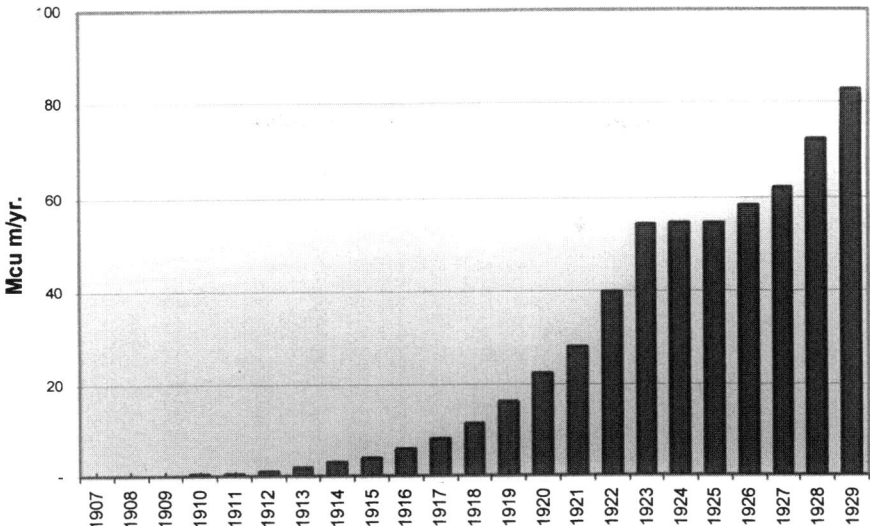

Figure 11. U.S. Annual Oxygen Production to the Depression (2, 3)

Mattias Frankl of Germany (1877-1947) (*41*) developed a regenerator with separate passages for the clean oxygen product (Figure 12).

In Frankl's system, two beds of fill were used: one was used in cleaning service while the other was being cleaned. The in-service bed received high-pressure air from the feed air compressor. This air flowed through the bed, cooling against the refrigeration stored in the bed. As the air cooled, first water, then carbon dioxide, condensed and froze on the fill. Free of contaminants, the feed air could be further cooled and processed without plugging the process equipment.

As the high-pressure air deposited water and carbon dioxide ices, the regenerator became plugged with ice. Before the air flow was completely blocked, the regenerator was "reversed." First, the high-pressure feed air was diverted to the second bed, which was now ice-free, having been previously cleaned. The now-fouled first bed was depressurized, and cold waste gas from the cryogenic distillation column was introduced into the cold end. This cold gas, roughly 80 percent of the feed air, proceeded through the bed, warming against the energy in the fill and evaporating the ices. Although the waste stream had less flow than the feed air, the lower pressure of the stream enabled it to evaporate all the water and carbon dioxide deposited by the feed air. The first bed, now completely clean and refrigerated, was ready to receive feed air on the next regenerator reversal. Because the oxygen product was required to be free of water and carbon dioxide, the product was segregated from the reversing streams in tubular passages imbedded in the regenerator's fill.

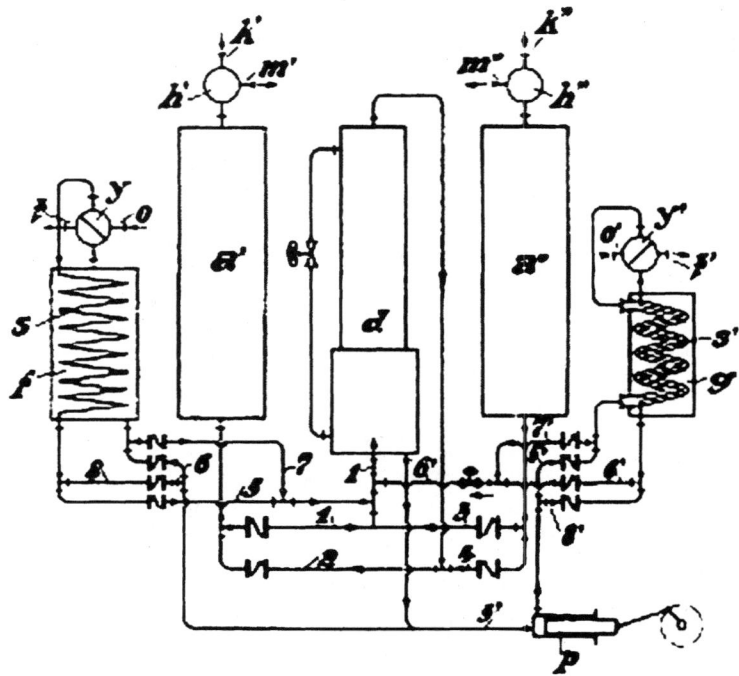

Figure 12. Process of Separating Gas Mixtures (42)

Once perfected, air separation companies licensed the Frankl technology and soon deployed it widely. Taking advantage of the greater evaporating power of the lower pressure waste stream, additional clean product passages could be added, enabling the production of modest quantities of high-purity nitrogen.

Another important innovation emerged during the late 1930s: Pjotr L. Kapitza (1894-1984) (*43*) perfected the turbo-expander for cryogenic service (*44*). The higher thermodynamic efficiency of the turbine allowed more energy-efficient production of refrigeration than the reciprocating expansion engines that were currently in use. This more efficient refrigeration production was manifest as lower feed air compressor head, and, consequently, lower energy input into the process. Lower energy input reduced product unit cost, further driving consumption.

More sophisticated air separation system designs, made possible by better property knowledge, better feed air clean-up systems, and improved energy efficiency, allowed the air separation industry to meet the growing demand for its products while reducing product costs. In the two decades between World War I and World War II, air separation plants grew significantly. By the early 1940s, "mass" plants capable of producing up to 500 tons per day of oxygen were in production (*45*).

Reduced product costs further drove demand. By 1941, U.S. production of oxygen was 228 million Nm^3 per year (Figure 13). The U.S. air separation industry was now a major contributor to economic growth, and it was poised for even more rapid expansion.

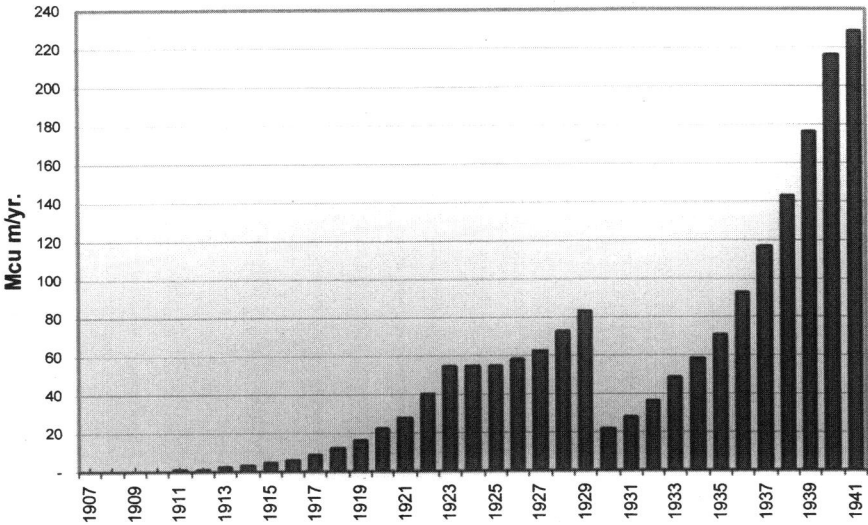

Figure 13. U.S. Annual Oxygen Production to World War II (2, 3)

Entrepreneurship Drives Technical Change

By the late 1930s, the profitable American air separation industry had a mixed structure consisting of large enterprises, such as Union Carbide and the General Air Reduction Company, along with many local firms. Products were sold in cylinders and as liquids; only modest volumes of product were sold as gas through pipelines.

The industry's profitability and mixed structure attracted new entrants. Among them was Leonard Pool (1906-1975) (*46*). Early business experiences led him to the industrial gases industry. As he learned the business, he envisioned an opportunity for a small gas oxygen plant to replace delivered cylinder and liquid oxygen.

Determined to make his dream a reality, Pool raised capital and formed Air Products Inc. in 1940 (*47*). He and his colleagues then developed a compact, effective cryogenic distillation plant optimized for the production of gaseous oxygen. But keeping the fledgling firm solvent was a struggle.

Military needs driven by the United States' participation in World War II created the opportunity Pool needed. Rapidly developing technology allowed aircraft to fly high enough to require air crews to need on-board oxygen. To supply this oxygen, liquid was carried aboard the aircraft and evaporated on demand to supply the air crew.

However, to supply the liquid, aircraft support facilities, including those in remote locations and at sea, needed sources of oxygen. Air Products' small air separation plants soon were adapted to this service and supplied to the U.S. military.

When the war ended, the military market was dramatically reduced. But Air Products had a capable oxygen plant, now fully mature, well-tested, and in large-scale production.

Pool realized he could compete with his larger, well-established competitors by leasing his plants to customers for a long-term lease, a practice analogous to the long-established leasing of liquid tanks and their associated infrastructure. Using this business model, the small plant could compete with delivered liquid oxygen, even when the customer paid for the electric energy required for operating the plant. This practice of placing the plant on the customer's site, running it with customer power, and supplying its output to the customer at a fixed lease regardless of the customer's demand, established the on-site plant concept.

Once again, an entrepreneur added to the American air separation industry's business. By "selling the milk, not the cow," Pool added another means of effectively selling industrial gases (*48*).

Post-War Market Needs and the Demands for Scale

Industrial gas companies would soon benefit from a confluence of technological innovations. These innovations increased the demand for air gases, created a dramatic increase in the production of gases, and placed the industry on a new growth trajectory.

Devastated by war, the Old World's industry lay in ruin, just as post-war rebuilding of the combatant's infrastructure became a social priority. American steel producers shouldered the load, and the country's war-driven expansion continued.

Soon after the war, innovations in steel making began to affect the demand for oxygen. The basic oxygen furnace began to be adopted on a large scale in the United States. Compared to the then conventional open-hearth furnace, a basic oxygen furnace, using oxygen in place of air, allowed pig iron to be refined into steel more efficiently and with better quality. This innovation ignited an unprecedented demand for oxygen.

Open-hearth technology uses only oxygen present in the blast air; no air separation is necessary. But to supply oxygen for basic oxygen furnaces, large air separation facilities are required. Today, modern basic oxygen steelmaking uses almost two tons of oxygen per ton of steel.

In 1955, approximately 90 percent of U.S. steel was produced in open-hearth furnaces. Twenty years later, more than 65 percent of U.S. steel was produced in basic oxygen furnaces. By 1985, less than 10 percent of U.S. steel was produced in open-hearth furnaces (*49*). Demand for oxygen increased accordingly.

Hundreds and even thousands of tons per day of oxygen production capacities were required at steel mills. Since delivered liquid oxygen was insufficient to meet those needs, it was imperative to build plants at customer sites; thus, the on-site business model became the norm.

Driven by civilian demand and reconstruction needs, the chemical industry also expanded. This increased demand for industrial gases: oxygen for capacity increases and nitrogen for inerting. Soon, on-site plants were required at chemical complexes, too.

The new cryogenic gas plants at customer sites led to a new paradigm. For modest additions of capital, incremental liquids production capabilities could be added to on-site plants. Often, this placed low-cost liquid production closer to liquids customers, thereby lowering the cost of delivery to serve them. This decentralization of liquids production enabled meaningful expansion of the liquids business, further driving industry growth.

Rapid expansion accelerated growth for all industry players. Demand for oxygen and other air gases grew faster than the U.S. Gross Domestic Product (GDP) (Figure 14). Union Carbide and Air Reduction (now renamed Airco) benefited. Air Products emerged from start-up status and became well-established. Smaller players like Big Three Industries of Texas became regional powers, and Liquid Carbonic, the leading carbon dioxide supplier, entered the air gases business.

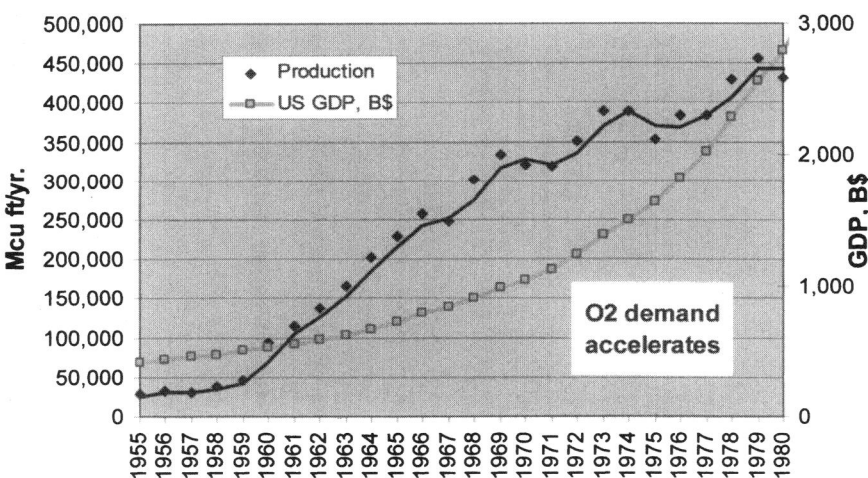

Figure 14. U.S. GDP and Oxygen Production (2, 3)

Innovation Responds to Market Needs

As demand escalated, new technologies emerged to enable larger plant sizes. In the mid-1950s, the brazed aluminum heat exchanger was developed. This technology allowed thousands of aluminum parts to be simultaneously assembled by stacking them and then dipping the stack into a salt bath.

This technology was near revolutionary. At first, only small heat exchangers were available. These were hybridized with regenerators to accommodate clean products. Later, as manufacturers grew more confident, sizes increased. By the mid-1960s, regenerators were replaced by brazed aluminum heat exchanger batteries in the latest plant designs.

The feed air clean-up services provided by the regenerators also were provided by the heat exchangers. Like their predecessor regenerators, heat exchangers removed feed contaminants by freezing and could then be reversed to restore performance.

In addition, brazed aluminum heat exchangers could be very complex, accommodating many streams, which allowed them to manage both single- and two-phase flow and permitted fluid removal or introduction anywhere along their length. This ability to manage complex heat transfer schemes inexpensively removed constraints from process designers and enabled more thermally integrated process schemes. This significantly improved separation efficiencies.

Driven by increasing demand for large volumes of nitrogen, air separation plant designers also began to search for ways of circumventing the nitrogen production limits inherent in regenerator- and reversing-heat-exchanger-equipped systems. Both technologies limited nitrogen production because adequate flows of low-pressure gas for ice clean-up were needed. For example, in an oxygen-only plant, almost 80 percent of the high-pressure feed air is available at low pressure to clean the off-line unit.

Adding the delivery of clean nitrogen as a product reduces the amount of low pressure gas available for cleaning. The physics of the clean-up process thus limits the amount of nitrogen it is possible to produce.

A solution emerged in the mid-1950s. Robert Milton and his colleagues, working at Union Carbide's Tonawanda, N.Y., R&D facility, developed synthetic zeolite molecular sieves. Improved and adapted to air separation service, this technology allowed expanded nitrogen production for air separation units.

Properly designed zeolites adsorbed both water and carbon dioxide to low enough levels to permit the processing of feed air without risk of ice deposition in the process equipment. By using two beds in the familiar "one in service, one cleaning up" system, a continuous supply of clean feed air could be maintained. The off-line bed could be cleaned using much less waste gas by heating the waste gas to a modest temperature. By reducing the waste gas requirements, more nitrogen was made available as product. Ultimately, this zeolite-based prepurification system permitted cryogenic air separation units to produce nitrogen and oxygen in the ratio of 3:1 (*50*).

The Markets Demand More

New demands for industrial gases continued to develop. Relentless demands from the steel industry for improved quality required more heat-treating applications to switch from combustion-generated inert gases to cryogenic nitrogen.

Electric arc furnaces became economically interesting. These systems used post-consumer ferrous scrap metal and recycled it into useable products of increasingly high quality.

Recycling of steel became one of the most successful environmental accomplishments of our time and one of the least acknowledged. The extensive scrap yards of the 1950s and 1960s, with their endless rows of derelict vehicles, disappeared as the mini mills reused the steel.

Applications soon were developed using oxygen in arc furnaces to save energy, enhance furnace productivity, and improve product quality. As mini steel mills, equipped with electric-arc-furnaces, appeared across America, they were accompanied by air separation units.

In addition, emerging environmental concerns created opportunities for air separation gases. Federal legislation and funding to improve the nation's water quality created a new market for oxygen. Oxygen-enhanced wastewater treatment technology was developed and deployed. Small air separation plant technology, optimized for this market, was developed and placed at wastewater treatment facilities around the United States.

Further, federal interest in reducing oxides of nitrogen created another opportunity. High-temperature industrial combustion processes using air as an oxidizer, such as those used in glass making, created objectionable quantities of oxides of nitrogen. Substituting oxygen for air as the oxidizer limited the nitrogen available in the high-temperature process, thus reducing or eliminating the emissions.

The air separation industry responded by creating application-specific air separation plants to supply the growing demand. In the glass industry, for example, plants were produced to serve oxygen-based glass production. Switching from air-blown to oxy-fuel processes allowed container glass producers to increase production while remaining below emission caps. Operating these high-temperature processes in an environmentally acceptable manner had an additional advantage of reducing fuel consumption.

62

Molecular Sieves and Pressure Swing Adsorption Oxygen Production

As applications for oxygen grew, a new oxygen production technology emerged. In the early 1950s, a team again led by Robert Milton and Donald W. Breck, working at Union Carbide Corporation's Tonawanda, N.Y., laboratories, developed another commercially useful application for synthetic zeolites.

Milton and Breck discovered that these engineered materials could be produced with pore structures capable of selectively adsorbing nitrogen from air. Exploitation of this phenomenon soon led to pressure swing adsorption (PSA) air separation plants for the production of oxygen at purities between 90 and 94 percent.

Production of oxygen using PSA and its successor, vacuum pressure swing adsorption (VPSA), began with small, specialty applications, but has grown steadily. Important improvements in zeolite technology, process design, and equipment over the last 30 years have contributed to this growth. These improvements have made possible larger and more efficient plants (51).

Today, VPSA oxygen plant sizes range up to 250 tons per day production capacity and typically deliver 90 percent oxygen. Production costs are about 75 to 80 percent of the costs of producing high-purity, cryogenic oxygen. Most plants are leased to customers, further validating the classic lease model pioneered by Leonard Pool.

For applications where energy efficiency and/or environmental concerns drive oxygen demand, VPSA-produced oxygen is attractive. Today, VPSA oxygen accounts for about 5 percent of the world oxygen air separation capacity and is the world's second largest source of industrial-scale oxygen. (Figure 15)

PSA production of oxygen created another market by effectively meeting the healthcare industry's need for an inexpensive, convenient gas source for oxygen therapy. Small, two-bed PSA machines are used to produce 3 to 7 lpm of 90 percent oxygen for oxygen therapy applications. These units are compact and reliable, and are used in both home and institutional settings. Patients are served without the inconvenience of managing high-pressure cylinders or liquid oxygen Dewars. Today, approximately 1 million patients in the United States are served by this growing market (52).

Ultra-High-Purity Gases

The extraordinary growth of the electronics industry created demand for an entirely new category of industrial gases: large volume, ultra-high-purity (UHP) atmospheric gases to produce state-of-the-art semiconductor devices. These gases must be extraordinarily pure, with contaminant concentrations in single-digit parts per billion. At first, the demand was met with trucked in-

Figure 15. A Modern VPSA Oxygen Plant. (Courtesy of PRAXAIR Technology, Inc.)

liquid, but as the scale of semiconductor production soared during the 1980s and 1990s, on-site production of gases, particularly nitrogen, became necessary.

To meet the demand both for plant size and purity, new nitrogen separation innovations were required (53). This led to the development of very powerful distillation units capable of driving oxygen, argon, and carbon monoxide concentrations to subparts per billion in nitrogen. Water, carbon dioxide, and other trace hydrocarbons were managed by molecular-sieve-based prepurification units.

Hydrogen removal to required concentrations required innovation. Air feed to air separation plants always contain trace, but significant quantities, of hydrogen. Any hydrogen in the feed air is concentrated in the nitrogen by distillation, which exacerbates the problem. Since part per billion hydrogen levels were required, another unit operation was necessary to remove the hydrogen.

One practical approach was to add a catalytic reactor upstream of the prepurifer. In this system, any feed-air-borne hydrogen reacts with oxygen in the feed air to make water. This water could be removed in the downstream standard prepurifer.

A second approach was to pass the product nitrogen, containing virtually all the hydrogen present in the feed air, over a bed of transition metals, typically nickel. Hydrogen removal was accomplished by chemisorption. Dual beds were required since, when expended, the bed must be regenerated off-line.

Yet another challenge posed by the electronics industry was particulate control. Particulate contamination is especially disruptive to the production of semiconductor devices. Consequently, semiconductor fabs demand very low levels of particulates in the process gases. Specifications of less than three particles per cubic foot of gas, less than 0.01 microns in diameter, are typical.

Rising to this challenge, filter makers and their industrial gases partners developed and tested sophisticated filtration systems capable of delivering gases to the required specifications. Today, providing these particulate levels in large volume gas flows is routine.

Accompanying these advances were essential developments in analytic and particulate measurement systems. These systems were required to test the innovations demanded by the makers and to validate the performance of these innovations in the field. Working with analytic instrumentation makers and particulate measurement hardware manufacturers, the industrial gases industry was able to dramatically advance the state of the art on measurement systems (54).

Today, real-time, parts per billion trace gas and particulate contamination measurements are provided to customers' manufacturing information systems. This allows customers to understand the purity of their process gases cubic meter by cubic meter. Such information is essential to process quality control and has contributed to the effectiveness of semiconductor manufacturing.

Today, major semiconductor fabs have leased on-site air separation facilities that provide UHP nitrogen and, often, UHP oxygen, with impurity levels of single-digit parts per billion of the contaminants of interest. Some are very large units, producing nitrogen in 1000 tons per day quantities.

Innovation Continues

Important advances in the last 20 years have made industrial gas production more efficient and, consequently, have made industrial gases available at lower costs.

A particularly important innovation during this period was the adoption of structured packing for cryogenic air separation. Previously, cryogenic distillation was performed using sieve trays and their associated accessories. By substituting much lower pressure drop structured packing for sieve trays, energy consumption was reduced 3 to 4 percent.

Improvements in compressor efficiency have also had a significant effect. In an interesting example of technology cross-fertilization, much of the improvement in compressor efficiency has been driven by advances in computational fluid dynamic modeling.

Process innovations also played a significant role. Process cycles to efficiently improve the production of both high- and low-purity oxygen continue to be developed. Other process innovations exploit the refrigeration resources inherent in cryogenic distillation plants to produce less costly liquid products.

Heat exchangers are essential to low-cost cryogenic air gas products. Improvements in brazed aluminum heat exchanger technologies have increased their pressure ratings, sizes, and performance, which have enabled meaningful reductions in air separations costs.

Even more important have been improvements in main condenser technology. Developments of improved surfaces and main condenser architecture have lead to significant reductions in air separation power.

These innovations have reduced energy costs to produce oxygen by almost 20 percent over the last 20 years (Figure 16).

These have led to today's safe and efficient cryogenic air separation plant (Figure 17).

The Next Century

Entering its second century, the American industrial gas industry will continue to serve the U.S. economy. Its growth will be ensured as it responds to emerging social needs as it has in the past. At the start of the 21st century, some of these needs are clear: energy efficiency, energy supply, and environmental responsibility.

Demands for energy-efficient process technologies will lead to more substitution of oxygen for air in combustion processes. In addition to its efficiency virtues, such substitution reduces air pollution. By removing nitrogen from high-temperature process streams, emissions of oxides of nitrogen are greatly reduced. Furthermore, this substitution enables more feasible carbon capture and sequestration by removing nitrogen from combustion waste streams.

Exploitation of lower value (or remote) hydrocarbon resources will be enabled by industrial gases products and technologies. Large volumes of carbon dioxide and nitrogen will be required for enhanced oil recovery. Cryogenic liquefaction, transport, storage, and distribution technologies will facilitate widespread use of liquid natural gas. Oxygen will be widely used for gasification of hydrocarbon resources for the production of syngas. Reformed hydrogen and hydrogen from other sources will continue to be in demand for upgrading lower-value hydrocarbon resources and for the creation of more environmentally friendly reformulated gasoline.

In addition, the industrial gases industry will support the growing demand for efficient, environmentally responsible electricity. By substituting oxygen for air in powdered coal burning power plants, carbon dioxide and other pollutants can be captured, facilitating carbon dioxide sequestration.

66

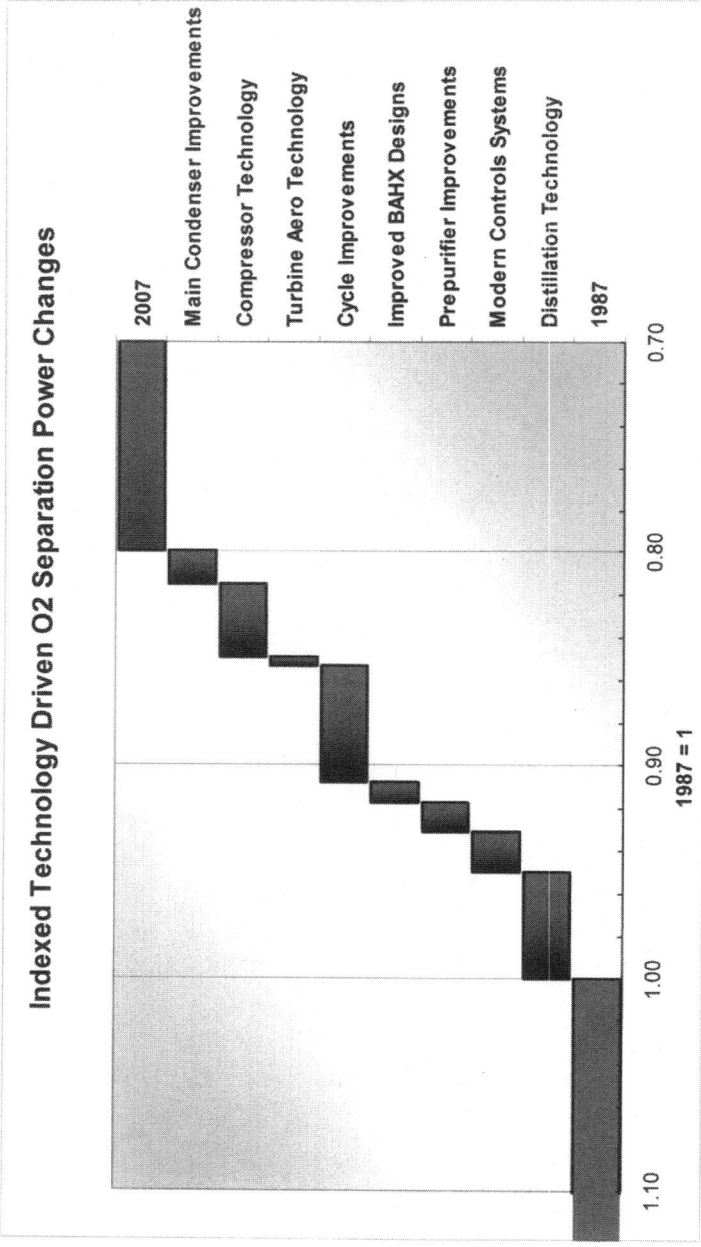

Figure 16. Indexed Technology-Driven Oxygen Separation Power Changes

Figure 17. A Modern Cryogenic Air Separation Plant. (Courtesy of PRAXAIR Technology, Inc.)

High-temperature superconducting power systems will be supported by liquid-nitrogen-based cryogenic refrigeration systems. These systems will facilitate the collection of renewable electric energy. Also, they will be essential to strengthening the nation's electric transmission and distribution grids. This will be especially important if the use of "plug and play" hybrid automobiles becomes widespread, as electric energy demand will dramatically increase.

As it has in the past century, the chemists and chemical engineers of ACS's Industrial and Engineering Chemistry Division will continue to add to the story of innovation and enterprise that is the history of the American industrial gases industry. They and the industry they participate in will grow as they rise to meet the needs of 21st-century society.

Acknowledgments

Many thanks to the many people who contributed to the preparation of this paper. It builds on the work of many predecessors and colleagues. Special thanks to Clem Demmin and his fellow retirees at the Praxair Heritage Center who have contributed their recollections and who have preserved the invaluable records referred to so often here. Thanks also to Crystal Megaridis of the Praxair library system and her staff. The facts they provided stand; any mistakes in their interpretation are mine.

Particular thanks to Mark Ackley, Neil Prosser and Bill Slye of the Praxair Technology Center for their help. Special thanks to Kristin Bojanowski of Praxair for applying her editorial talents to this piece.

References

1. Kirschner, E. M. Growth of Top 50 Chemicals Slowed in 1995 from Very High 1994 Rate. *Chemical & Engineering News*, April 8, 1996.
2. Moeller, W. The Story of the Linde Air Products Company. Manuscript, Praxair Heritage Center, Tonawanda, NY; January 23, 1939; p 5.
3. U.S. Bureau of Commerce; *Current Industrial Reports: Industrial Gases*, MQ325C and predecessor series; *U.S. Statistical Abstract*, various years; http://www.census.gov/mcd/.
4. Greenwood, H. C. *Industrial Gases*; D. Van Nostrand Co.: New York, NY, 1919; p 96.
5. Ebbe, A. *History of Industrial Gases*; Kluwer Academic/Plenum Publishers; New York, NY, 2003; p 111.
6. Greenwood, H. C. *Industrial Gases*; D. Van Nostrand Co.: New York, NY, 1919; pp 96-97.
7. Sloane, T. O. *Liquid Air and the Liquefaction of Gases – Theory, History, Biography, Practical Applications Manufacture*; Norman W. Henley & Co.: New York, NY, 1900; p 299, et seq.
8. Linde, K. U.S. Patent 727,650, 1903.
9. Sloane, T. O. Liquid Air and the Liquefaction of Gases – Theory, History, Biography, Practical Applications Manufacture; Norman W. Henley & Co.: New York, NY, 1900; p 320.
10. Charles E. Tripler Dead; New York Times; June 23, 1906.
11. Tripler, C. E. Liquid Air – The Newest Wonder of Science. The Cosmopolitan. June 1898, Vol. XXV, No. 2.
12. Johnson, G. H. Liquefied Air and Its Uses. Frank Leslie's Popular Monthly; 1899.
13 A Liquid Air Company; New York Times; November 5, 1901.
14. Ebbe, A. History of Industrial Gases; Kluwer Academic/Plenum Publishers; New York, NY, 2003; p 150.
15. ibid; p 111.
16. Greenwood, H. C. Industrial Gases; D. Van Nostrand Co.: New York, NY, 1919; p 68.
17. Waks, Fabienne. 100 years of Inspiration, the Air Liquide Adventure; Les Editions Textuel: Paris, 2002; p 14.

18. Greenwood, H. C. Industrial Gases; D. Van Nostrand Co.: New York, NY, 1919; p 76.

19. 125 Years of Linde, A Chronicle; p 25; www.linde.com/.../linde/like35lindecom.nsf/repositorybyalias/pdf_ch_chron icle/$file/chronicle_e%5B1%5D.pdf.

20. Greenwood, H. C. Industrial Gases; D. Van Nostrand Co.: New York, NY; 1919; p 82.

21. Foerg, W. History of Cryogenics: The Epoch of the Pioneers for the Beginning to the Years 1911. International Journal of Refrigeration 2; 2005; pp 283-292.

22. Ruhemann, M. The Separation of Gases; Oxford at the Clarendon Press: 1949, p 158.

23. De Baufre, W. L. U.S. Patent 1,864,585, 1932.

24. Cecil Lightfoot was the son of T.B. Lightfoot, who became a director of the British Oxygen Company.

25. Moeller, W. The Story of the Linde Air Products Company. Manuscript, Praxair Heritage Center, Tonawanda, NY; January 23, 1939; p 4.

26. The Linde Story Seventy Five Years in the Making; p 13.

27. Moeller, W. The Story of the Linde Air Products Company. Manuscript, Praxair Heritage Center, Tonawanda, NY; January 23, 1939; p 15.

28. Rockefeller Riches Back Air Nitrate Co.; *New York Times*, May 25, 1916.

29. Praxair Historical Outline; Manuscript; Praxair Heritage Center Collection, Tonawanda, NY, undated.

30. Moeller, W. The Story of the Linde Air Products Company. Manuscript, Praxair Heritage Center, Tonawanda, NY; January 23, 1939; p 5, et seq.

31. Union Carbide Historical Sketch; Manuscript, Praxair Heritage Center, Tonawanda, NY; undated; p 2.

32. 125 Years of Linde, A Chronicle; p 38; http://www.linde.com/international/web/linde/like35lindecom.nsf/repository byalias/pdf_ch_chronicle/$file/chronicle_e[1].pdf.

33. Heylandt, P. U.S. Patent 1,414,359, 1922.

34. Heylandt, P. U.S. Patent 1,464,319, 1923.

35. Early Memories; Editor, Demmin, H. C.; Manuscript No. 4c, Praxair Heritage Center, Tonawanda, NY; May 19, 2006.

36. Leo Dana Dies at 94; New York Times; August 23, 1990.

37. Dana, L. U.S. Patent 1,943,059, 1934.

38. Dana, L. U.S. Patent 1,950,353, 1934.

39. Dana, L. U.S. Patent 2,396,459, 1946.

40. Moeller, W. The Story of the Linde Air Products Company. Manuscript, Praxair Heritage Center, Tonawanda, NY; January 23, 1939; p 6, et seq.

41. 125 Years of Linde, A Chronicle; p 38; http://www.linde.com/international/web/linde/like35lindecom.nsf/repository byalias/pdf_ch_chronicle/$file/chronicle_e[1].pdf.

42. Frankl, M. U.S. Patent 1,945,634, 1934.

43. http://nobelprize.org/nobel_prizes/physics/laureates/1978/kapitsa-bio.html.
44. Ebbe, A. History of Industrial Gases; Kluwer Academic/Plenum Publishers; New York, NY, 2003; p 69.
45. Theophilos, N. A Brief History of Praxair Cryogenic Air Separation; Praxair Heritage Center, Tonawanda, NY; January 2007.
46. Butrica, A. J. Out of Thin Air, A History of Air Products 1940-1990; Praeger Publishers: 1990; p 3.
47. ibid, p 27.
48. ibid, p 13.
49. The Making, Shaping and Treating of Steel; 10th Edition; The AISE Steel Foundation, 1985; p 34.
50. Wilson, K.B.; Smith, A.R.; Theobald, A.; Air Purification for Cryogenic Air Separation Units; IOMA Broadcaster, January – February 1984; pp 15-20.
51. Sherman, J. D. Synthetic zeolites and other microporous oxide molecular sieves. Proceedings of the National Academy of Sciences. 1999, Vol. 96, Issue 7; pp 3471-3478.
52. Stegmaier, J. P. Long Term Ambulatory Oxygen Therapy Systems, a Work in Progress; RT Magazine (http://www.rtmagazine.com/issues/articles/2008-01_03.asp), January 2008.
53. Cheung, H.; Royal, J. Efficiently Produce Ultrahigh Purity Nitrogen On-Site. Chemical Engineering Progress. October 1991, vol. 87, No. 10; pp 64-71.
54. Goddard, J.; Malczewski, M.; Peres, G. Delivering Tomorrow's Ultrahigh Purity Nitrogen Today. *Microcontamination 91 Proceedings*; Cannon Communications; 1991: pp 181-205.

Chapter 4

Polyethylene: An Account of Scientific Discovery and Industrial Innovations

Rajen M. Patel, Pradeep Jain, Bruce Story, and Steve Chum

Polyolefins Research, The Dow Chemical Company, Freeport, TX 77541

Introduction

Polyethylene is the highest volume plastic available today. Global consumption of polyethylene was about 150 billion pounds (67.8 million metric tons) in 2006 and is forecast to grow to about 185 billion pounds (82.9 million metric tons) in 2010[1]. Polyethylene demand, total capacity and percent operating rates from 1995 to 2010 are shown in Figure 1[1]. Polyethylene is composed of mainly carbon and hydrogen (with some notable exceptions such as ethylene vinyl acetate copolymer, acid copolymers, etc.) which can be combined in number of ways. Various polyethylene molecular architectures have been commercialized over last 70 years to make different types of polyethylene. These various molecular architectures can be grouped into ten major types of polyethylene:

- LDPE, low-density polyethylene
- EVA, ethylene vinyl acetate copolymer
- Acid copolymers such as ethylene acrylic acid (EAA) or ethylene methacrylic acid (EMA) copolymers
- Ionomers
- HDPE, high-density polyethylene
- UHMWPE, ultra-high-molecular-weight high-density polyethylene
- LLDPE, linear low-density polyethylene
- VLDPE or ULDPE, very-low or ultra-low-density polyethylene

72

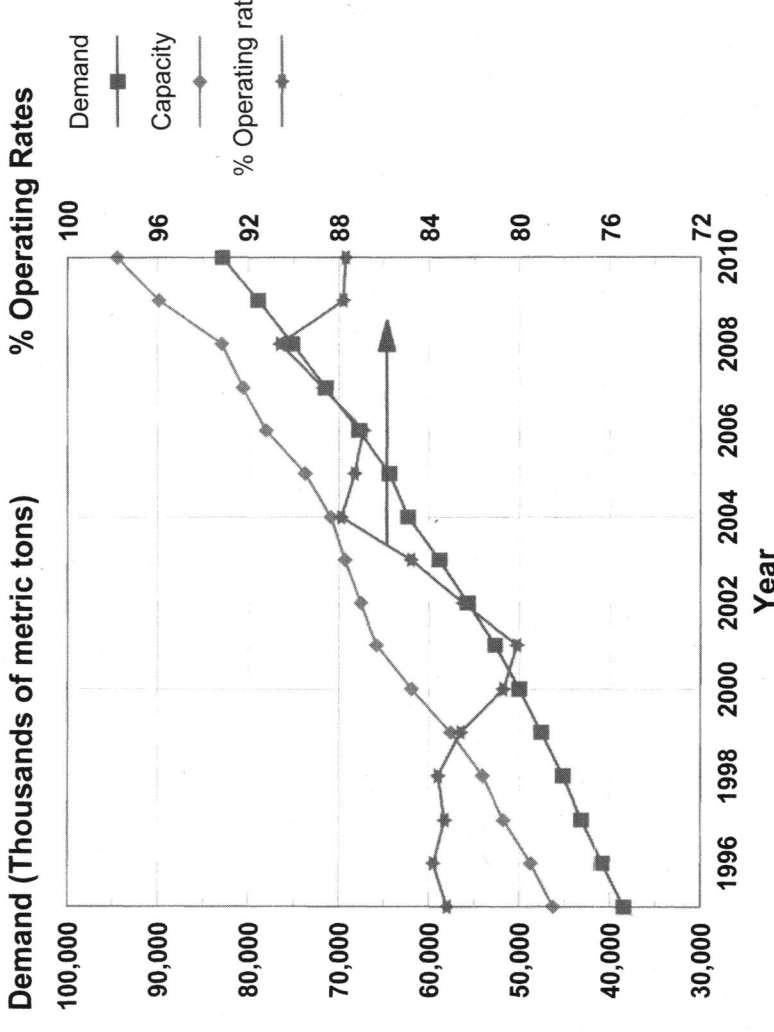

Figure 1. Polyethylene demand, capacity, and percent operating rates from 1995 to 2010. Data provided by Dr. Balaji Singh of Chemical Market Resources Inc., Texas.

- Homogeneous polyethylene (metallocene catalyzed)
- Olefin block copolymer (e.g. INFUSE[TM1] from The Dow Chemical Company)

There are other minor types of polyethylenes such as ethylene carbon monoxide (ECO) copolymers which are photodegradable (used to make six-pack loop carrier), ethylene ethyl acrylate (EEA) and ethylene n-butyl acrylate (EnBA), cyclic olefin copolymers (COC) made from ethylene and norbornene, chlorinated polyethylene (CPE), etc. Most of the polyethylene types listed above have a polymer backbone consisting of carbon and hydrogen with different types of branches coming off the backbone. The branches range from simple alkyl groups in LDPE, LLDPE, VLDPE etc. to ester groups in EVA to acid groups in EAA or EMA. The degree and type of branches control the degree of crystallinity in polyethylene by introducing defects in a regular chain architecture, thereby affecting solid state properties of polyethylenes[2, 3]. The ability to control types and degree of branching via incorporation of different comonomers allows one to make polyethylene from non-polar to polar, from stiff and rigid plastics to elastomers, from having a high melting point to a low melting point, etc. Hence, a wide range of properties are obtained from the above mentioned polyethylene types, making polyethylene one of the most versatile plastics. HDPE, LLDPE and LDPE are three major types of polyethylenes and their demand from 1995 to 2010 is shown in Figure 2. Chemical formula of major types of polyethylene are shown in Figure 3. The LLDPE chemical formula shown in Figure 3 is for an ethylene/1-octene copolymer.

The highest global consumption of any plastic coupled with the many distinct types of commercially available polyethylenes are testament to the rich history of major innovations in products, processes and breadth of applications of polyethylene. This chapter will give a historical perspective of these innovations in polyethylene including a breadth of product applications of polyethylene and the impact of metallocene polyethylenes commercialized in last 15 years. A very recent innovation of olefin block copolymers by The Dow Chemical Company will be described and some remarks will be made on future product innovations and trends.

Polyethylene History: Major Product and Process Innovations

The history of polyethylene started with its synthesis by German chemist Hans Von Pechmann who prepared polyethylene by accident in 1898 by

---

[1] Trademark of The Dow Chemical Company

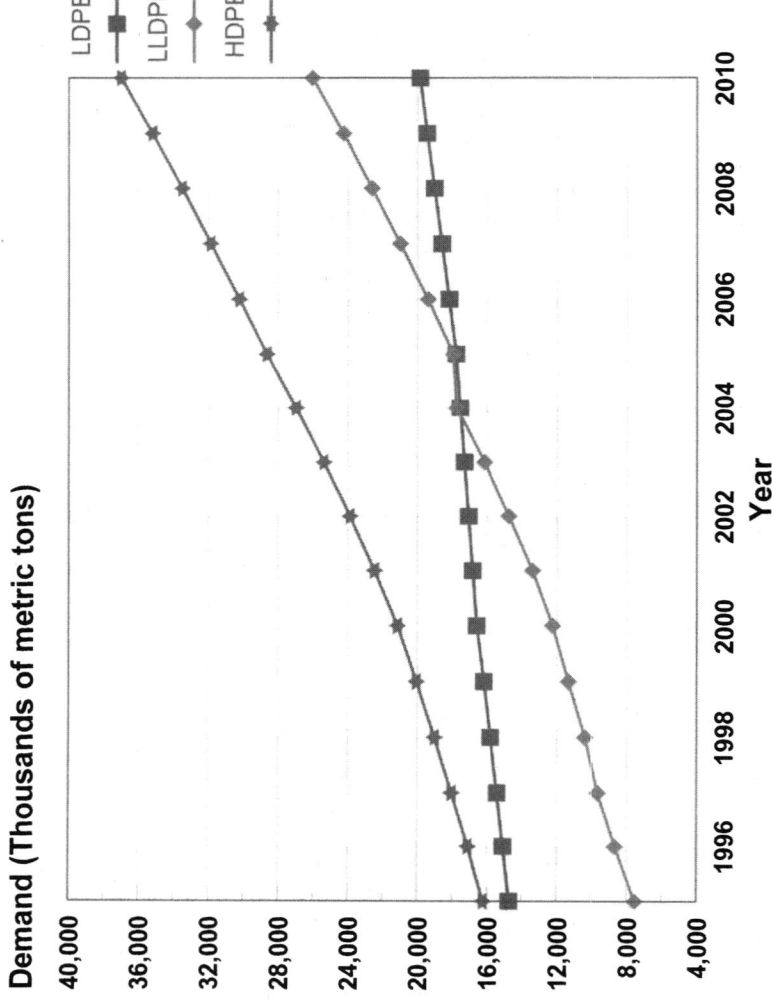

Figure 2. LDPE, LLDPE, and HDPE demand from 1995 to 2010. Data provided by Dr. Balaji Singh of Chemical Market Resources Inc., Texas.

High Pressure Low Density Polyethylene (LDPE)

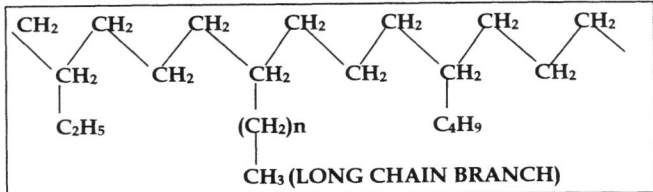

Ethylene Vinyl Acetate (EVA) Copolymer

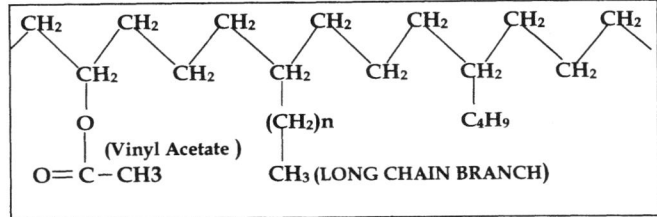

Ethylene Acrylic Acid (EAA) Copolymer

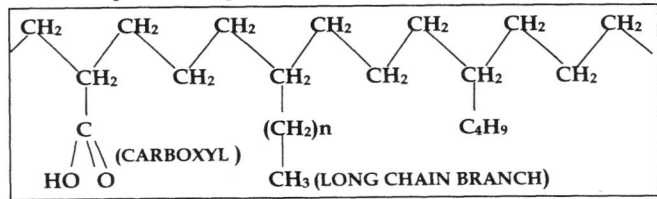

High Density Polyethylene (HDPE)

Linear Low Density Polyethylene (LLDPE)

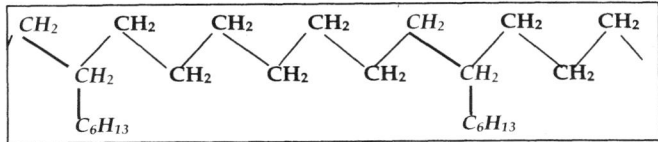

Figure 3. Chemical formula of major types of polyethylene resins. LLDPE is an ethylene/1-octene copolymer.

decomposition of diazomethane.[4] Bamberger and Tschirner[5] later characterized the white substance Pechmann created and found it to contain long sequences of -CH_2- groups and termed it polymethylene. However, the history of polyethylene product and process innovations really started with the accidental discovery and subsequent commercialization of high-pressure low-density polyethylene (LDPE) resin by Imperial Chemical Industries (ICI) in England. LDPE was accidentally discovered at ICI by Fawcett and Gibson as a waxy solid powdery substance found while investigating high-pressure and high-temperature reactions of ethylene and benzaldehyde. LDPE was later made commercially practical by Perrin. The researchers realized after months of experimentation that oxygen had to be present to initiate the conversion of ethylene to polyethylene under high pressure. The presence of oxygen leads to the formation of free-radical initiators (peroxides) that subsequently decompose to start the free-radical polymerization of LDPE. ICI filed a series of patents to claim the inventions, claiming continuous production of LDPE under pressures greater than 500 atmospheres and temperatures greater than 100°C, preferably in the presence of 0.01% to 5% oxygen.[6-8] In 1937 a 10-ton/year pilot plant was opened followed by a 100-ton/year autoclave process plant that started on the first day of World War II (in contrast, current world-scale LDPE tubular process plants are in the 200,000 to 350,000 ton/year capacity range, 2000 to 3500 times larger than the first ICI autoclave process plant). Gearing up to full scale production using high-pressure autoclave reactors was a major chemical engineering challenge that concluded with the release of ALKATHENE LDPE in 1939 using the high-pressure autoclave process. Polymerization of ethylene to polyethylene is a highly exothermic reaction, releasing approximately 1450 BTU/lb or 3370 J/g of ethylene. Removal of the exothermic heat of polymerization determines the percent conversion of ethylene to polyethylene per pass. The autoclave reactor is a continuous stirred-tank reactor (CSTR) with an agitator designed to promote good mixing. Autoclave reactors are essentially adiabatic and rely on back mixing of the hot reactants with the cold incoming ethylene feed stream to remove heat of polymerization and to keep the reaction stable. The electrical insulating properties of LDPE (high dielectric strength) were soon realized and availability of LDPE for use as a light-weight compact insulator allowed the Allied forces to use airborne radar, which gave them an enormous advantage in air warfare during World War II. BASF subsequently produced LDPE using a tubular high-pressure reactor, a non-adiabatic reactor that allows heat removal via a coolant in addition to heat removal via cold ethylene feed. Thus, tubular reactors allow higher conversion of ethylene to polyethylene compared to autoclave reactors. Rapid development of high pressure autoclave and tubular process technologies soon followed. In the tubular process polymerization takes place in a long tube (>1000 m) at high pressures. The reactor stream in a tubular reactor typically follows a plug-flow pattern. The free radical initiator is introduced at different zones along the tube to optimize process productivity and properties of the LDPE that is produced.

The reaction temperature profile along the tube is controlled by the amount and nature of the initiators. The heat of the polymerization reaction is partly removed by cooling the tubular reactor wall with hot water (hot water prevents fouling of the reactor wall by preventing crystallization of LDPE) with a temperature anywhere from 120°C to 180°C depending on the plant and location along the tube.[9] The hot water is kept under pressure so that it remains liquid at those temperatures. Early LDPE process technologies used the oxygen in air as a source of free radicals initiators. Organic peroxides are now predominantly used as a source of free radicals. At the high reactor pressures, the polymerization step is very rapid. DuPont, Union Carbide Corporation, ICI and BASF played an important role in improving LDPE process technology.

After World War II, several other commercial applications of LDPE were realized. LDPE resins exhibited excellent processability and melt strength due to the presence of a high degree of tree-like long-chain branching. High melt strength allowed manufacturing of blown films at high output rates. LDPE could be used to make high-clarity, high-gloss thin film for display packaging. LDPE could be injection molded, extrusion coated, extruded into a pipe, sheet and other profiles etc. In extrusion coating, the desired low neck-in (change in width of the web) characteristics of LDPE resin helped to minimize edge trim wastage. Also, the LDPE extrusion coating grades, especially those made in autoclave reactors, could be coated onto a substrate at very high line speeds with low neck-in. LDPE resins, however, exhibited low modulus and exhibited melting points in the vicinity of 110°C leading to a low softening point and an upper service temperature of about 90°C. The root cause of the low modulus and low melting point was the formation of in-situ short chain branches.[10] These branches lower the degree of crystallinity due to introduction of defects along the polymer chain, decreasing the modulus and lowering the melting point due to formation of thinner crystallites. LDPE also exhibited poor environmental stress crack resistance (ESCR) in presence of certain liquids. Due to the limitations in reactor pressure and temperature range, LDPE resins could be commercially produced only in the density range of about 0.915 to 0.930 g/cm3, corresponding to a weight percent degree of crystallinity from about 43% to 53%. Extremely high pressure is needed for making higher density LDPE, while lower pressures and higher reactor temperatures are needed to make lower density LDPE. Both of these extremes are either uneconomical or have undesirable product property implications. A further small reduction in density can be achieved by adding a small amount of comonomer like propylene or 1-butene (both of which also act as a chain transfer agent reducing molecular weight). Thus, the narrow range of commercially available densities (crystallinity) also limited commercial applications of LDPE resins.

DuPont, in the late fifties, introduced ethylene vinyl acetate (EVA) copolymers as specialty low-crystallinity copolymers. EVA copolymers are produced by introducing vinyl acetate comonomer in the high-pressure process.

As more vinyl acetate comonomer is introduced in the reactor, and hence gets incorporated in the polymer, the degree of crystallinity and the melting point of the resulting EVA copolymer decreases. However, clarity, flexibility, and toughness of the resulting EVA copolymer increase. These copolymers found uses in specialty applications such as sealants, ice bags, bottle cap liners and hot-melt adhesives. Acid copolymers such as ethylene acrylic acid (EAA) or ethylene methacrylic acid (EMAA) are also polymerized using the high-pressure process. These acid copolymers have excellent adhesion to aluminum foil and are used extensively as a tie layer for multi-layer packaging involving aluminum foil or metallized film packaging.

DuPont introduced SURLYN[TM2] ionomers in the early sixties. Surlyn is a family of semi-crystalline ethylene methacrylic acid (EMAA) copolymers produced using the high-pressure process, in which part of the methacrylic acid is neutralized with metal ions such as zinc or sodium. Inclusion of a few mole percent ionic groups along the backbone has a significant effect on the morphology and properties of the resulting polymer. The resulting polymer structure is comprised of three regions: amorphous polymer, crystalline polymer, and ionic clusters. The ionic clusters act as reversible physical crosslinks and provide superior abrasion resistance, transparency, scratch and mar resistance, low-temperature impact resistance, and compatibility/adhesion to metal, polyamides, polyesters and polyolefins. Ionomer applications include sealants, tie-layers, skin packaging, polymer modification, golf-ball covers and bowling-pin covers.

One important process limitation of LDPE and EVA was the need of pressures in excess of 30,000 psi in their manufacture. This required thick-walled autoclave and tubular reactors, and large compressors. This in turn made LDPE and EVA plants highly capital intensive and also required high maintenance and high energy costs, especially to compress ethylene to such very high pressures. Hence, a large research effort was undertaken to enable polymerization of ethylene at lower pressures to improve the process economics. This need was the true "catalyst" that spurred key fundamental inventions in the area of polyolefin catalysis. The resulting new polymerization processes stimulated many important innovations to fully exploit these catalysts inventions and in the process revolutionalized the polyethylene industry, starting in the early fifties! This rich history of catalyst inventions and process/product innovations will be detailed in the subsequent sections.

In 1953, Karl Ziegler, of the Max Planck institute in Germany, made the crucial discovery that titanium and zirconium halides with aluminum alkyls produced high molecular weight linear polyethylene (HDPE) from ethylene at atmospheric pressure and room temperature.[11, 12] The polymer they produced exhibited a higher density, higher melting point, and higher stiffness compared to high-pressure LDPE due to its linear nature, i.e. a lack of short and long chain

---

[2] Trademark of E.I. DuPont deNemours

branching. More importantly, the polymer could be produced at much milder process conditions compared to LDPE. Ziegler shared the Nobel Prize for his discovery with Giulio Natta, who discovered that polypropylene can also be produced using the same catalyst. These catalysts, in general, are referred to as Ziegler-Natta (Z-N) catalysts.

Around the same time frame, Hogan and Banks at Phillips Petroleum were working on improving the octane level of gasoline by passing natural gas liquid components through silica/alumina supported chrome oxide catalyst. They noticed white powder plugging up the catalyst bed. The white powder was identified as crystalline polypropylene. Related research using ethylene led to HDPE using the catalyst at relatively low pressures.[13] Both Hoechst and Phillips Petroleum commercialized HDPE products in 1956.

By the end of 1950s, both the Ziegler-Natta and the Phillips chrome oxide catalyst were being used to produce HDPE commercially. Both these catalysts underwent a series of improvements for improved process economics and product differentiation. An important catalyst breakthrough in the late sixties was high-yield $MgCl_2$-supported Z-N catalyst. HDPE resins made from ethylene monomer exhibited high stiffness and high melting point and hence, high heat resistance. The initial application of HDPE resin was to make the Hula-Hoop toy. However, linear HDPE resins were prone to environmental stress cracking. To improve environmental stress crack resistance (ESCR), a small amount of other alpha-olefin comonomers such as propylene, 1-butene, 1-hexene or 1-octene was incorporated into the polymer backbone. These comonomers introduced short-chain branches along the backbone reducing the degree of crystallinity and improving ESCR via increasing tie-chain concentration. Tie-chains are amorphous, rubbery molecules connecting crystallites. It was also found early on that higher alpha-olefins such as 1-hexene and 1-octene were much more effective in improving ESCR compared to propylene and 1-butene at the same density (crystallinity).

Various process innovations were needed to make HDPE commercially. Phillips developed a loop slurry process to make HDPE[14]. HDPE resins made using Phillips chrome oxide catalyst in the loop slurry process soon found acceptance in making blow-molded bottles. The very broad molecular weight distribution (MWD) of the loop slurry uni-modal HDPE resins gave the desired processability, swell characteristics, and ESCR for the blow-molded bottles application. Hoechst developed a cascade stirred slurry tank process to make bi-modal HDPE. Bi-modal HDPE resins are significantly superior in mechanical properties (ESCR and impact toughness) compared to the uni-modal HDPE resins and are being increasingly used in film, pipe and blow-molding applications.[15] This is due to higher tie-chain concentration in bi-modal resins than uni-modal resins at a given density and melt index (processability). Higher tie-chain concentration is achieved by incorporating alpha-olefin comonomer into very high molecular weight fraction in one reactor (making a very high molecular weight, lower density fraction) and incorporating low or preferably no comonomer into the very low molecular weight fraction in the other reactor (making a very low molecular weight, high density fraction). This is shown conceptually in Figure 4. Note that tie-chain concentration is a very useful concept to understand and rationalize mechanical properties of polyethylenes. However, the tie-chain concentration cannot be quantified experimentally and theoretical models have been developed to predict relative tie-chain concentration as a function of molecular weight and alpha-olefin incorporation.[16, 17]

Union Carbide Corporation developed a gas-phase process to make HDPE. In the gas-phase process, ethylene is catalyzed into HDPE in a fluidized-bed reactor consisting of catalyst-polymer particles. The polymerization occurs at the interface between the solid catalyst and the polymer matrix, which is swollen with monomers during polymerization. Ethylene is then easily separated from the HDPE particles which are then converted into pellets using an extrusion step. The first commercial gas-phase polymerization plant for making HDPE using a fluidized-bed reactor was constructed by Union Carbide in 1968 at Seadrift, Texas. Union Carbide also developed the dual reactor UNIPOL-II[TM3] gas phase process in the 1980s to make bi-modal HDPE resins with superior mechanical properties compared to UNIPOL single-reactor HDPE resins. A world scale UNIPOL-II plant was constructed in 1996 to make differentiated bi-modal HDPE resins.

In the mid-sixties, The Dow Chemical Company developed a solution process to make HDPE. In this process, HDPE is made at a high temperature in an inert solvent and the polymer thus made is dissolved in the inert solvent. The inert solvent is later evaporated and recycled to retrieve the polymer. The solution process typically cannot be used to make the very high molecular weight HDPE resins needed for blow-molding, pipe and high modulus film (grocery sacks) applications. Hence, initial applications of HDPE produced via solution process were in the area of injection molding and roto-molding.

Depending upon the degree of short-chain branching, polyethylene was initially classified as HDPE or medium density polyethylene (MDPE). However, initially only a limited amount of alpha-olefin comonomer could be incorporated in the polymer backbone in a commercially viable process due to catalyst limitations. There was a strong commercial incentive to develop low-pressure polyethylene resins having similar density range as high-pressure LDPE resin (0.915 to 0.930 g/cm^3) by incorporating higher levels of comonomer, primarily due to the lower capital cost of the low-pressure process. These resins were termed linear-low-density-polyethylene (LLDPE) to reflect the lack of long-chain branching in these resins compared to high-pressure LDPE resins. DuPont, Canada was first to commercialize and market Ziegler-

---

[3] Trademark of The Union Carbide Corporation

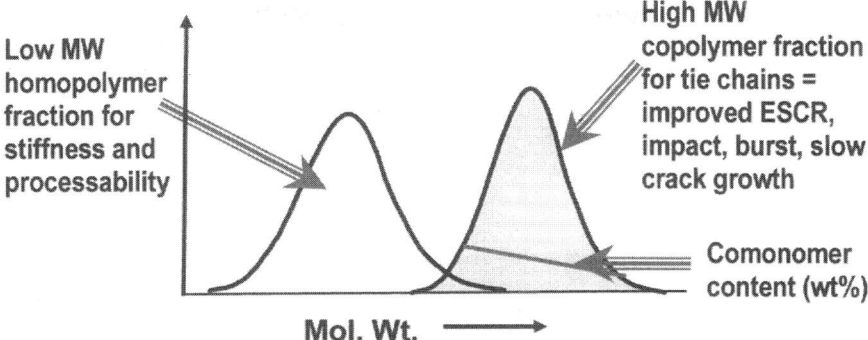

Low MW homopolymer fraction for stiffness and processability

High MW copolymer fraction for tie chains = improved ESCR, impact, burst, slow crack growth

Comonomer content (wt%)

Mol. Wt.

Figure 4. Bi-modal HDPE resins designed for optimum stiffness, toughness and ESCR balance via optimizing tie-chain concentration

Natta (Z-N) catalyzed LLDPE resins in the early 1960s utilizing catalyst developments from the U.S. parent company in combination with a low-pressure solution-polymerization process. The new LLDPE resins exhibited superior mechanical properties compared to LDPE resins in film applications and the early LLDPE resins were used by DuPont to make proprietary films. This first commercial introduction of LLDPE resins by DuPont took place rather quietly and its potential was not appreciated initially. This was mainly due to poor extrusion processability of the new LLDPE resins compared to LDPE resins. The rheological properties of the new LLDPE were very different from those of LDPE resins. A paper by Phillips Petroleum[18] on improved toughness, ESCR and low-temperature properties of LLDPE resins in telephone-cable jacketing in 1970 caught the attention of Union Carbide which at that time was a leading supplier of cable jacket materials based on LDPE and provided impetus for its research in LLDPE[19]. This led to a major breakthrough by Union Carbide with adaptation of its fluidized-bed gas-phase HDPE process to make LLDPEs. The catalyst and equipment design used for making HDPE could not be used to make LLDPE resins due to the potential for particle agglomeration in a fluidized bed because of the lower melting point of LLDPE resins. New low-temperature and low-pressure catalysts were developed by Union Carbide to enable production of LLDPE resins.[20] Union Carbide extended their commercial HDPE gas-phase process to make Chrome oxide (CrO) and Z-N catalyzed LLDPE resins in 1975. The Union Carbide gas-phase process, called the UNIPOL[TM4] process, was made available for worldwide licensing in 1977, which accelerated commercialization of LLDPE resins. In the late 1970s, against the backdrop of rising energy costs, the UNIPOL gas-phase process was

---

[4] Trademark of Univation Technologies, LLC

touted to be much less energy-intensive. The availability of the UNIPOL process, with its capital cost and energy advantage and superior product properties, caused high-pressure LDPE producers to suspend their plans for future investments in the high-pressure process. Later, in the early 1980s, British Petroleum announced its own Innovene™ gas-phase process for LLDPE. Looking back, in 1980 world polyethylene capacity was dominated by high-pressure processes for LDPE and by slurry processes for HDPE. Together they accounted for about 95% of the total capacity in 1980.[21] Since then the gas-phase UNIPOL process has become a dominant process for producing HDPE and LLDPE resins due to its adoption by major producers such as ExxonMobil Chemical and SABIC. Today, about 25% of the world's polyethylene is produced using the UNIPOL gas-phase process. Note that HDPE and LLDPE resins must be considered one product from the supply point of view (i.e. made on the same solution, gas or slurry train) but they remain two quite separate products from the demand point of view. Initially, gas-phase LLDPE resins were made with 1-butene as the comonomer. Later on, due to improved mechanical properties compared to 1-butene-based LLDPE, 1-hexene-based LLDPE resins were commercialized by Union Carbide under the trade name of TUFLIN™[5] using the gas-phase UNIPOL process.

In 1978, The Dow Chemical Company commercialized LLDPE based on 1-octene comonomer using their proprietary solution process, under the trade name of DOWLEX™[6]. Octene based Z-N LLDPE resins, made using the solution process, exhibited superior film mechanical properties such as dart impact and tear compared to the 1-butene based gas-phase Z-N LLDPE resins introduced by Union Carbide. DOWLEX resins soon were widely accepted as premium LLDPE resins. This led other suppliers to develop their own LLDPE resins using higher alpha-olefins as comonomers (e.g. 1-hexene, 4-methyl 1-pentene). However, octene based DOWLEX resins, made using Dow's solution process, still had performance advantages and remained the premium LLDPE resins. High efficiency Z-N catalyst on $MgCl_2$ support enabled economic production of both HDPE and LLDPE resins in all these processes due to lower catalyst cost and elimination of the catalyst de-ashing step (removal step).

Z-N LLDPE resins became commercially attractive because of the use of the low pressure process resulting in lower capital and operating cost, and more importantly, due to their substantially improved toughness (dart impact and tear, ESCR, tensile strength) properties and higher melting point compared to high pressure LDPE resins. LLDPE resins were quickly utilized as a blend component with LDPE resins to improve toughness, optics, and to allow down-gauging while achieving desired processability. Thus, LLDPE resins extended the versatility of LDPE through the use of blending. Higher toughness and ESCR coupled with a faster molding cycle due to the higher melting and

---

[5] Trademark of The Union Carbide Corporation
[6] Trademark of The Dow Chemical Company

crystallization temperature of the LLDPE resins made them very attractive for injection molding applications, displacing LDPE resins. The improved toughness of LLDPE resins along with exceptional draw-down capability, even at fractional melt indices, due to low extensional viscosity allowed down-gauging compared to LDPE resins and led to their wide acceptance in film and packaging applications. However, LLDPE resins were harder to process due to their higher shear viscosity and exhibited lower melt strength compared to LDPE resins leading to bubble instability at higher output rates. More energy is required to extrude LLDPE than LDPE and the machinery has to be modified or replaced to handle LLDPE resins or LLDPE rich blends. The processing deficiencies of LLDPE resins were systematically addressed using lower (L/D) screws, new screw designs such as the barrier-flight screw, improved die designs and cooling-air ring designs, and the use of wider die gaps for extrusion and film processes. This was achieved by cooperation among equipment suppliers, film processors and resin suppliers, led by Union Carbide. Pure LLDPE film also did not exhibit cross direction (CD) shrinkage and could not be used in collation shrink film or in pallet shrink hoods by itself. Blending with LDPE, frequently leading to LDPE-rich blends, allowed one to obtain desired CD shrinkage for those applications while obtaining improved toughness due to LLDPE. Pure LLDPE resins could not be used in extrusion coating applications due to high neck-in and tendency of draw resonance (web instability). Still today, autoclave LDPE resins dominate the extrusion coating market.

The discovery of fluoropolymer processing aids[22] by DuPont, to eliminate melt fracture during processing of LLDPE resins, allowed significantly higher output rates with narrow die gaps during fabrication[23] and allowed the use of lower melt index LLDPE resins to improve melt strength and to even further improve toughness of LLDPE resins. Development in anti-oxidants and stabilization technologies ensured optimum LLDPE product quality. Key film and packaging applications of LLDPE include sealant layer, pallet stretch film, collation shrink, greenhouse film, silage film, mulch film and heavy duty shipping sacks.

VLDPE (or ULDPE) resins, having lower density (< 0.915 g/cm^3) were subsequently developed by incorporating even higher levels of alpha-olefin comonomer in the copolymer. These resins exhibited improved puncture, impact, tear, ESCR, low temperature toughness, heat seal and hot-tack properties compared to EVA resins. VLDPE/ULDPE resins found acceptance in flexible packaging, flexible tubing, sealants, low-temperature packaging, barrier shrink bagging, and as a cling layer in stretch film.

One key feature or limitation of Z-N catalyzed LLDPE and VLDPE resins is the non-uniform or broad composition distribution and broad molecular weight distribution (MWD or polydispersity index of ~ 4 to 5) of the resins. This is due to multi-site nature of Z-N catalyst with differences in each site's ability to incorporate alpha-olefin comonomer. Catalyst sites in Z-N catalyst that readily incorporate alpha-olefin comonomer tend to produce lower molecular weight chains. Catalyst sites that do not readily incorporate alpha-olefin tend to produce higher molecular weight chains. Thus, Z-N catalyzed LLDPE resins are molecular blends of high molecular weight lightly short-chain branched molecules, lower molecular weight highly short-chain branched molecules and everything in between! Hence, LLDPE and VLDPE resins are classified as heterogeneous polyethylenes. Note that HDPE and LLDPE resins produced using chrome oxide (CrO) catalysts exhibit much broader molecular weight distribution (MWD ~ 10 to 20) than the MWD (~ 4 to 5) of Z-N catalyzed HDPE and LLDPE resins Very broad MWD Chrome oxide (CrO)-catalyzed LLDPE resins exhibit very good melt strength during film blowing compared to the Z-N catalyzed LLDPE resins. However, thin film (less than 2 mil or 50 micron thick) made from Chrome oxide-catalyzed LLDPE resins typically exhibit lower machine direction (MD) Elmendorf tear compared to thin films made using Z-N catalyzed LLDPE resins, because of the high level of molecular orientation in the MD[24].

The theoretical relative tie-chain concentration model developed by Patel et al.[16] clearly shows that, at a given molecular weight, both the lightly-branched (high density) and the highly-branched (very low density) fractions have relatively low tie-chain concentrations and that it is the medium-branched fraction that has the highest relative tie-chain concentration at a given molecular weight. Irrespective of short-chain branch content (i.e. density), the predicted relative tie-chain concentration increases with increasing molecular weight though the highest increase is also predicted for medium branched fractions. Thus, the broad composition and molecular weight distribution of Z-N catalyzed LLDPE/VLDPE resins leads to less than optimum relative tie-chain concentration at a given density and melt index and hence, lower dart impact toughness. However, films made from Z-N catalyzed LLDPE/VLDPE resins exhibit very good Elmendorf tear strength. The high-pressure LDPE resins have even lower relative tie-chain concentration compared to LLDPE due to presence of a very high level of long-chain branching. The presence of a very high level of long-chain branching (tree-like) in LDPE leads to compact molecular coils with a lower radius of gyration in the molten state, thereby reducing overlap of the molecular coils and hence, reducing tie-chain concentration.[17]

Another key limitation of Z-N catalysts is the inability to incorporate very high levels of alpha-olefin comonomer such as 1-butene, 1-hexene, or 1-octene to make the density less than about 0.885 g/cm^3. The lowest density Z-N catalyzed VLDPE resin commercially available today (FLEXOMER^{TM} from The Dow Chemical Company) has a target density of 0.885 g/cm^3 (approximately 20 wt% crystallinity). Due to this, Z-N catalyzed VLDPE resins cannot be used in applications requiring very low modulus, low shore A hardness (density less than 0.885 g/cm^3). Note that logarithm of modulus of polyethylene resins is related to density (degree of crystallinity)[25, 26].

Significant research efforts were undertaken to commercially produce MDPE/LLDPE/VLDPE resins with narrow composition and molecular weight distributions. Such resins are classified as homogeneous polyethylene. Elston of DuPont Canada first synthesized homogeneous polyethylene resins and a new composition of matter patent for homogeneous polyethylene was awarded to Elston of DuPont Canada in 1972, foreshadowing metallocene-based polyethylene.[27] He used vanadium catalysts in a solution process that acted like a single-site catalyst to make homogeneous polyethylene. Elston taught many of the key advantages of homogeneous polyethylene such as improved optics and impact strength. Note that vanadium catalysts were an outgrowth of Z-N catalyst research. Ethylene-propylene rubbers were commercialized using vanadium catalysts followed by EPDM in the early 1960s.

Impact of Metallocene Polyethylene and Recent Advances

Metallocene catalysts for ethylene polymerization were first described in 1957 by Breslow et al. of Hercules.[28] They patented a dicyclopentadienyl chromium-metal alkyl catalyst for ethylene polymerization. The major draw back of the early metallocene catalysts was the low catalyst efficiency. During the mid 1970s several groups discovered by accident that addition of trace levels of water to metallocene catalyst systems containing $AlMe_3$ led to improved catalyst efficiency. Kaminsky's and Sinn's research led to the discovery of methylalumoxanes as the key activator of dicyclopendadienyl (bisCp) metallocene catalysts for highly improved efficiency to enable com- mercialization of homogeneous polyethylene.[29, 30] Metallocene catalyst research exploded in the 1980s with major plastics producers patenting variety of metallocene catalysts and co-catalysts to make homogeneous polyethylenes with wide range of molecular weight and density. Notably, Welborn and Ewen[31] of ExxonMobil Chemical patented catalysts comprising derivatives of mono-, bi- and tricyclopendadienyl coordination complexes with a transition metal and an alumoxane to produce homogeneous polyethylene of controlled and high molecular weight at conventional polymerization temperatures. Also, Stevens and Neithamer[32] of The Dow Chemical Company and Canich[33] of ExxonMobil Chemical patented monocyclopentadienyl or substituted monocyclopentadienyl metal complexes, termed as "constrained geometry catalysts" (CGC), to produce homogeneous polyethylenes. Turner et al.[34] of ExxonMobil Chemical patented cationic dicyclopentadienyl (bisCp) systems.

Another breakthrough obtained by Lai et al. of The Dow Chemical Company is described in a new composition of matter patent[35] and its continuation in part patent[36] pertaining to homogeneous polyethylene resins containing small and controlled amounts of long chain branching. These resins were made using constrained geometry catalysts (CGC) in a continuous solution process. These resins are termed as "substantially linear polyethylenes" to differentiate them from linear homogeneous polyethylene resins containing no long chain branching. Substantially linear resins exhibit improved processability during extrusion, as measured using the I_{10}/I_2 ratio (ratio of melt index measured using 10 kg weight to melt index measured using 2.16 kg weight, both at 190°C) or other shear rheology parameters, even at narrow molecular weight distribution due to the presence of long chain branching. The inventors (Lai, Wilson, Stevens, Knight and Chum of The Dow Chemical Company) of U.S. patent # 5,272,236 were selected as the winners of the National Inventor of the Year award in 1994. This prestigious award is given by the Intellectual Property Owners (IPO) association. Senators DeConcini and Hatch presented the award to the inventors. Incidentally, this was the first time a chemical company (The Dow Chemical Company) received this prestigious award.

A review of the most cited U.S. patents in the polyolefins field as of August 2006, gives an indication of the most important patents in this fast-developing field. The two patents by Lai et al. (U.S. # 5,272,236 and 5,278,272) are the two most cited patents in the field of polyolefins (553 and 445 citations, respectively) and these two patents survived opposition worldwide[37]. The third most cited patent (389 citations) in the polyolefins area is by Stevens and Neithamer[32] of The Dow Chemical Company, describing constrained geometry catalyst systems for producing homogeneous polyethylene. The fourth and fifth most cited U.S. patents in this field are ExxonMobil Chemical's U.S. # 5,198,401 by Turner, Hlatky, and Eckman with 372 citations and U.S. # 5,153,157 by Hlatky and Turner with 306 citations, claiming cationic bis-cyclopentadienyl metallocene catalyst systems. Of the 25 most cited polyolefin patents, two were the polyethylene composition of matter patents (substantially linear polyethylene containing long chain branching), two were condensed-mode gas-phase polymerization process patents and the other 21 were catalyst patents! This clearly shows that catalysts are the key enabler of polyolefin innovation.

Metallocene catalysts allowed commercial production of homogeneous polyethylene resins over a broad density range of 0.855 g/cm^3 to 0.965 g/cm^3, thus overcoming the key limitations of Z-N catalysts. These homogeneous polyethylene resins exhibit a broad range of morphology (from lamellar morphology at high crystallinity to granular morphology at low crystallinity) and solid state properties (from necking and cold drawing at high crystallinity to uniform drawing and high elastic recovery at low crystallinity)[38]. Homogeneous polyethylene random copolymer resins produced by metallocene catalyst are known as polyolefin elastomers (POE) (density < 0.885 g/cm^3) and polyolefin plastomers (POP) (density 0.885 to 0.910 g/cm^3) due to their unique microstructure and properties. ExxonMobil Chemical commercialized POE and

POP resins in the early 1990s using bis-cp metallocene catalyst in a high-pressure autoclave process in Louisiana, U.S.A.[39] These polymers, due to their low melting points, are more suited to be made using a solution process where the polymer is dissolved in an inert solvent during polymerization. The Dow Chemical Company, being a wide user and innovator of the solution process, rapidly commercialized POE and POP resins made using constrained geometry catalyst (CGC) in their solution process in early nineties. Note that the solution process offers great flexibility to make products with a very broad density range, from about 0.857 g/cm^3 to 0.97 g/cm^3 on the same production train – a much broader range than that can be made on a gas phase or slurry processes. The solution process also allows a much faster grade change resulting in lower amounts of off-grade products.

A key feature of metallocene catalyzed homogeneous polyethylene is narrow composition (intermolecular) and molecular weight distribution.[25] The narrow composition distribution of metallocene catalyzed POP versus Z-N catalyzed LLDPE is readily observed in temperature-rising elution fractionation (TREF) profiles as shown in Figure 5. In Figure 5, the x-axis is the elution temperature and the y-axis is the weight percent of polymer eluting at a given temperature. The TREF technique is described in detail by Wild.[40] The narrow composition distribution (a.k.a. short-chain-branch distribution, SCBD) leads to a lower melting point compared to heterogeneous Z-N catalyzed LLDPE/VLDPE resins at the same density. This is illustrated in Figure 6. This is due to narrow intermolecular distribution of the comonomer in POP and POE resins leading to narrow distribution of crystallite sizes compared to Z-N catalyzed LLDPE, where a small fraction of the chains with highest density (low comonomer incorporated) exhibits the highest melting point. The lower melting point of homogeneous polyethylene (POP) at the same density is advantageous in key applications such as sealants and shrink film. Before 1990s, EVAs and ionomers were the primary choice for sealant layers due to their low melting point and low heat-seal-initiation temperatures. However, EVA resins exhibit very low hot-tack strength due to the presence of high levels of long-chain branching and can have taste and odor issues. Ionomers exhibit somewhat higher hot-tack strength compared to EVA resins due to presence of ionic domains and were predominantly used where higher hot-tack strength was needed such as in vertical form fill and seal (VFFS) machines. However, ionomers are more expensive resins than EVA. Introduction of metallocene POPs in 1990s exhibiting lower melting point allowed the packaging industry to take advantage of their lower heat-seal and hot-tack initiation temperature. The lower heat-seal and hot-tack initiation temperature allows faster packaging line speeds and hence, improved productivity. Metallocene POP resins also exhibit significantly higher hot-tack strength compared to EVA. This is illustrated in Figure 7 for AFFINITY POPs of various melt index (MI, measured at 190°C using 2.16 kg load) and densities in gm/cm^3. Hot-tack strength of AFFINITY

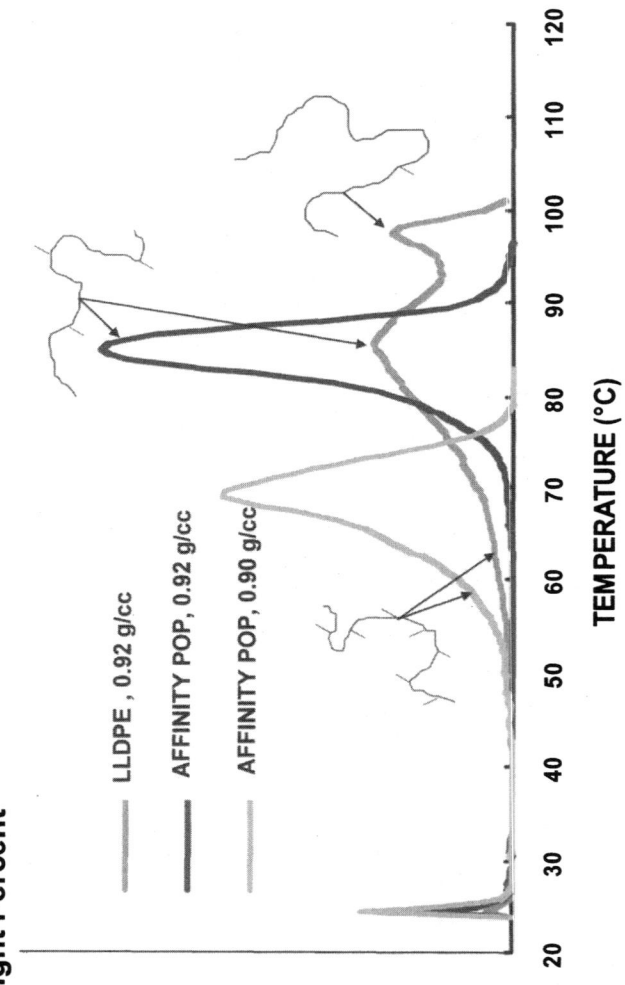

Figure 5. Composition distribution measured using TREF profile of Z-N LLDPE and metallocene catalyzed POP resins.

Heat Flow (Watts/gram)

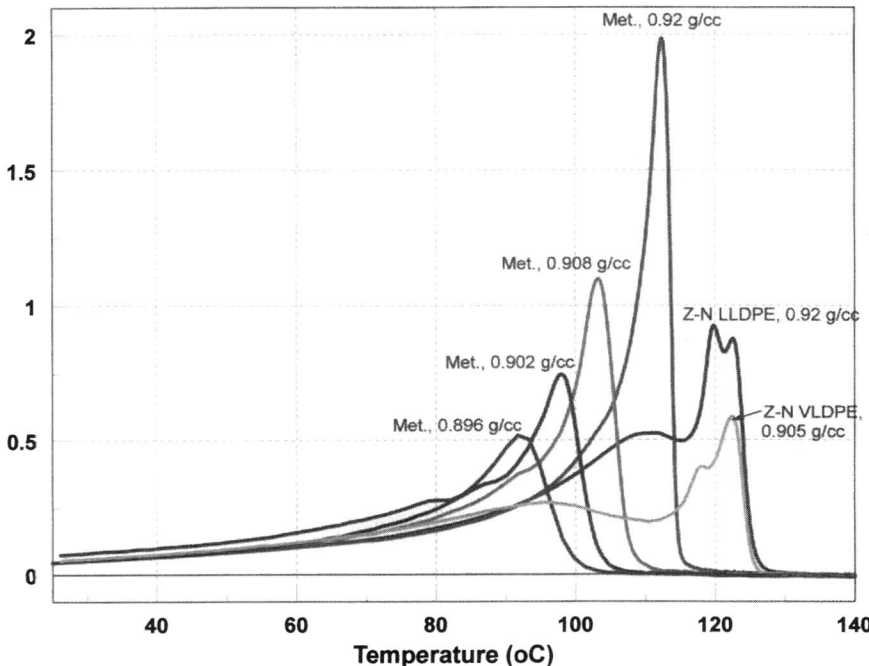

Figure 6. DSC melting curves of Z-N catalyzed LLDPE/VLDPE and metallocene catalyzed resins. (Reproduced with permission from reference 25. Copyright 2002 Taylor and Francis Group, LLC.)

POPs are also compared with various EVA resins in the same figure. Hence, metallocene POPs became a popular choice of resin in the 1990s and beyond, for sealant applications requiring low initiation temperatures as well as high hot-tack strength (e.g. in VFFS machines), and better taste and odor performance. Rapid growth of multi-layer packaging applications in last 15 years has led to significant growth in usage of metallocene POP resins as sealants.

In the last ten years, many manufacturers have also introduced metallocene-catalyzed LLDPE resins at densities above 0.91 g/cm^3. These resins exhibit improved dart impact and puncture, and improved optics (especially when blended with LDPE), compared to standard Z-N LLDPE resins. Univation Technologies, LLC, jointly owned by The Dow Chemical Company and ExxonMobil Chemical, has become a major worldwide licensor of gas-phase metallocene resins.

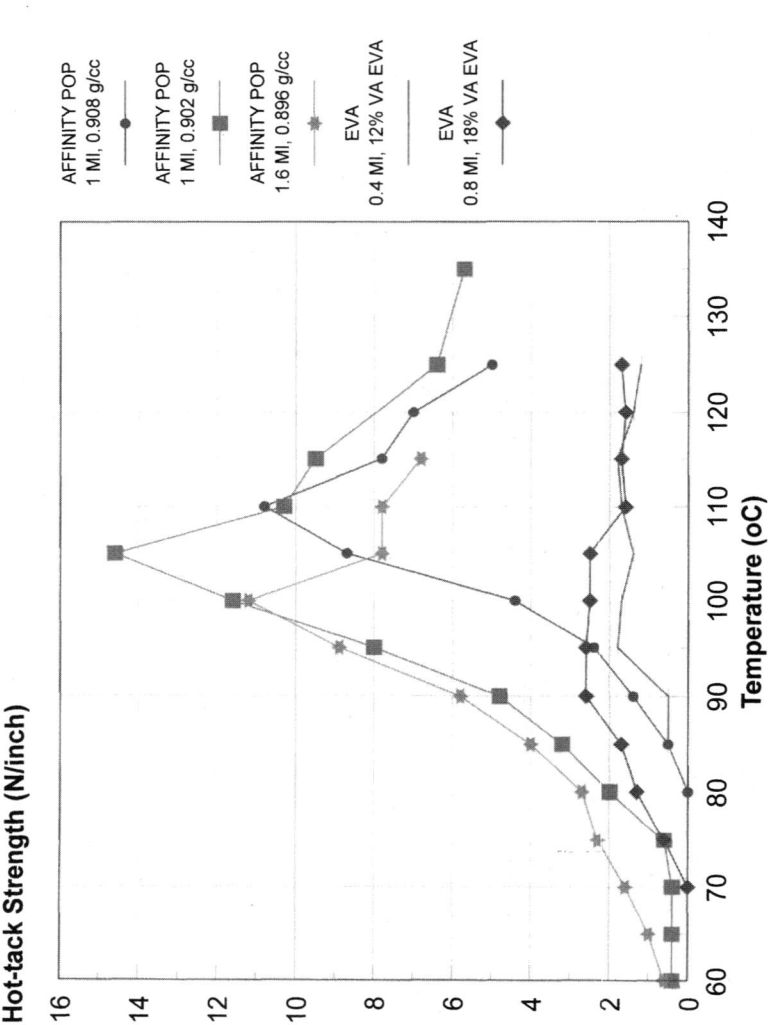

Figure 7. Hot-tack strength of AFFINITY POP and EVA resins as a function of seal-bar temperature. Nylon/EAA/Sealant (1/1/1.5 mil) blown co-ex film was used. (Reproduced with permission from reference 25. Copyright 2002 Taylor and Francis Group, LLC.)

Another key feature of the metallocene catalyst is their ability to incorporate a very high level of alpha-olefin comonomer to make very-low-density or even completely amorphous polyethylenes. This allowed commercial production of very low density polyethylene elastomers (POE, density less than 0.885 g/cm^3). Such low density POE resins exhibit very low modulus and low Shore A hardness for enhanced flexibility and soft touch, very low glass-transition temperature, and high elastic recovery. Note that in the case of HDPE and ethylene co-polymers (and in general for semi-crystalline polymers having glass transition temperature below room temperature), modulus increases with degree of crystallinity.[26] Before the 1990s, ethylene-propylene rubber (EPR) and ethylene-propylene-diene monomer (EPDM) resins were predominantly used as impact modifiers for polypropylene resins. Since the 1990s, metallocene POE resins with density less than 0.87 g/cm^3 have found wide commercial acceptance as impact modifiers for polypropylene to make thermoplastic olefins (TPO) primarily used in automotive applications. POE resins made using 1-octene as a comonomer exhibit a lower glass transition temperature (Tg ~ -55°C as measured by DSC) and better compatibility with polypropylene resins compared to EPR and EPDM resins, leading to improved low-temperature toughness and stiffness balance in TPO resins. Availability of POE resins in pellet form vs. bales for EPDM and EPR grades is also advantageous for compounding with polypropylene. With the rapid growth of TPO resins in the last 15 years, especially in automotive applications such as bumper fascia and instrument panels, the use of POE resins, especially those with 1-octene as a comonomer, for impact modification of polypropylene has increased significantly. POE resins have also found commercial acceptance for soft and flexible goods, adhesives, cling layer in stretch films, and for elastic films and fibers.[41] Very high melt index POE resins (melt index in the range of 500 to 1000) have found commercial acceptance in hot melt adhesive applications for case and carton sealing.

Typical POE resins are either linear (no long chain branching) or substantially linear due to their low levels of long-chain branching. However, such POE resins exhibit poor melt strength (sag resistance) and have a narrow processing window for applications such as thermoforming, blow molding, and profile extrusion. To improve the melt strength performance of POEs, The Dow Chemical Company recently commercialized high melt strength ENGAGE[TM7] POE resins. High melt strength is achieved via incorporation of a higher level of long-chain branching. High melt strength ENGAGE POE resins exhibit excellent sag resistance, especially when blended with high melt strength polypropylene resins for thermoforming and blow molding applications. These blends also exhibit the lower gloss desired for auto interior applications.

---

[7] Trademark of The Dow Chemical Company

In case of the metallocene-catalyzed random copolymer of ethylene and alpha-olefin (POP and POE), incorporating more comonomer along the polymer backbone reduces density and crystallinity and hence, increases flexibility and softness. However, as the density is decreased, the melting point, crystallization peak temperature and heat resistance decrease and cycle times in injection molding increase. These deficiencies have limited the use of POEs in applications where heat resistance, high temperature compression set, and faster cycle times are desired.

The most recent advance in polyethylene is by The Dow Chemical Company, and it has overcome most of the deficiencies of POEs via the introduction of olefin block copolymers (OBC) the under the trade name of INFUSE[TM8]. OBCs are made from same raw materials (ethylene and alpha-olefin comonomers) arranged into alternating "soft" and "hard" blocks (Figure 8). Soft blocks contain a high level of alpha-olefin comonomer and have low density, low crystallinity and low melting point. Hard blocks contain almost no or very low level of alpha-olefin comonomer and have high density, high crystallinity, high melting point and crystallization temperature. The soft blocks deliver flexibility and the hard blocks deliver improved heat resistance, compression set at 70°C and faster cycle time via high crystallization temperature. Hence, customers get a flexibility and softness similar to those of POEs but with improved heat resistance, elastic recovery, compression set at 70°C, abrasion resistance, and faster cycle/set-up times.

The catalytic system used to make OBCs uses a chain-shuttling agent (CSA) to shuttle or transfer growing chains between two distinct catalysts with different comonomer (alpha-olefin) selectivity.[42] This is shown in Figure 9. Synthesis of olefin block polymer via chain shuttling requires the chain transfer to be reversible. OBCs are produced in a continuous solution polymerization process more economically favorable than the batch processes employed to make styrenic block copolymers.

DSC melting curves of OBCs are compared with those of POE random copolymers in Figure 10. It can be seen that at the same density, OBCs exhibit a much higher melting point compared to POP and POE random copolymers. DSC crystallization curves of OBCs are compared with POE random copolymers in Figure 11. Again, it can be seen that at the same density, OBCs exhibit a much higher crystallization peak temperature compared to POP and POE random copolymers. This results in much faster cycle time for OBCs than for POEs in injection molding applications and much faster set-up time in profile extrusion applications. OBCs are targeted for elastic film, elastic fibers, bottle cap liners, profile extrusions, soft and flexible goods, soft touch over-molding and adhesives applications. OBCs will expand the competitive space

---

[8] Trademark of The Dow Chemical Company

Random Copolymers

- Adding more comonomer lowers the polymer's density and crystallinity while increasing flexibility.
- However, the melt temperature, crystallization temperature, and heat resistance also drop as density is lowered.

Less comonomer and higher density

More comonomer and lower density

Block Copolymers

- OBCs use same raw materials arranged into alternating "soft" and "hard" blocks.
- The soft blocks deliver flexibility and the hard blocks deliver heat resistance.
- The customer gets flexibility similar to random copolymers (e.g., AFFINITY™ POPs, ENGAGE™ POEs) but with improved heat resistance, elastic recovery, compression set, and cycle times.

Soft blocks

Hard blocks

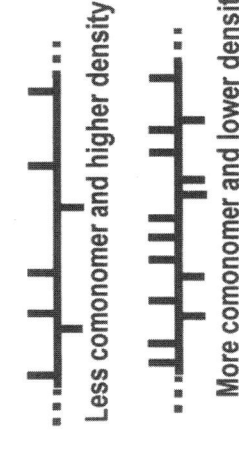
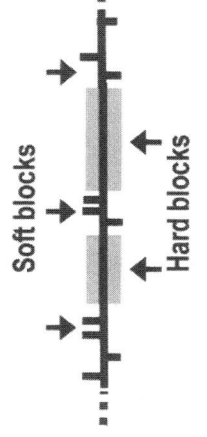

Figure 8. Random copolymer versus Olefin Block Copolymer (OBC).

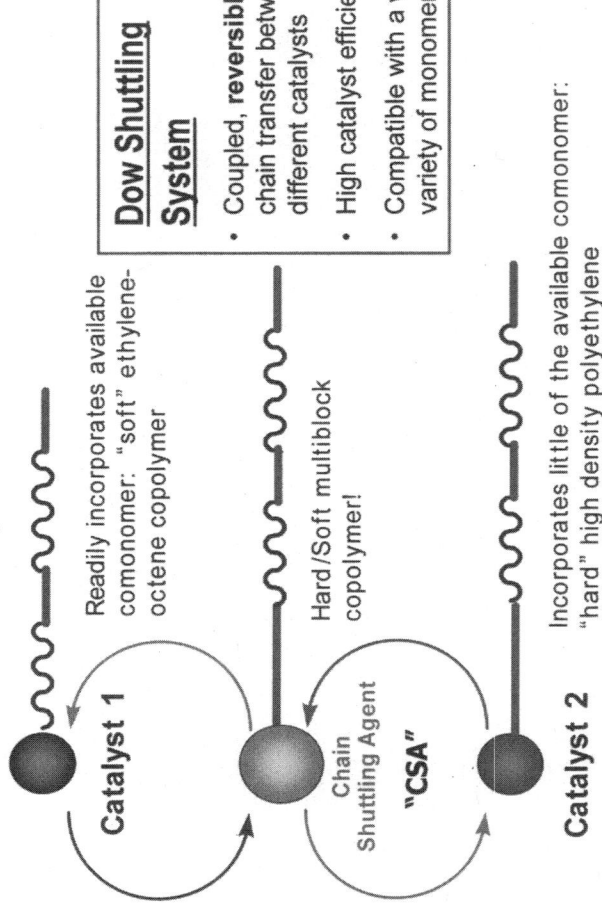

Dow Shuttling System

- Coupled, reversible chain transfer between 2 different catalysts
- High catalyst efficiency
- Compatible with a wide variety of monomers

Readily incorporates available comonomer: "soft" ethylene-octene copolymer

Hard/Soft multiblock copolymer!

Chain Shuttling Agent **"CSA"**

Incorporates little of the available comonomer: "hard" high density polyethylene

Catalyst 1

Catalyst 2

Figure 9. Catalytic block technology used to make OBCs.

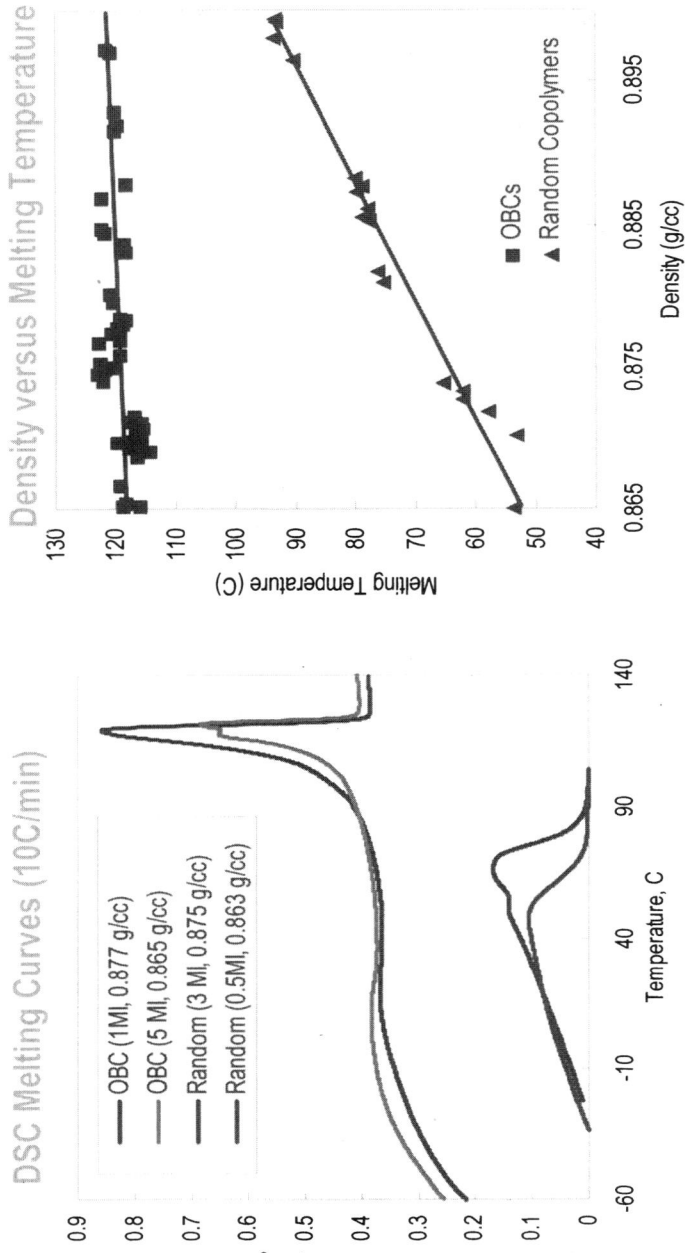

Figure 10. DSC melting curves and melting point versus density of OBCs and POE random copolymers.

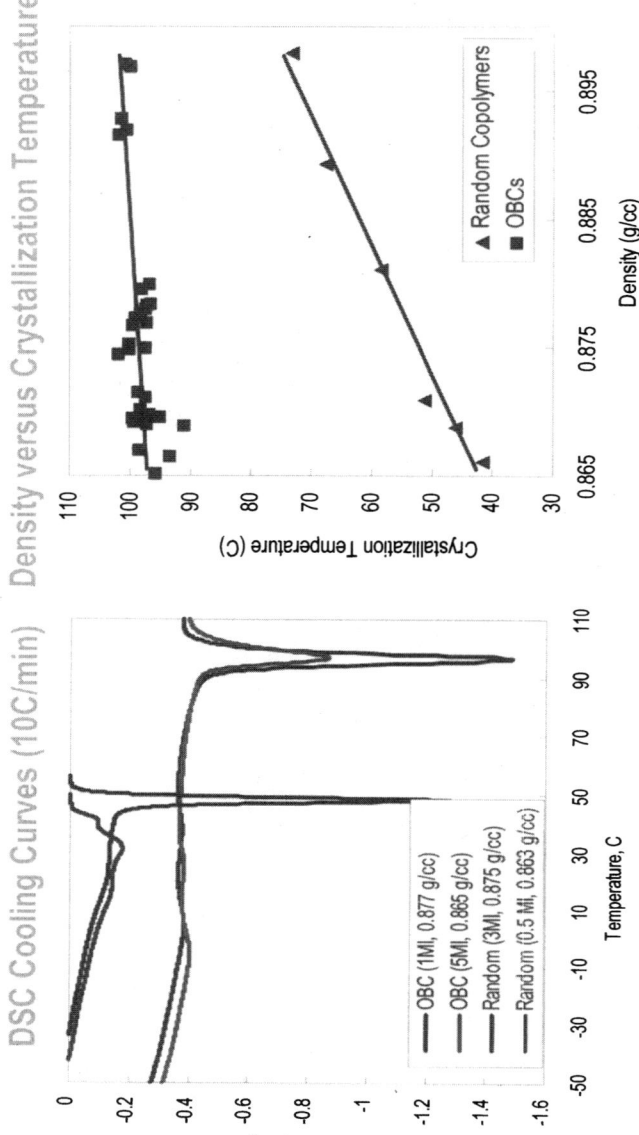

Figure 11. DSC crystallization curves and crystallization peak temperature versus density of OBCs and POE random copolymers.

for polyolefin elastomers against a range of flexible materials and could compete with high value elastomers such as SBS, SIS, SEBS, TPVs, etc.

In the last fifteen years, ethylene copolymers other than those mentioned above have been produced using metallocene catalysts. These include ethylene-styrene interpolymers, ethylene-norbornene copolymers (a.k.a cyclic olefin copolymers, COC) and EPDM (NORDEL™ IP from The Dow Chemical Company).

Future Trends

The polyolefin industry continues to capitalize on metallocene catalyst chemistry and rapidly developing "post-metallocene" chemistry for new product developments in the area of ethylene copolymers and propylene copolymers. Currently polar ethylene copolymers such as EVA, EAA, EMA, etc. can only be produced using a high-pressure free-radical polymerization process. These polar copolymers contain a very high degree of long-chain branching that is detrimental to many product properties such as toughness, hot-tack strength, etc. Polar comonomers are poisons for traditional Z-N, chrome oxide, as well as metallocene catalysts and hence cannot be polymerized with ethylene using such catalysts. Efforts have been made in last ten years and are ongoing to develop catalyst systems that can incorporate polar comonomers, resulting in linear polar ethylene copolymers, with advantageous properties, using standard low-pressure processes.

High-throughput catalyst-screening techniques, pioneered by Symyx Corporation, have allowed rapid development of new catalyst systems for reducing cost and for developing novel polyolefins. These techniques allow catalyst research that is faster and less expensive without sacrificing the quality of the data. Such high-throughput R&D processes will continue to be used to accelerate new product developments. Traditional Z-N catalysts are continually being improved for improved process economics and product properties.

New commercial uses of polyethylene such as polyolefin dispersions, apparel elastic fibers, artificial grass turf, and wood plastic composites, will continue to expand the applications of polyethylene resins. For example, The Dow Chemical Company recently introduced HYPOD™ polyolefins high-solids waterborne dispersions, made using mechanical dispersion process technology, to provide converters an opportunity to bring olefinic performance benefits to latex applications. There is a continuing drive on the part of major polyolefin producers to get access to low-cost feed stocks in the Middle East through joint ventures and partnerships. A huge increase in production capacity in the Middle East, to take advantage of low-cost feed stocks like natural gas and naphtha, will make the Middle East a major producer of polyolefins. There is a trend towards increasing the size of polyethylene production plants to take advantage of economics of scale. Advances in reactor technologies such as super-condensed mode in a gas-phase process using supported metallocene catalyst[43] as well as post-reactor technologies have allowed the introduction of mega size trains to make commodity polyethylenes.

Finally, sustainability will become a significant driver in making and using plastics including polyethylene. There will be more emphasis on producing less plastics waste by practicing the three Rs; reduce, reuse and recycle. There will also be an emphasis on making plastics including polyethylene from renewable resources. Currently ethylene is produced using naphtha or natural gas, both of which are petroleum products. The Dow Chemical Company, one of the largest global producer of polyethylene, and Crystalsev, one of Brazil's largest ethanol producers have recently announced plans for a world-scale facility to manufacture polyethylene from sugar cane. The new facility will use ethanol derived from sugar cane, an annually renewable resource, to produce ethylene. It is estimated that the new process will produce significantly less CO_2 compared to the traditional polyethylene manufacturing process.

Concluding Remarks

Polyethylene constitutes a large and diverse family of large-volume commodity and differentiated resins exhibiting a very wide range of properties, from rigid plastics to elastomers, from non-polar to polar, etc. Summary of density, degree of crystallinity, melting point, and year of commercial introduction of major types of polyethylene resins is shown in Figure 12. This very wide range of properties is accomplished by molecular design, starting primarily with carbon and hydrogen and enabled by catalysts. Our industrial world would indeed be much poorer without polyethylene. The long history of polyethylene with many remarkable innovations along the way clearly tells us that catalysts are key enablers of innovations and market success in polyethylene. Catalyst innovations coupled with process and post reaction (e.g. oxygen tailoring) process innovations, enabling advantaged cost/performance balance and low environmental impact, have led to high growth rates in applications and the usage of polyethylene, especially in flexible and rigid packaging, and durable and non-durable applications. An excellent example of a durable application of polyethylene is the use of HDPE pipes for gas transport where pipes are designed to last at least 50 years. Polyethylene provides excellent toughness such as impact/puncture and tear resistance in non-durable applications such as packaging, trash liners and stretch film. This allows down-gauging resulting in material cost saving. Traditional packaging materials such as metal, glass, paper, etc. are continuously being replaced by polyethylene due to cost, performance and environmental factors. Advances in metallocene

POLYETHYLENE TECHNOLOGY				

Type	Density (g/cm^3)	Melting Point (°C)	% Crystallinity	Year
LDPE	0.915 - 0.93	106 - 120	40 - 60	1939
EVA	0.93 - 0.95*	40 - 105	5 - 40	1955
HDPE	0.94 - 0.965	125 -135	65 - 80	1955
LLDPE	0.915 - 0.94	120 - 125	40 - 60	1975
VLDPE	0.885 - 0.915	118 -122	25 - 40	1983
Metallocene	0.857 - 0.965	40 - 135	5 - 80	1991

* Density of EVA is higher due to bulky vinyl acetate group. Hence, EVA density can not be compared to other polyethylenes.

Figure 12. Summary of density, melting point, degree of crystallinity, and year of commercial introduction of major types of polyethylene resins. (Reproduced with permission from reference 25. Copyright 2002 Taylor and Francis Group, LLC.)

catalyst technology, such as the development of constrained geometry catalysts in the last fifteen years, have greatly expanded the property envelop of ethylene copolymers and allowed polyethylene manufactures to deliver new products with improved performance to customers. The authors of this chapter have witnessed the metallocene and now the post-metallocene revolution in polyethylene over the last fifteen years.

There are many references to academicians such as Kaminsky, Jordan, Bercaw, Mark, Baer, and Hiltner in the polyolefins patent art and literature linking their pioneering research to industrial innovations. In last 15 years, the prestigious ACS award for industrial academic cooperative research was given to The Dow Chemical Company and ExxonMobil Chemical for long term collaborations with universities in the polyethylene field[37]. Continued collaboration between academia and industry is necessary for further innovations and developments. This review shows the evolution of technology from one research group to another across time. Significant improvements have been made in last 25 years to catalysts, processes, plant capacity and polyethylene performance. There will be many more developments in polyethylene catalyst, especially post-metallocene catalysts, processes, applications as we all seek to achieve new levels of performance and the patents will record these developments.

References

1. Data provided by Dr. Balaji Singh of the Chemical Market Resources Inc., Texas, U.S.A.
2. Peacock, A. J., *Handbook of Polyethylene - Structures, Properties, and Applications*. Marcel Dekker, Inc.: New York, 2000.
3. Vasile, C., *Handbook of Polyolefins, second edition*. Marcel Dekker, Inc.: New York, 2000.
4. Von Pechmann, H., *Berichte* **1898**, *31*, 2643.
5. Bamberger, E.; Tschirner, F., *Chem. Ber.* **1900**, *33*, 955.
6. Fawcett, E. W.; Gibson, R. O.; Perrin, M. W. US 2,153,553, 1939.
7. Fawcett, E. W.; Gibson, R. O.; Perrin, M. W.; Paton, J. G.; Williams, E. G. GB 471,590, 1936.
8. Perrin, M. W.; Paton, J. G.; Williams, E. G. US 2,188,465, 1940.
9. Zoller, W. WO 02/100907, 2002.
10. Fox, J. J.; Martin, A. E., *Tran. Faraday Soc.* **1940**, *36*, 897.
11. Ziegler, K. DE 878,560, 1953.
12. Ziegler, K. BE 533,362, 1955.
13. Hogan, J. P.; Banks, R. L. US 2,825,721, 1958.
14. Hogan, J. P.; Norwood, D. D.; Ayres, C. A., Phillips Petroleum Company loop reactor polyethylene technology. *J. Appl. Polym. Sci.: Appl. Polym. Symp.* **1981**, *36*, 49-60.
15. Galli, P.; Vecellio, G., Polyolefins: the most promising large-volume materials for the 21st century. *J. Polym. Sci., Part A: Polym. Chem.* **2004**, *42*, (3), 396-415.
16. Patel, R. M.; Sehanobish, K.; Jain, P.; Chum, S. P.; Knight, G. W., Theoretical prediction of tie-chain concentration and its characterization using postyield response. *J. Appl. Polym. Sci.* **1996**, *60*, (5), 749-58.
17. Seguela, R., Critical review of the molecular topology of semicrystalline polymers: The origin and assessment of intercrystalline tie molecules and chain entanglements. *J. Polym. Sci., Part B: Polym. Phys.* **2005**, *43*, (14), 1729-1748.
18. Levett, C. T.; Pritchard, J. E.; Martinovich, R. J., New low-density polyethylene. *SPE Journal* **1970**, *26*, (6), 40-3.
19. Staub, R. B., Would we be better off without LLDPE? In *LLDPE in Europe - World Perspectives and Developments*, Madrid, Spain, 1986; p 1/1.

20. Xie, T.; McAuley, K. B.; Hsu, J. C. C.; Bacon, D. W., Gas Phase Ethylene Polymerization: Production Processes, Polymer Properties, and Reactor Modeling. *Ind. Eng. Chem. Res.* **1994**, *33*, (3), 449-79.
21. McMaster, L. P., The gas phase process. In *Polyethylene - The 1990s and beyond*, London, 1992; p S3A/1.
22. Blatz, P. S. Extrudable composition consisting of a polyolefin and a fluorocarbon polymer. US Patent #3,125,547, 1964.
23. Achilleos, E. C.; Georgiou, G.; Hatzikiriakos, S. G., Role of processing aids in the extrusion of molten polymers. *J. Vinyl Addit. Technol. FIELD Full Journal Title:Journal of Vinyl & Additive Technology* **2002**, *8*, (1), 7-24.
24. Sukhadia, A. M., Trade-offs in blown film LLDPE type resins from chromium, metallocene and ziegler-natta catalysts. *Journal of Plastic Film & Sheeting* **2000**, *16*, (1), 54-70.
25. Patel, R. M.; Chum, P. S., Structure, Properties and Applications of Polyolefins Produced by Single-Site Catalyst Technology. In *Encyclopedia of Chemical Processing and Design*, 2002; Vol. 69, Supplement 1, pp 231-262.
26. Sehanobish, K.; Patel, R. M.; Croft, B. A.; Chu, S. P.; Kao, C. I., Effect of chain microstructure on modulus of ethylene-α-olefin copolymers. *J. Appl. Polym. Sci.* **1994**, *51*, (5), 887-94.
27. Elston, C. T. US 3,645,992, 1972.
28. Breslow, D. S.; Long, W. P. US 3,013,002, 1961.
29. Kaminsky, W.; Hahnsen, H.; Kiilper, K.; Woldt, R. US 4,542,199, 1985.
30. Sinn, H. W.; Kaminsky, W. O.; Vollmer, H. C.; Woldt, R. US 4,404,344, 1983.
31. Welborn, H. C.; Ewen, J. A. US 5,324,800, 1994.
32. Stevens, J. C.; Neithamer, D. R. US 5,064,802, 1991.
33. Canich, J. M. US 5,026,798, 1991.
34. Turner, H. W.; Hlatky, G. G.; Eckman, R. R. US 5,198,401, 1991.
35. Lai, S. Y.; Wilson, J. R.; Knight, G. W.; Stevens, J. C.; Chum, P. S. US 5,272,236, 1993.
36. Lai, S. Y.; Wilson, J. R.; Knight, G. W.; Stevens, J. C. US 5,278,272, 1994.
37. Story, B. A., Polyolefin patents encourage innovation. In *Flexpo 2006*, Galveston, Texas, 2006.
38. Bensason, S.; Minick, J.; Moet, A.; Chum, S.; Hiltner, A.; Baer, E., Classification of homogeneous ethylene-octene copolymers based on comonomer content. *J. Polym. Sci., Part B: Polym. Phys.* **1996**, *34*, (7), 1301-15.
39. Hemmer, J. L., New high pressure polyolefins based on EXXPOL[TM] technology In *Polyethylene - The 1990s and beyond*, London, 1992; p S2A/3.
40. Wild, L., Temperature rising elution fractionation. *Adv. Polym. Sci.* **1991**, *98*, (Sep. Tech. Thermodyn. Liq. Cryst. Polym.), 1-47.

41. Obijeski, T. J.; Huff, G. M.; Maugans, R. A.; McKinney, O. K. US 5,472,775, 1995.
42. Arriola, D. J.; Carnahan, E. M.; Hustad, P. D.; Kuhlman, R. L.; Wenzel, T. T., Catalytic Production of Olefin Block Copolymers via Chain Shuttling Polymerization. *Science* **2006**, *312*, (5774), 714-719.
43. DeChellis, M. L.; Griffin, J. R.; Muhle, M. E. US 5,405,922, 1995.

Chapter 5

The History of Petroleum Cracking in the 20th Century

Alan W. Peters[1,2], William H. Flank[3], and Burtron H. Davis[4,*]

[1]W. R. Grace Company, 7379 State Route 32, Columbia, MD 21044
[2]Deceased
[3]Department of Chemistry and Environmental Science, Pace University,
861 Bedford Road, Pleasantville, NY 10570
[4]Center for Applied Energy Research, University of Kentucky,
2540 Research Park Drive, Lexington, KY 40511

An overview of the development of thermal and catalytic cracking for petroleum refining is presented. The introduction of research departments in a petroleum company led to the development of the Burton Process for thermal cracking to enhance the yield of transportation fuels demand caused by the rapid increase in the number of autos on the highways. As other companies developed processes to compete with the Burton Process, competition led to a number of legal actions that are briefly detailed. As these court actions were reaching a climax, the introduction of catalytic cracking by Eugene Houdry reduced the need for thermal cracking processes. Houdry's efforts required outstanding advances in process control, process engineering and catalysis. The discovery that the transport of finely divided solids by a gas resembled fluid flow allowed Standard Oil (New Jersey) to develop the fluid catalytic cracking (FCC) process. The entry of the U.S. into W.W.II and the urgent need for high octane aviation fuel permitted the development of FCC at an astoundingly rapid rate where, supported by government guarantees, a commercial sized reactor could quickly become the pilot plant for the development of ever larger and improved plants. The need for catalysts with higher activity led to the replacement of natural clays by high surface area amorphous silica alumina catalysts during the 1930-1940s. These catalysts were in turn replaced in the 1960s by the synthetic and natural zeolite catalysts. The revolutionary advances during the century - thermal cracking, fixed-bed catalytic cracking, fluid-bed catalytic cracking, synthetic silica-alumina catalysts and synthetic zeolite catalysts - were superimposed on a background of many evolutionary advances.

Processes initially developed for the "destructive distillation" of coal tars involved thermal cracking (*1*). As petroleum crude became a source of distillate products, the initial cracking processes for petroleum were "borrowed" from the coal tar technology base. However, when these were applied to petroleum, they were not very efficient.

At the turn of this century, the introduction of the auto quickly transformed the refining industry from one based on the production of products for illumination to products suitable for transportation fuels.

In viewing the history of petroleum cracking, the introduction of technological advances far outdistances the development of scientific understanding of the processes. In this review, emphasis will be placed upon the technology and the individuals involved in the technological advances. There have been many instances where the introduction of advances in petroleum cracking, both thermal and catalytic, has not been clear-cut, and this has resulted in much litigation. In this review, coverage of these legal battles will be brief, because of the limitations of space and the abilities of the authors.

Thermal Cracking Processes

Back in 1901 R. E. Olds sold 425 Oldsmobiles, representing the first commercially successful U.S.-made automobile. By 1910, there were over half a million cars registered, and the appetite for gasoline hasn't faltered since. Petroleum refiners, who up until then were primarily interested in kerosene and lubricants, saw a problem coming. The yield of gasoline from crude oil in 1910 was only about 13%, and was obtained by simple distillation (*2*).

Burton Process

The Burton process is usually viewed as the first great advance in petroleum cracking over those adapted from coal tar cracking. The major advance introduced in this process was the feature of conducting the cracking at elevated pressures and the significant increase in gasoline yield (*1*).

William M. Burton was an early product of the Ph.D. program in chemistry that was started by Professor Ira Remsen at The Johns Hopkins University. Burton was employed by Standard Oil of Indiana (Indiana Standard) in 1890 and was probably the first Ph.D. employed in the petroleum refining industry. He rapidly rose in the administrative ranks and was soon Vice President for refining. At Indiana Standard, he put together a small group of formally trained scientists and engineers. Initially this team investigated systematically a number of approaches to cracking that were suggested from the literature and from earlier industrial experience (*1*). One approach involved testing a variety of catalysts. They found that aluminum chloride gave the best catalytic results, but cost eliminated it from further consideration. About 1910 the team made the decisions that ultimately led to the Burton Process: (1) limit the process to converting gas oil rather than the entire crude and (2) utilize distillation under pressure. Because high pressure equipment was not readily available at that time, safety considerations caused Burton to order suspension of the pressure work. It was only after he learned of an incident where a petroleum fraction heavier than gas oil had been distilled with steam at pressures of 50 psi without an explosion that he agreed to resume the high-pressure work (*1*). Allowed to again work at high pressures, the group gradually demonstrated the advantage of the process. Working with a 50-gallon still, the advantage of pressure cracking was demonstrated by Indiana Standard workers. They learned how to control the temperature and pressure so that construction of the first commercial plant was completed and started operation in January, 1913 (*1*). Burton's still was operated at 95 psi and 750°F, and just about doubled gasoline yield (*2*). Trial and error eventually led them to procedures to define the feed rate to the furnace. Controlling pressure was a major problem that was solved by Dr. Robert Humphreys, a Hopkins chemist that Burton hired in 1900. He solved this problem by locating the pressure-control valve beyond the condensing equipment, and thereby removed much of the pressure fluctuations. The fractionating towers (dephlegmators) had to be improved to obtain reflux for further cracking. Humphreys overcame this problem by designing a more selective intermediate partial condenser that was a fan-like arrangement of air-cooled pipes. Also, Humphrey observed that the solid coke formed in the oil rather than on the metal surfaces. To take advantage of this observation, he devised a false bottom of grids to trap the coke before it settled to the bottom of the still. The false tray led to a more constant cracking rate, the elimination of "hot-spots" that forced shut-down for repairs and cleaning, and better removal of coke.

A schematic of the Burton process that incorporates advances introduced by Humphreys and the Lewis-Cooke bubble tower is shown in Figure 1. This process operated on the batch principle. Within a short time, many Indiana Standard employees made improvements to the design of the first process. However, all of these advances were nearly for naught. The directors of the

106

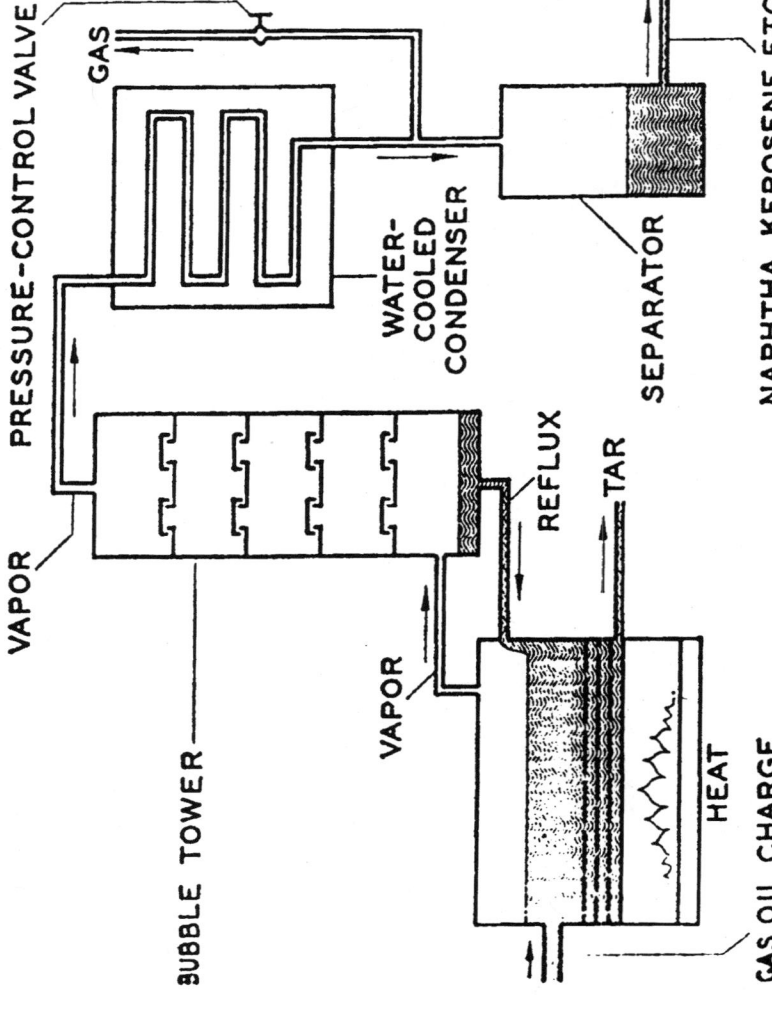

Figure 1. The Lewis-Cooke bubble tower and staged distillation, added to the Burton still. (Reproduced with permission from reference 1. Copyright 1963 Northwestern University Press.)

parent Standard Company rejected Burton's request for funds to build 100 commercial stills, each with a capacity of 8,000 gallons (*1*). When anti-trust reasons led the U.S. Supreme Court, as a result of the 1911 Standard Oil Trust divestiture decision, to dissolve the parent company, Indiana Standard became independent and approved the funds to construct the pressure stills. They also began licensing the Burton patent to competitors for a reasonable royalty, a novel approach at the time. Burton eventually became president of what eventually became Amoco Oil (*2*).

After a few years of commercial operation with the Burton Process, E. M. Clark, superintendent of one of the refineries, made a very important improvement. Using gravity to transport heavy liquid since a pump was not available that could pump heavy hot petroleum, Clark developed a modification to allow semi-continuous processing, resulting in a significant improvement in transportation fuel yield (Figure 2).

As word spread of the success of the Burton process, demand for licensing developed from many refiners. By 1920, Indiana Standard was reaping "the greatest windfall in the history of petroleum refining", at least up to that time (*1*).

In describing the development of his work, Burton indicated that he followed the advice given to him by his undergraduate and graduate instructors: Professor Ira Remsen of The Johns Hopkins University taught that "...the best preparation for a career in technical chemistry is thorough training in the pure sciences..." and Professor E. W. Manley counseled that "The best incentive for research work is the work itself" (*3*). It is obvious that Professor Manley's outlook could not be sold easily to today's refinery management.

The development of the process would not have been possible without making significant improvements in the hardware (*1*). Electrical welding was introduced a decade after the Burton process. Thus, the first stills had to be constructed of riveted steel plates, and they frequently burst at the seams. They also tended to leak at the seams and the escaping vapors would ignite; fortunately the coke deposits that formed with time on-stream tended to plug the leaks as the still was used. Pumps that could handle hot oil safely were still to be developed.

The Burton process was an outstanding success, and eventually an outstanding failure. Enos (*4*) estimates that the total cost of research and development of the process was about $200,000 and that during the first ten years the process returned to Indiana Standard more than $100,000,000 from its own refining operations and from royalties. The developers were chemists, and were not impressed by the limitations of such processing. The company, with its monopoly on cracking, had "a goose laying golden eggs" so why should it spend funds on the Burton process when there were so many other pressing concerns? For example, the anti-trust breakup of the old Standard Company left them without crude oil reserves, and money was needed to develop such

108

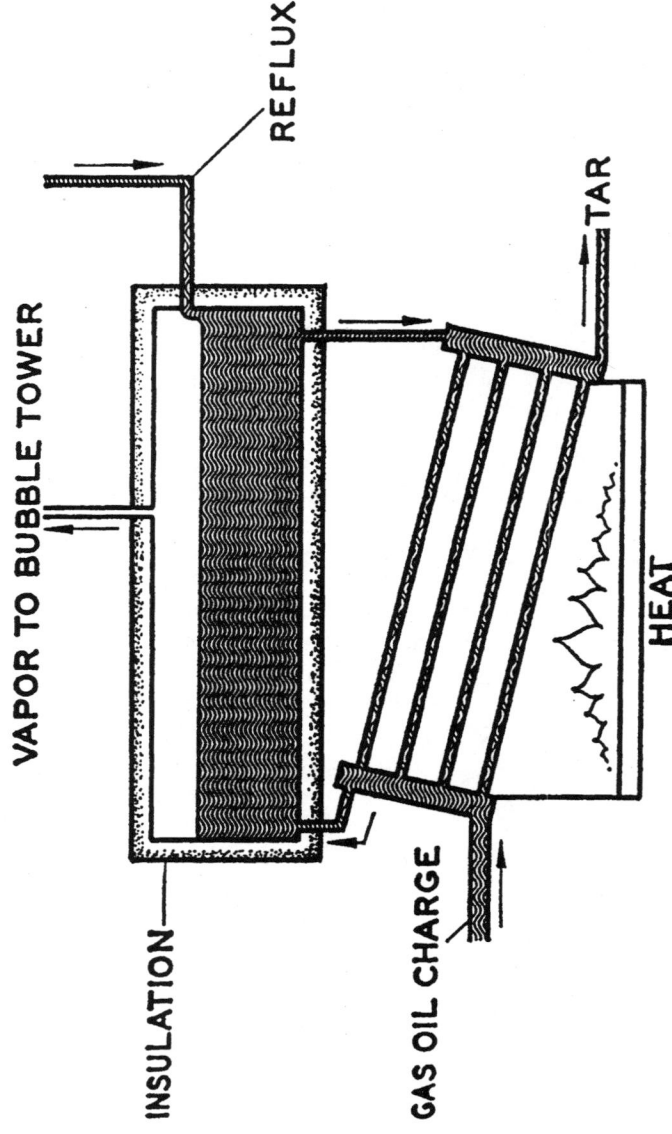

Figure 2. The Burton-Clark pressure still. (Reproduced with permission from reference 1. Copyright 1963 Northwestern University Press.)

reserves. In addition, like many successful persons, Burton did not see the need to improve his process, and did not appear to have even encouraged improvements from within, such as the one accomplished by Clark. The decline in the use of the Burton Process is evident by considering the refining capacity for the various processes that were operating in 1931 (Table 1).

Table 1. Cracking plants in the United States in 1931

Type of process	No. of units	Total capacity, bbl/day	Average capacity
Burton	793	164,249	207
Dubbs	185	252,250	1,366
Cross	150	245,800	1,638
Tube and Tank	118	385,460	3,266
Holmes-Manley	115	233,900	2,033
Jenkins	46	66,150	1,438
de Florez	6	13,550	2,258
Gyro	20	16,000	800
Isom	115	179,150	1,557
Others	320	394,272	1,232

SOURCE: Data from reference 18.

Potential competing processes existed in 1912, and more arose very quickly. Two dominant alternatives to the Burton process were the Dubbs and the Holmes-Manley thermal processes and two similar catalytic processes advanced by the Texas Company (Texaco) and by Gulf Refining (Gulf and Texaco are now incorporated into Chevron).

Dubbs Process - Universal Oil Products (UOP)

Dubbs encountered problems in processing crude from the Santa Maria oil field in California and his solutions evolved into the Dubbs Process (*1*). This crude was prone to emulsification during ordinary distillation. Dubbs invented and utilized a tube still that operated under autogenous pressure (the pressure generated by the vapors during heating). Jesse Dubbs' commitment to his petroleum profession may be gleamed from the name of his eldest son, Carbon Petroleum (CP) Dubbs. While the distillation technique that Dubbs developed had solved the emulsion problem, apparently the senior Dubbs did not recognize the role that pressure could play in crude cracking, since this was not covered in his two initial patents filed in 1909 and 1911 (*5,6*). Apparently the emergence of the Burton process caused Dubbs to make a number of amendments to cover cracking and to broaden his patent coverage. He presumably was encouraged in this direction by Frank Belnap of UOP (*1*).

UOP traces its origin to the meat processing industry. J. Ogden Armour, a very wealthy meat packer, was apparently at odds with the petroleum giant, John D. Rockefeller (*7*). Frank Belnap, as a young patent attorney, approached Armour to finance the legal, experimental and commercial development of pending patent applications by Jesse A. Dubbs. Armor already was part owner of an asphalt production plant located near Independence, Kansas, the Standard Asphalt Refining Co. (Sacco). It produced asphalt by air-blowing crude oil from the Santa Maria oil fields, the same crude Dubbs used at his refinery at Orcutt, California to produce heavy fuel oil for ocean-going vessels. A new company, National Hydrocarbon Company, was formed June 17, 1914, and the name changed in August 1915 to Universal Oil Products Co. An Armour lieutenant, R. J. Dunham, was named president of the National Hydrocarbon Company but was soon replaced by Hiram J. Halle, a financial wizard. At Independence, Missouri, a small building was constructed to house the labs and offices. To ensure secrecy, a high fence surrounded the building.

The oldest son, CP Dubbs, took over the experimental development of his father's clean circulation thermal cracking process (*1,8*). On February 15, 1917, Dr. Gustav (Gus) Egloff was hired by UOP to direct research and development. Egloff, an extremely energetic and talented man with an encyclopedic mind, soon become known as "Gasoline Gus" (*7*). During a patent suit in a St. Louis courtroom, the opposing lawyer asked, "Is it fair to state that your study [about oil emulsions] has been really thorough and exhaustive?" (*9*). In answer, Egloff spoke about emulsions for 21 days, providing a detailed explanation of the role emulsions play in petroleum production and refining (*10*). His testimony was widely reported in U.S. newspapers and became a classic in training potential lawyers to understand that you do not ask a question in court if you do not know what the answer is going to be.

Egloff, despite his small stature, was very athletic. As a youth he was attracted to cycling, winning short races in the old Madison Square Garden in

New York City and 100-mile events on the roads (*10*). At Cornell, when he weighed 125 pounds, he won an intercollegiate wrestling championship. A very private person at home (Who's Who did not even mention his wife), he was outstanding at obtaining publicity for his and his company's scientific and technical accomplishments.

Egloff normally awoke at about 5:00 AM, had a quick breakfast of orange juice and coffee, and then walked the four miles from his apartment home to his office (*11*). Arriving at the office before others allowed him to plan his day's activities without interruption. During the day, he did not just sit in his office - he visited the laboratory to learn first-hand about the work in progress. He not only learned, he also taught. Vladimir Haensel, Ipatieff's first graduate student at Northwestern University during the 1930s, worked at UOP in the summer (*9*). Passing through the laboratory devoted to naphtha reforming studies, Egloff commented that, since the man who directed Haensel's work was on vacation, Haensel should be challenged to develop a catalyst which would not undergo aging by carbon deposition. Needless to say, even though he worked the rest of the summer on his new project, Hansel failed to accomplish this. However, about 15 years later Haensel developed the bifunctional chlorided platinum-alumina naphtha reforming catalyst that transformed naphtha reforming, and made UOP financially secure. Nearly 50 years later this was the dominant reason for Haensel receiving the National Academy of Engineering Charles Stark Draper Prize with its $450,000 award (*12*).

Egloff could be impulsive. "At a scientists' dinner, a friend admired the new golden-brown tie Gus wore. Without a word Gus took it off and handed it to his friend as a gift, then made his address tieless (*10*)."

Egloff was attached to things old. "At a chemist's banquet, his friends took his old hat, filled it with water, ash trays and hard rolls, carried it in as an exhibit, and then presented him with a new hat. Gus still continued to wear the old one" (*10*).

Egloff was outstanding at recognizing and recruiting scientific talent. Vladimir Ipatieff had been a general in the Russian army and had utilized the knowledge he obtained in his artillery training to develop high pressure reactors (*13*). He developed a research program using these reactors that made him a well-known international figure. As Stalin progressed with his "purges," in 1929 a close friend, E. I. Spitalsky, was arrested and five of Ipatieff's former students and co-workers were arrested and shot without a trial. After receiving a secret warning that his arrest was planned, Ipatieff and his wife were leaving Russia to attend a scientific meeting in Germany. Crossing the border into Poland, Ipatieff turned to his wife and said, "Take a good look at your country, Barbara, as we are leaving it for good." Egloff met Ipatieff at a meeting in Germany and invited him to join UOP. They encountered U.S. immigration problems, but Egloff overcame this by arranging the only solution that could be accomplished quickly: have the president of Northwestern University make Ipatieff a faculty member. Thus, at age 65, Ipatieff began his third career as a professor at Northwestern University in what has now become the Ipatieff Laboratory for Catalysis and as the head of a research group of about 10 persons at UOP. At UOP, because of his "inspiring leadership, practical experience, and canny foresight in assigning problems," Ipatieff made many significant discoveries (13). The first of these was silicophosphoric acid catalyst for oligomerization and alkylation, and this is one of the catalysts used for these reactions even today. Egloff also brought Hans Tropsch (co-discoverer of the Fischer-Tropsch reaction of converting syngas to hydrocarbons) to the U.S. but, when Tropsch learned of his fatal illness, he returned to Germany.

Egloff was a visionary (10). Looking to the future he noted that, to reduce weight, football players' pants should be made of nylon, padding should be made of foam-rubber, and the jerseys of fiber-glass. Even decades in advance of the green revolution, Egloff stated that, "There is no fume in an industrial plant that is not controllable." He foresaw airplanes crossing the ocean at 1,000 miles an hour, gasoline giving 50 miles to the gallon, puncture-proof tires that are made from petroleum and good for 100,000 miles, and synthetic textiles made from petroleum. The Concord crossed the ocean at speeds exceeding 1,000 mph, autos can be purchased that provide greater than 50 miles per gallon, most textiles contain synthetics, and the tires on a 1996 Lumina have attained more than 100,000 miles of use. However, Egloff's view that tiny sea animals are at work making deposits of petroleum faster than man can bring it out of the ground remains to be proven.

The three basic components of the Dubbs process as it ultimately evolved were: (1) operating at the self-generating pressure, (2) continuous distillation and (3) clean recirculation (1). The first component was included in Dubbs' first patent on cracking (5). The second component was developed only after work at UOP and after acquiring the patent rights of M. J. Trumble (14) from Shell Oil Co. in late 1919. Trumble provided for the continuous withdrawal of the heavy unvaporized oil that contained most of the coke-forming materials and by continuously adding feed to the loop at a rate to replace the amount of unconverted material that was withdrawn (Figure 3). This allowed the oil circulating in the closed ring to be maintained at a more uniform temperature, pressure and composition than the earlier processes. The third component was a result of CP Dubbs' work. In addition to removing a tar stream containing the coke formers, a second fractionator allowed the desired product to distill overhead while condensing the tar-free unconverted fraction that was recycled to mix with fresh oil for further cracking (Figure 4) (15). During 1917-1918 UOP worked on the development of the process in utmost secrecy.

Progress was rapid in developing the continuous cracking process and on July 29, 1919, 21 people representing 14 companies gathered in Independence to observe a demonstration run (1). After 10 days of operation, the tubes were still free of carbon and the demonstration induced 8 refiners to reach licensing agreements with UOP. Compared to the competing non-continuous Burton

113

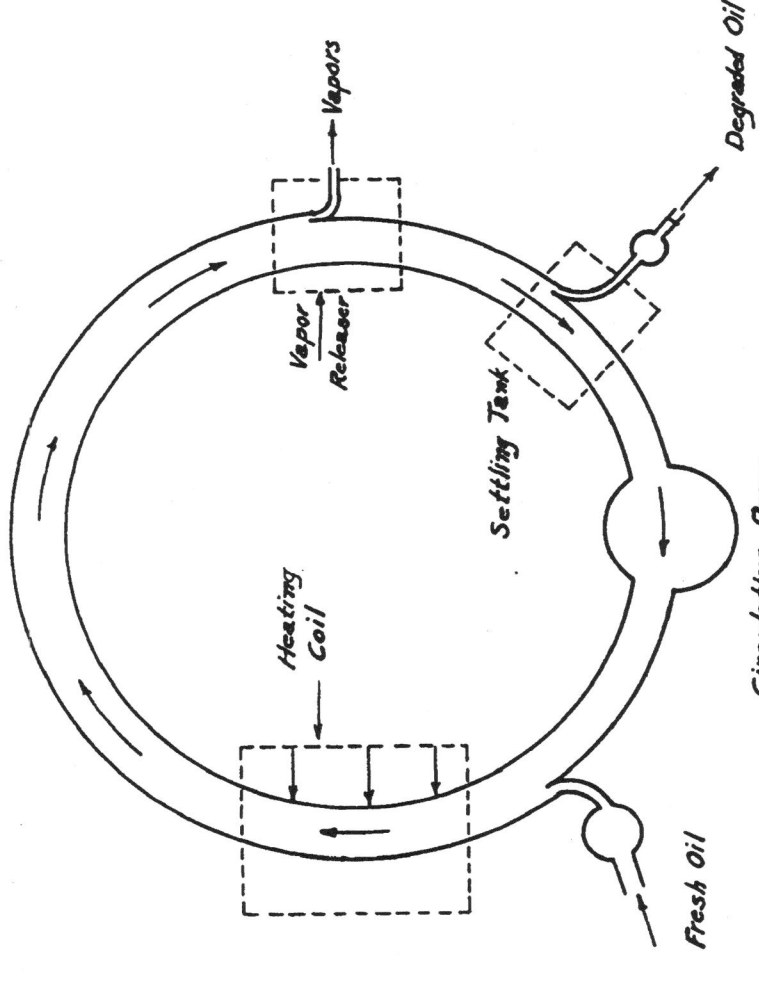

Figure 3. Trumble's "closed ring." Schematic diagram from patent 1,281,884. (reference 14)

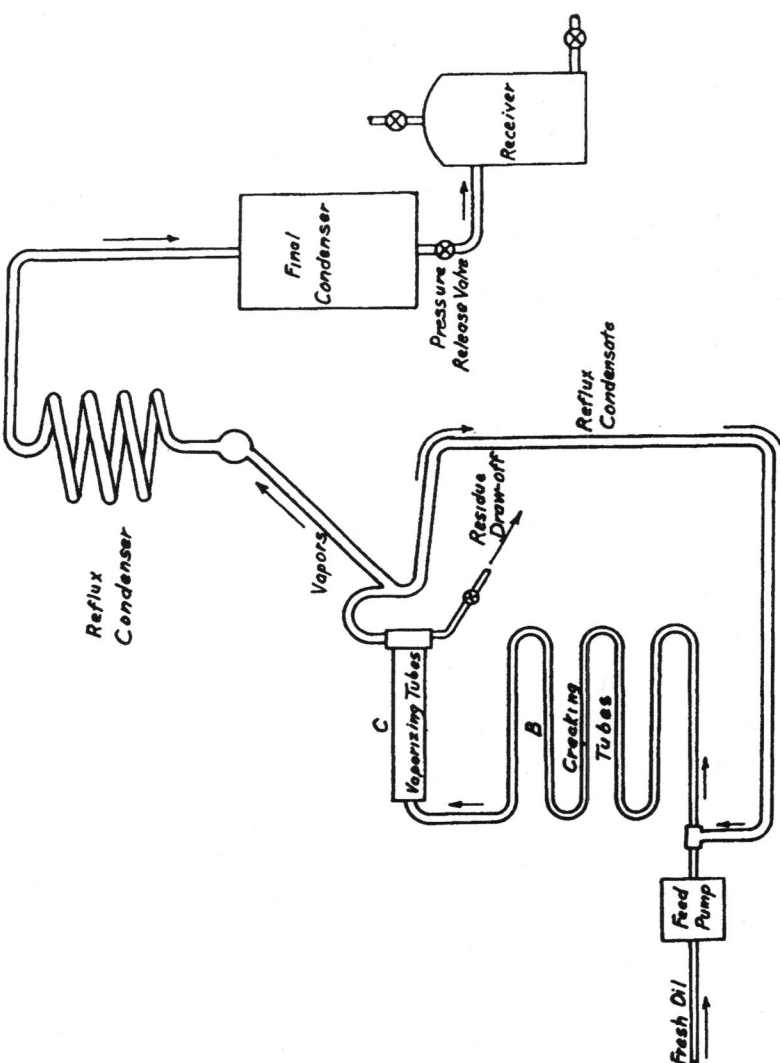

Figure 4. The principle of "clean recirculation" as taught in Dubbs' patent 1,392,629. (reference 15)

process, as improved by Clark, the Dubbs process, now owned by UOP, was continuous, produced a plentiful supply of reflux for dilution, could crack topped crudes as well as gas oil and similar fractions, and the continuous, even flow through the heating zones facilitated the removal of coke-forming materials. These features offered much longer run times between shut-down and allowed a significant increase in gasoline yield. The licensing arrangements with UOP were much more liberal in fees and they guaranteed much better performance than those of Indiana Standard; an additional feature was that UOP would allow the users of the Dubbs process to practice any improvements that occurred during the life of the contract.

The Texas Company acquired American patent rights to W. A. Hall's British patents on cracking that featured low pressure and the absence of steam. Texaco managers, R. C. Holmes and F. T. Manley, together with a group of chemists that included G. W. Gray, developed the Holmes-Manley process to the point where, by 1920, it could compete with the Burton process. In the following decade, the Holmes-Manley process became established as a leading commercial cracking process (*16*).

Standard Oil Co. of New Jersey (Jersey Standard) initially utilized the Burton process. Using the rights to patents obtained by Ellis and their experience gained in operating the Burton process, they developed their Tube and Tank process. Carlton Ellis was an inventor who started a company and located his plant adjacent to Jersey Standard's refinery in Bayonne, New Jersey, in order to convert the propylene that was present in the gases from thermal cracking to isopropanol. During WWI the company provided acetone, derived from isopropanol, as an additive to boost the performance of airplane fuel. However, when a source of aromatics became available, it was apparent that the sale of acetone would decline. Ellis approached Jersey Standard, whose management noted that the patents held by Ellis that involved distillation under pressure may predate those of Cross. This led Jersey Standard to acquire the rights to Ellis' patent. Starting at low pressures, Jersey Standard's personnel gradually increased the pressure to 350 psi, then to 450 and 750 psi, and finally 1,000 psi as better welded chambers and better alloy tubes became available. This process, partly because of the size of Jersey Standard, soon became the largest volume cracking process.

Cross Process

The Cross process was an outgrowth of work by a University of Kansas chemistry professor, Walter M. Cross (*17*). Together with his brother, Roy, they began experiments on thermal cracking in 1913. During the years that followed they obtained several significant patents, especially for high pressure distillation. Eventually the brothers hit upon the idea of placing a reaction chamber between the heated tubes and the evaporator (Figures 5 and 6) (*18*). The brothers licensed the first plant utilizing their concepts to the Indian Refining Company at Lawrenceville, IL in 1921, and within another year six other companies, including Pure Oil, had installed units.

"In August 1922 the Standard Oil Company of New Jersey brought suit against the Pure Oil Company, operating the Cross process, for infringement of the Ellis patents. The case was never brought to trial and was dismissed following an agreement between Jersey Standard and the Gasoline Products Company, a company formed by the Cross brothers, entered into in 1923. In 1915 the New Jersey Company had entered into an agreement with the Standard Oil Company of Indiana, owner of the Burton patents, and later the Texas Company was included. In 1925, the United States Government brought suit against this group that was involved in these cross licensing agreements. In this suit the government unsuccessfully sought to show that the patents relied upon by the defendants were not valid in that they were anticipated by prior art (*18*)."

UOP, once they had strong coverage with the Dubbs patents, brought suit in 1920 against Indiana Standard. This would involve a significant sum of money, since infringement would involve not only Indiana Standard's own operations but also its licensees. Both UOP and Indiana Standard showed great willingness to provide financial support for litigation and in 1930, ten years later, the case seemed no closer to settlement than when it started (*4*). By this time Shell and Standard Oil of California were paying UOP over six million dollars in royalties a year. Deciding that it would be cheaper to buy UOP than to continue paying royalties, they set out to do so, and on January 6, 1931 Shell, Union Oil Company and Standard Oil Company of California completed the purchase of UOP for 25 million dollars. Actually Shell paid ten million and Standard five million. The rest was provided by other oil companies, including three million apiece by Standard of Indiana, Standard of New Jersey and The Texas Company, who was allowed three million for adding their patents to the pool.

The Patent Club

The beginning of legal problems in the ownership of rights to thermal cracking may be considered to begin on August 7, 1916 when UOP filed a suit against Indiana Standard. UOP claimed that Indiana Standard infringed upon their basic patent (*19*) when it operated the Burton process. This suit was prolonged and resulted in testimony being taken before a special master at numerous times and places so that by 1927 it was estimated the record of these consisted of more than 30,000 pages (*19*). Indiana Standard filed a counter suit against UOP on September 29, 1923 in Illinois to the effect that the Dubbs process infringed on patents by Burton, Clark and Humphreys. UOP, on

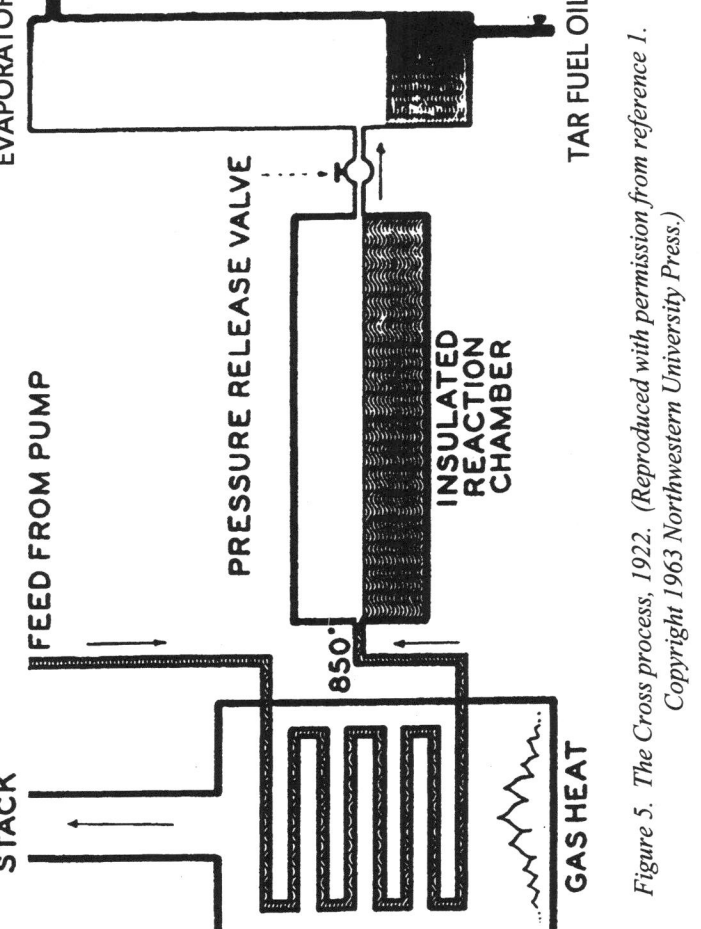

Figure 5. The Cross process, 1922. (Reproduced with permission from reference 1. Copyright 1963 Northwestern University Press.)

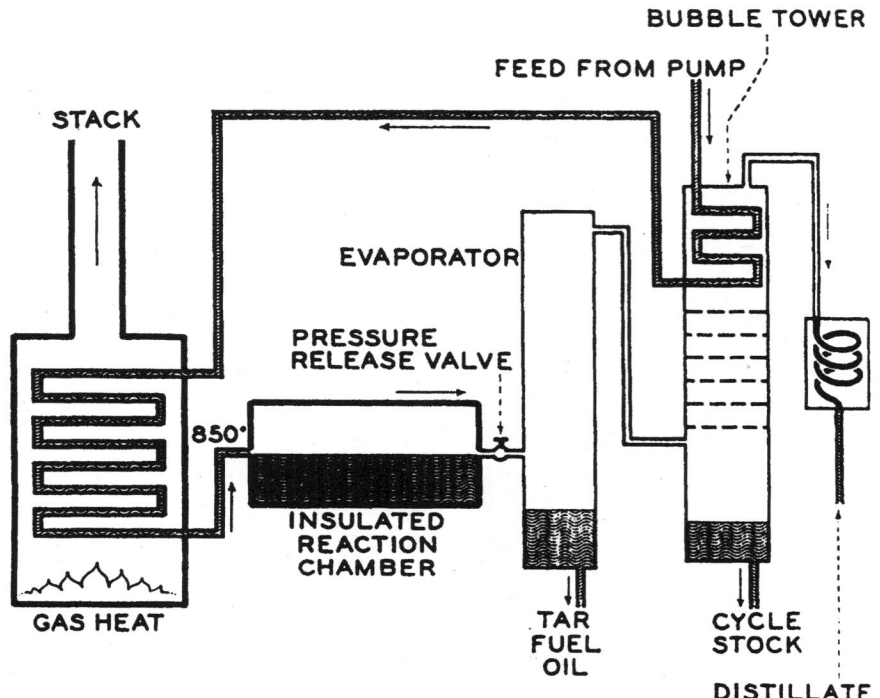

Figure 6. The Cross process with bubble tower, 1924. (Reproduced with permission from reference 1. Copyright 1963 Northwestern University Press.)

January 28, 1926, brought suit in Indiana claiming that Indiana Standard had infringed on many of UOP's patents. These suits continued until settled out of court in 1931.

During the early years of this litigation, the Texas Company and Indiana Standard contested rights involving patents obtained by Adams and Burton (*19*). These two companies settled their dispute by pooling their interests in these patents. Jersey Standard and the Texas Company also settled a dispute over patent rights in 1923 by pooling their interests. As noted earlier, Jersey Standard had brought suit against the Pure Oil Company, charging that Pure infringed a patent by Ellis covering the Tube and Tank process by operating the Cross process under license from the Gasoline Products Company. About 1923, Jersey Standard and Gasoline Products Company also settled by pooling their interests. The result of these pooling arrangements brought together four strong companies - Jersey and Indiana Standard, The Texas Company and Gasoline

Products Company. Shortly thereafter the M. W. Kellogg Company joined these four companies and the five became known as the "Patent Club."

The Patent Club was a strong group with patent pooling arrangements and certain agreements concerning market territory, controlled by the licensing process. The U.S. government filed suit in the federal district court in Chicago in 1924, claiming that it constituted a combination in restraint of trade that violated the anti-trust laws. A number of patents held by the Patent Club members were also attacked. During three years, a special master, appointed by the court, heard testimony and then ruled in favor of the defendants, the Patent Club. The court, in 1929, did not accept the master's finding regarding the patent pooling arrangements, and did not rule regarding the validity of the patents. The Patent Club appealed the decision of the district court to the U.S. Supreme Court. The Supreme Court, in April, 1931, reversed the lower court's decision with regard to the agreements between the companies and licensees, and the Patent Club suit ended.

Thus, by 1930, there were two dominant groups involved in liquid phase cracking processes: the Patent Club and UOP. UOP was therefore a small fish in a very large pond filled with many large fish. Undeterred, UOP had filed suits against many of these larger organizations, including, the Standards of Indiana, New Jersey and New York, Gulf Refining Company and Gasoline Products Company as well as various organizations licensing processes from these companies. In turn, many of these organizations filed suits against UOP. In spite of these attacks, UOP was able to continue to interest refiners in using the Dubbs process and the proceeds allowed them to continue to wage a strong legal battle.

The refiners had continued to expand their scientific and engineering operations, as illustrated by the example of Indiana Standard (Figure 7). However, refiners were still dominated by a management that relied on practical experience in a refinery; i.e., people who had advanced through the ranks by acquiring experience in all facets of the refining operations. UOP, on the other hand, was a small organization whose only product was processes they convinced others to use. Thus, UOP's management took advantage of the available scientific and engineering advances to make their processes the best that were available to the refiner, and especially to the smaller refiners. UOP also rapidly built up a vast catalogue of patent coverage. Up to 1930, in spite of the Patent Club, UOP had made some headway in attracting larger refiners. For example, California Standard had taken a license for the Dubbs process in 1926. Shell, through its subsidiaries, had also license agreements with UOP. In 1931, California Standard and Shell Union Oil Company bought UOP for $30 million. This purchase also involved other companies who had brought legal action against UOP. Thus, with this purchase, all legal action was dropped and the "Peace of 1931" was thought to end the conflicts concerning cracking patents.

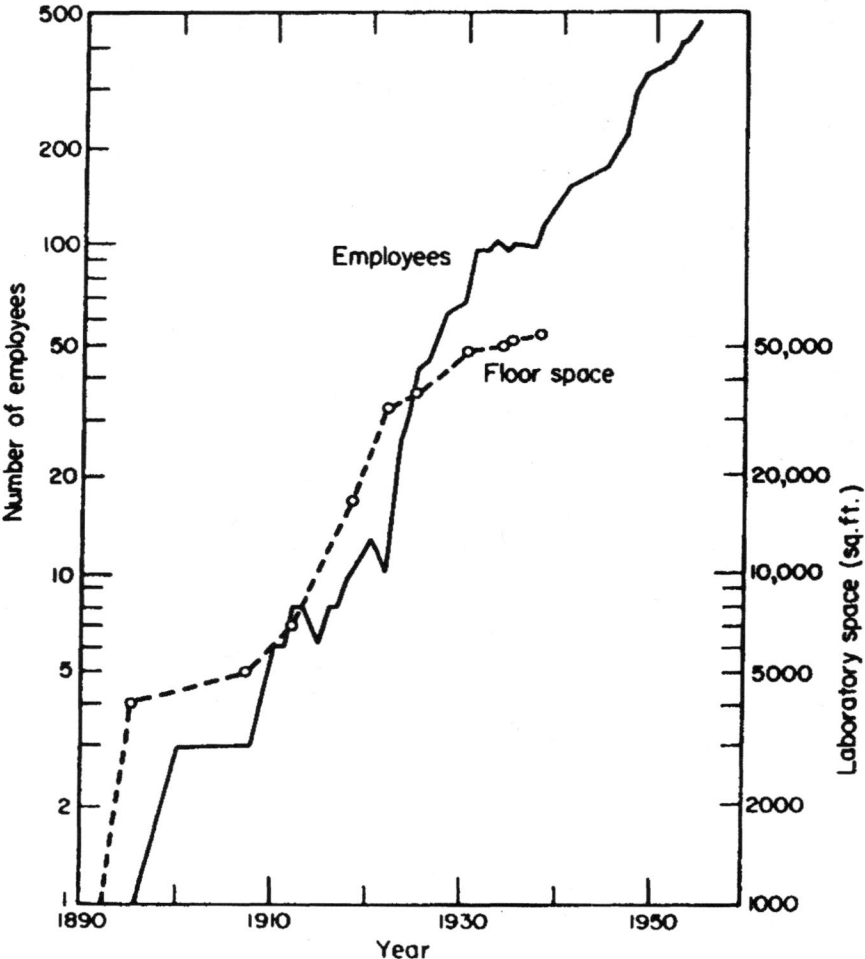

Figure 7. Growth of Whiting research laboratory, Standard Oil Co. (Indiana), 1890-1955. (Reproduced with permission from reference 8. Copyright 1962 Massachusetts Institute of Technology.)

As McKnight (*19*) details, the Peace of 1931 was short-lived. First, conflict developed between those companies not included in the treaty and the treaty members, UOP and the Patent Club members. Soon conflict arose among the treaty partners themselves. McKnight (*19*) indicates that many refiners held the view that cracking was a natural evolution of the refining art and that patents covering thermal cracking were not valid. Officials of the Skelly Oil Company advanced this view when they were made a defendant in a suit brought by UOP contending that Skelly's use of the Jenkins process infringed upon many of their patents, including those they acquired from Trumble and Dubbs. Eventually the suit involved only the Trumble patent, and the Federal District Court at Wilmington, Delaware held this patent to be valid and to be infringed by Skelly. Upon appeal, the decision was upheld. When the U.S. brought suit against the Patent Club, American Refining Company terminated royalty payments to the Gasoline Products Company on the basis that if they continued to make payments, they would be included in the suit. Gasoline Products Company obtained a judgment against American, indicating that the latter company had infringed Cross's U.S. Patent 1,423,500. This decision not only increased the presumption of the validity of the Cross patent but also strengthened the position of all of the royalty-collecting companies.

In the 1930s, the Winkler-Koch Engineering Company designed a cracking process based upon the best and latest knowledge available to the art at that time (*19*). This organization had no patent position and contended that their process was not based upon any patent, whether valid or not. Operators of the Winkler-Koch process were soon attacked in court. Initially, Winkler-Koch successfully defended their position, and early court rulings were in their favor. However, UOP brought suit against Winkler-Koch and the Root Refining company, initially claiming that the process infringed many UOP patents. Eventually, the case turned on a single Dubbs patent (1,392,629) and one by Egloff (1,437,593). In a decision on April 27, 1934, the court held both of these patents to be valid and infringed. The Circuit Court of Appeals of the Third Circuit upheld the decision.

In winning and/or settling many suits in their favor, UOP was becoming a major player in the cracking process business. Gradually even the members of the Peace of 1931 accord developed animosity toward UOP. Again, UOP found itself being attacked in court by several of the Standard group, The Texas Company, Gasoline Products and others. The situation became so bad that in several instances one of the contestants would be actively participating in the defense of another refiner against UOP. As the trials were approaching the decision stage, a far-reaching settlement was attained. On December 1, 1937, an agreement was reached whereby UOP obtained non-exclusive rights under patents owned by The Texas Company, the Standard Oil Companies of New Jersey and Indiana, Gasoline Products Company, Atlantic Refining Company and Gulf Oil Corporation (*19,20*). The "Treaty of 1937" cost UOP a substantial payment, but the amount was not disclosed (*20*). This treaty permitted both groups to continue doing what they had been doing without fear of litigation. While UOP had to pay a substantial settlement cost, they obtained rights to the Behimer patents owned by The Texas Company, and this would extend the "monopoly" held by UOP by virtue of the Dubbs patent to at least 1949 when the Behimer patent rights would become available to the public.

After more than 20 years of legal battles over cracking patents, it appeared that peace would reign from 1938 onward. However, the revolutionary advance of Houdry catalytic cracking was now at hand and the rights covering thermal cracking would rapidly decline in value. Peace was obtained at a great cost, but once won could not be enjoyed. Catalytic cracking made the objects of the battles obsolete, and soon to be nearly worthless. The legal battles were now to be joined in a new area, but by many of the same players.

Catalytic Cracking

Aluminum Chloride "Homogeneous" Catalyst

By 1914, George W. Gray, employed by The Texas Company, had developed a process known as AlCl3, because it was based upon anhydrous aluminum chloride as a catalyst (*1*). Thus, 5 wt.% of the catalyst was heated with dry petroleum distillate and 15 to 60% gasoline fraction was obtained, depending upon the source of the feedstock. The Texas Company did not commercialize the process because of the high cost of the catalyst and their inability to recover and recycle the anhydrous catalyst.

Dr. Almer M. McAfee, a Texan who graduated from the University of Texas in 1908 and from Columbia University with a Ph.D. in chemistry in 1911, joined The Texas Company in 1912 and a year later changed employment to the Gulf Oil Corp. McAfee was not enamored by thermal cracking, as the following illustrates (*21*):

"The conversion of high boiling petroleum oils into lower boiling oils of greater commercial value is an old problem....The inventors have rung the permutations on this simple idea; they heat under pressure and they heat under vacuum; they heat in the presence of gases; they heat in the presence of catalysts; they heat in tubes and they heat in boilers, etc., etc. It is safe to say that in 99 percent of the methods which have been proposed for converting high boiling oils into lower boiling oils, 'cracking' by heat is involved. *Sometimes it is disguised in ornate language; sometimes it lurks behind intricate apparatus, but it is always there (emphasis added).*"

At this early date, McAfee recognized that the preferred hydrogenated oil products could not be obtained without rejection of carbon. McAfee thought that the discovery of Friedel and Crafts (22) should be applicable to effect catalytic cracking. He found that with proper control of the vapors leaving the batch reactor and with sufficient time of contact, nearly all of the oil could be cracked to produce a high yield of transportation fuel. McAfee made improvements using bauxite ore as a cheaper source of aluminum chloride and in treating the aluminum chloride/hydrocarbon mixture that remained after cracking with chlorine to recover the aluminum chloride for recycle (23). While the McAfee process was operated on a commercial scale for a short period by Gulf and thus became the first commercial catalytic cracking method, it was not widely accepted and even Gulf soon abandoned it in favor of thermal cracking. McAfee retired from Gulf after thirty eight years of service. His son eventually became president of Gulf Oil.

The petroleum refiners were becoming the "milk-cow" for the legal profession. After lengthy litigation by Texas Oil and Gulf over two patents and who infringed in the aluminum chloride cracking process, the case was finally settled in favor of Gulf Refining in 1928 (1). The victory came long after the process had been retired.

Houdry - the Process and the Man

Catalytic cracking is usually considered to begin with the work of Eugene Houdry, and his process has been described as the largest single advance in the development of refining processes. However, the superiority of catalytic cracking over the thermal processes was recognized by many prior to Houdry's work. The problem was that catalysts could not be utilized effectively because of their rapid activity deterioration. Thus, the critical advance introduced by Houdry was a process which would allow the restoration of the initial catalytic activity to an aged catalyst.

Eugene Houdry is as unique an individual as his cracking process was. Born in France, his father operated a very successful steel fabrication business (24). Graduating from college in 1911, he not only finished first in his class because of his scholastic work but was also captain and halfback on his school's soccer team, which won the French national championship in 1910. Following his college training, he joined the family business as an engineer. He was soon drafted into the military and, as a lieutenant, he took part in the first battle of WWI in which tanks were used. On April 16, 1917, he was seriously wounded in the battle of Juvincourt; for his actions in this battle he was subsequently awarded the Croix de Guerre and made a chevalier in the Legion of Honor. Following the war he joined the family business again. In spite of his business success, he maintained a strong interest in the automobile, and especially the operation of its engine. During 1922, he visited the U.S. to attend the

Indianapolis 500 race and to visit Ford Motor Company. At Ford he became convinced that further improvements in the manufacture and use of autos could only come as a result of improvements in the operation of the motor, and motor operation was then limited by the properties of the fuel. Returning to France, he set out to develop a superior motor fuel as well as to provide France with an internal source of transportation fuel. This latter goal required the liquefaction of lignite, the only abundant fossil fuel found within France.

Houdry and his father were initially guided in this effort by the results of an Italian group, working under the direction of a French chemist, E. A. Prudhomme (4,8). The Italian group used nickel and cobalt catalysts to convert carbon monoxide and hydrogen to hydrocarbons that were claimed to be an excellent motor fuel. Houdry bought into the group, but as time went on it was evident that this was not going to be successful. According to E. Houdry, "Prudhomme always thought of catalysts as little animals. By putting a little gasoline with them, he thought that he could give them the right idea - help them along (4)." Others, including Jersey Standard, had looked into Prudhomme's results and concluded that an accurate mass balance indicated that the publicized yields were higher than actually obtained.

Rather than abandon the effort altogether, the son decided to change the direction to lignite liquefaction, to leave the family business and to devote himself to the effort full time. Following his life-long practice, Houdry would spend days at the laboratory, sleeping for brief periods on a cot, paralleling the Edison model. He was sufficiently successful in his lignite work that the French government supported the construction and operation of a large pilot plant. The studies included the testing of a wide variety of catalysts, usually selected following a true Edisonian approach. The results from the operation of the plant showed in 1929 that the process was technically sound but that it could not compete economically, and financial support from the government was terminated. In addition to his involvement in the research activities, Houdry had to be the "sales person" for the effort and this was a nearly full time job during the initial years.

Houdry had fortunately conducted studies of petroleum cracking in parallel with the lignite work. After three years of effort, at 3:00 AM in the morning one day during April 1927, Houdry suddenly noticed that one reactor, containing a silica-alumina clay, was converting a heavy, low-grade crude to high quality gasoline (1). By any standard, gasoline quality testing was crude (4). The initial judgment was by appearance. Three tests were also used. First, a rough molecular weight was obtained by burning a jet of the fuel: a blue or invisible flame meant that the gasoline contained too much of the low molecular weight compounds. A color test and gum forming properties were obtained by hanging a bottle on a line at the back of the lab. The crucial test was to determine the octane but there was only one motor in Europe at that time that could do this. Undeterred, Houdry used his sporty Bugatti racing car for the test. With each new batch of promising gasoline, Houdry would fuel up his car and head to a long hill located on a road not far from the lab; the better the gasoline the further up the hill Houdry's car would go prior to beginning to knock under the load of hill-climbing. With the success of the clay silica-alumina catalyst, Houdry began to publicize his process. Among the first to visit the Houdry labs were representatives of the Anglo-Persian, the Royal Dutch Shell and the Standard Oil Company (New Jersey). These companies sent crude samples to be processed but did not pursue this since they concluded that major design and construction problems remained to be solved. Not meeting with success in selling his process in Europe, Houdry tried to interest U.S. companies in his process. He met with success when he visited Vacuum (eventually Mobil Oil and now ExxonMobil), primarily because their European representative was impressed by the results coming from the Houdry lab and had "pre-sold" Vacuum management. In 1930 Houdry, encouraged by Vacuum, moved his laboratory to the U.S. and located in Paulsboro, NJ adjacent to a Vacuum refinery and research laboratory. His French team was joined by Vacuum personnel, and the results from their work were so promising that the Houdry Process Corporation was formed in 1931, with Houdry owning 2/3rds of the stock and Vacuum the remaining third. Soon Vacuum merged with Standard Oil (New York) to become Socony-Vacuum and the new management was not enthusiastic about continuing the funding for Houdry's work.

The search for funds began anew, and Houdry was able to convince Arthur E. Pew, a member of the family that owned controlling interest in Sun Oil, to provide support; for this Sun obtained half of Houdry's stock so that Sun, Socony-Vacuum and Houdry each owned one-third of the company. The laboratory moved across the Delaware River to Marcus Hook, Pa. (actually Linwood, Pa.), where Sun Oil was located (the Houdry Laboratory site has been designated as a National Historical Landmark by the American Chemical Society). In 1932 Houdry was able to announce his new process to the world. However, while he had a process, the valves, pumps and reactors needed for the process were not available. With much effort, by 1936, it was considered that the problems of the hardware had been overcome.

In spite of the decrease in demand for gasoline because of the depression, Mr. Pew soon decided that in the future there would be an increasing demand for higher quality gasoline. Owning the majority of Sun stock, once he decided that there was the need for catalytic cracking, he could make the decision to go forth with the construction of a new plant. The Sun Company approached Socony-Vacuum, who agreed to provide an amount of funding equal to that which Sun had spent in development during the past four years. It was also agreed that each company would build a commercial plant. Socony-Vacuum decided to build a small plant and to crack light gas oil before making the more expensive investment in a large plant. Because the small-sized equipment was available by converting a thermal reactor, Socony-Vacuum was able to bring on stream in June, 1936 a 2,000 barrel per day (BPD) plant in their Paulsboro, NJ refinery, the first commercial catalytic cracking unit. Sun built a larger 12,000

BPD plant which came on stream in April, 1937. The successful operation of these plants led Sun and Socony-Vacuum to build more than ten plants during the next four years, and for Houdry Process Corp. (HPC) to license to other companies. During the first two years of WWII about 90% of the aviation gasoline was obtained from catalytic cracking in 24 Houdry plants charging a total of 330,000 barrels a day. By 1944, there were 29 units in operation, with a capacity of 375,000 barrels a day, and by 1947 there were 37 licensed units.

The Houdry process operated at about 900°F and up to about 30 psi. Gasoline yields were increased almost a third to over 30%, and the product octane was in the low 80's compared to the low 70's for thermally cracked gasoline. However, the catalyst had to be regularly reactivated by burning off the coke and coke precursors. Initially this was done by frequently switching from a hydrocarbon vapor stream to a dilute air stream and back again, with a purge in between, each of these steps lasting about ten minutes, then by swing-case operation with up to four manifolded reactors, and later by moving the catalyst pellets from a reaction zone to a regeneration zone and back again, at first with buckets and later in a circulating moving bed system with compressed air (the Thermofor catalytic cracking or TCC process) or with flue gas and steam (the Houdriflow process). With catalyst to oil ratios ranging from one to as high as seven, there was a very considerable catalyst circulation rate.

Houdry and his team encountered many problems in developing the hardware for his process. One reason for this was that they were taking refining to a higher level and had to develop their own ways to do this. The limitations on heat transfer required the catalytic reactor tubes to have a small diameter. Thus, to develop a process with a large capacity required that many tubes be utilized. To accomplish this, Houdry placed many of these tubes in a reactor shell (known as case from the French). A horizontal cross section of one of these cases is shown in Figure 8 (25). An inlet tube (RT), for oil vapor and air, is surrounded by collector tubes (CT), for outlet vapors, and cooling tubes (KT and DT). Initially, the heat was removed by placing cooling tubes close to the reactor tube and allowing cooling by radiation. Water/steam was used to transfer heat and for cooling but this was not successful, due to corrosion. The water coolant was replaced by molten salt, as shown in the vertical cross section (Figure 9) (26). The collector tubes (CT) are welded to the salt-cooled tubes (DT). Fins are attached to the cooling tubes to facilitate heat transfer. The cooled surfaces are separated by only about 5/8 inch. Provision was made for changing catalyst but this normally required shut-down of about half of the cases and reloading catalyst required 4 to 7 weeks. If tubes had to be removed the down-time could extend for 3 months, and longer. The time sequence for a set of four cases is illustrated in Figure 10 and represents a truly amazing feat in that pre-computer control era. While the Houdry Process Corp. continued to make improvements upon the hardware, the basic design did not change.

What did Houdry accomplish with his fixed-bed catalytic cracking process? He introduced the first really successful catalytic cracking process (25). While

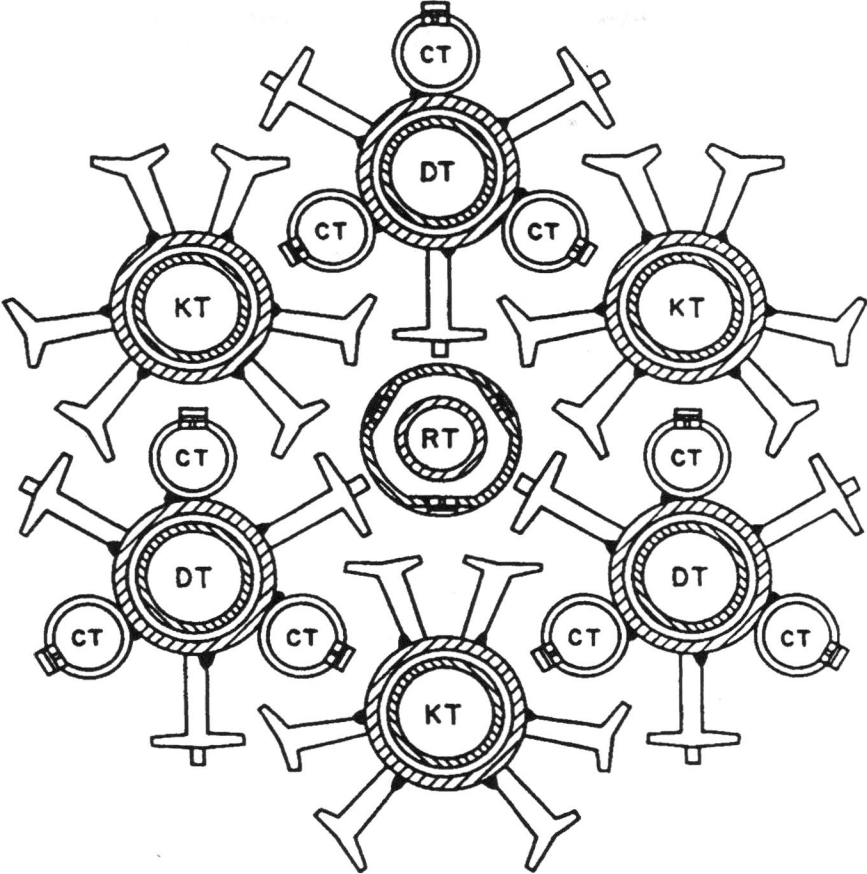

Figure 8. Horizontal cross section of Houdry catalyst case (early design). (Reproduced with permission from reference 32. Copyright 1954 Elsevier.)

the advances in catalyst formulation, first based upon naturally occurring clays and then synthetic silica-alumina, should not be overlooked, the Houdry accomplishments were foremost in the areas of superior advances in hardware and in process control. He introduced the first large-scale cracking process that practiced air regeneration. He developed the first process to employ automatic control of the cycles; this required the development of the cycle timer. He was the first to introduce and use high-temperature operating valves. He introduced the first use of large-scale gas turbine-driven compressors. In doing this, he was

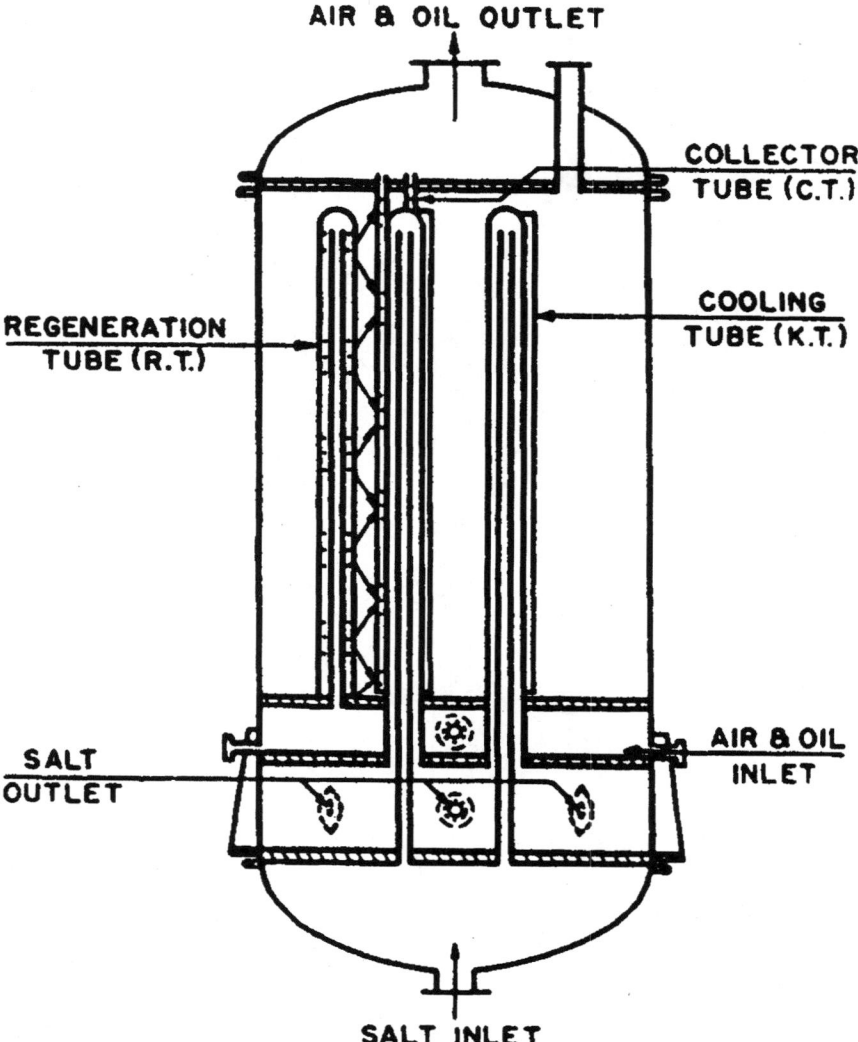

Figure 9. Vertical cross section of Houdry catalyst case (early design). (Reproduced with permission from reference 32. Copyright 1954 Elsevier.)

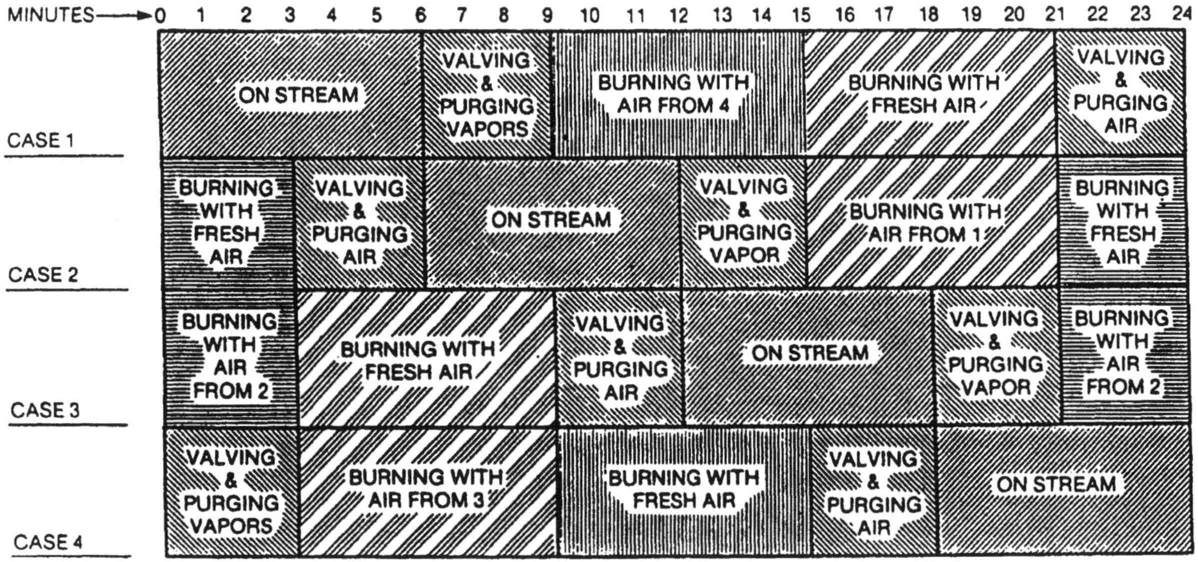

MINUTES——▶ 0 1 2 3 4 5 6 7 8 9 10 11 12 13 14 15 16 17 18 19 20 21 22 23 24

CASE 1

ON STREAM | VALVING & PURGING VAPORS | BURNING WITH AIR FROM 4 | BURNING WITH FRESH AIR | VALVING & PURGING AIR

CASE 2

BURNING WITH FRESH AIR | VALVING & PURGING AIR | ON STREAM | VALVING & PURGING VAPOR | BURNING WITH AIR FROM 1 | BURNING WITH FRESH AIR

CASE 3

BURNING WITH AIR FROM 2 | BURNING WITH FRESH AIR / | VALVING & PURGING AIR | ON STREAM | VALVING & PURGING VAPOR | BURNING WITH AIR FROM 2

CASE 4

VALVING & PURGING VAPORS | BURNING WITH AIR FROM 3 | BURNING WITH FRESH AIR | VALVING & PURGING AIR | ON STREAM

Figure 10. Four case Houdry unit on 24 minute cycle. (Reproduced with permission from reference 27. Copyright 1990 Freund Publishing House Ltd.)

forced to utilize expensive alloys and large, tall units necessitating lots of framework for support. In common with many initial, revolutionary advances, the developers were forced to improvise, and this resulted in complex and overly large units that were expensive to construct and complicated to operate.

Initially, in spite of its complexity, the Houdry process hardware worked satisfactorily. However, the steam cooling tubes were subjected to severe conditions so that after about two years corrosion was sufficiently severe to allow water to contact the catalyst. The steam reacted with the catalyst during regeneration, resulting in severe declines in activity. To overcome this problem, the water/steam fluid was replaced with molten salt to cool the tubes in the case and to act as an intermediate heat transfer agent from the case to the steam generators. This invention has been credited to Socony-Vacuum (27).

It frequently happens that the person who is the driving force in accomplishing a very difficult task does not have the ability, or sometimes the opportunity, to make the transition to manage a large successful organization. So it was with Houdry. In 1944, Socony-Vacuum brought in a person who was appointed as vice-president of Houdry Process Co. In 1948, Eugene Houdry terminated his work at Houdry, but retained his financial position, and formed Oxy-Cat, Inc., located in Radnor, PA. This company made catalysts to convert unburned fuel from internal combustion engines to carbon dioxide. He successfully accomplished this but was well ahead of his time. The only vehicles using this type of catalyst were those used in enclosed areas, and therefore the market was limited. The monolith noble metal catalyst that he developed for this purpose was the forerunner of today's automotive catalytic convertor. Houdry soon started a third company, this time to employ his views of catalytic conversions to extending life.

Houdry's lifestyle could serve as a model of an aristocratic man at the turn of this century. He settled in a house on City Line, the high society section of Philadelphia, with four employees serving as butler, maid, cook, etc. One evening during the first International Congress on Catalysis, held in Philadelphia in 1956 during the height of the US-USSR cold war, Houdry entertained the local organizers and the Russian guests. Charlie Plank, a co-inventor of the zeolite cracking catalysts, frequently recounted that, when one of the Russians arrived, he looked around the mansion and remarked, "So this is the house that Gene built!", displaying evidence that he was familiar with the Western children's story, "The House that Jack Built," in spite of the anti-western feelings advanced by the Russian leadership at that time. Houdry is reported to have stated, in response to an inquiry as to why he would have a mistress, "But how could any man treat his wife so badly by not having one?" In his later years, when it was necessary for him to go to the hospital for a minor treatment; he reserved a three bedroom suite: the room in the middle was for him, the one to his left was for his wife, and the one to his right was for his current mistress. Houdry also believed there were several ways to prolong one's life. Thus, during his latter years he slept with a tent over his bed so that he could increase the concentration of oxygen he breathed during the night by adding a stream of pure oxygen into the tent to mix with normal air.

UOP Cracking Process

In 1940, UOP workers wrote an encyclopedic volume covering catalysis (*28*). Egloff was an active writer, and many of his articles were included. Most of these were reviews of the literature or promotional articles of benefit to U.O.P. processes. Professor Herman Pines, for years a coworker of Vladimir Ipatieff at UOP, the person responsible for the development of many of the U.O.P. polymerization and alkylation processes, admired Egloff. To emphasize the extent of Egloff's writing, Pines stated that "Egloff is probably the only person who has written more books than he has read."

In their review of catalytic cracking, Berkman et al. (*29*) write, "yields of 85% gasoline of 81 octane have been obtained from Mid-continent gas oil by the U.O.P. process." This yield is based upon recycle operation, but at an extent that can be practiced commercially. Also included in the yield was polymerization of the cracked gases and the authors claimed that 95% of the cracked gases could be polymerized. They describe their plant as "...consisting essentially of a heater, catalyst reactor, and automatic controls for alternating the flow of oil through the furnace and reactors, and another section for reactivating the catalyst. The cycle is about 40 minutes in duration." They indicate that U.O.P. has catalytic cracking units in laboratory and commercial development stages. After nearly four pages of description of the U.O.P. process that is still in the development stage, they describe the Houdry process in two paragraphs covering one-half page, indicating that "A number of commercial [Houdry] units have been installed." The authors conclude by referring the reader to an Egloff paper, "The Catalytic Cracking of Aliphatic Hydrocarbons" (*29*). As the above indicates, U.O.P. was aggressively trying to enter the catalytic cracking business. In the 1930s, UOP was very active in research in polymerization (which includes what is known as alkylation and oligomerization today) and isomerization. By 1938, UOP had developed their own fixed-bed catalytic cracking process which utilized catalyst formed into the shape of pills, but this work was put aside in 1939 when the resources allocated to cracking were diverted to study fluid catalytic cracking (*4*).

Moving Catalyst Bed

Thermofor Catalytic Cracking – TCC

During the period in the 1930s when Socony-Vacuum was working with Houdry, other groups at Paulsboro were looking at other options for catalytic cracking. One that Socony-Vacuum eventually commercialized was the Thermofor Catalytic Cracking (TCC) process. Leonard Drake recalls that 30/60 mesh catalyst particles were initially used but that it was soon realized that larger and stronger catalyst spheres were needed (*30*). A method for the preparation of synthetic silica-alumina beads was developed by Milton Marisic, who has four of the basic patents on the preparation of the Socony-Vacuum bead catalyst. A measure of the security surrounding catalytic cracking at that time can be gleamed from the fact that Marisic had developed the small, laboratory scale synthesis and the technique of forming spheres by allowing the gel to fall into a heated organic liquid where the retained water evaporated and the gel gained rigidity. When the Socony-Vacuum engineers decided to scale up the synthesis, they set up a large unit in Building 8 and placed it off limits except to a few engineers. In this instance, the inventor of the process was not allowed to enter the building and Marisic, frustrated by this action, left the company.

A similar and amusing situation occurred some years later at the Houdry Labs when one of the present authors (Flank) wrote a memorandum for corporate senior management on a top-secret cracking catalyst project, and was not allowed to have a copy. When the lab director misplaced his copy and asked for help in retrieving the details, Flank dutifully produced his handwritten notes.

The synthetic silica-alumina bead catalyst developed at Socony-Vacuum had a distinct advantage because of its resistance to attrition. The bead also allowed for operating without the need of baffles in the reactor; therefore, more reactor space was occupied by catalyst and the operation without baffles was much simpler.

The technique of forming the large spheres by dropping the gel into a hot benzene-rich solvent was practiced. As the catalyst gel fell through the drum of solvent, it formed a spherical shape and lost water to become a hard sphere by the time it had fallen to the bottom of the drum. Years after this work was completed, Ed Rosinski noted that nearly all of the workers on the project had become ill with cancer. Rosinski notified Mobil management of his observations and his concerns about the issue. Mobil management contracted with a medical doctor to investigate the possibility that there was a connection between the lab work and the employees' concerns. For some reason, the doctor was able to publish the results of his study, directly linking the long-term exposure of the workers to benzene and their cancers. Rosinski noted that this publication did not make him a favorite of Mobil management for the next few years.

Socony-Vacuum utilized Thermofor kilns to burn off coke deposited on Fuller's earth during the filtration of lube oils (*31*). They adapted one of these kilns to introduce the first moving bed catalytic cracking process. The first semi-commercial 500 BPD (barrel per day) Thermofor Catalytic Cracking (TCC) unit went on stream in the Paulsboro refinery in 1941. It utilized bucket elevators to transport catalyst from the reactor to the regenerator. In 1943, Socony-Vacuum installed a 10,000 BPD TCC unit (*32*) at a subsidiary refinery,

Magnolia Oil Company in Beaumont. By the end of WWII the TCC capacity was nearly 300,000 BPD. The first units utilized countercurrent flow of reactant and catalyst in both the reactor and the regenerator. Two major improvements were soon made in the process: a change to concurrent flow in the reactor, and the replacement of bucket lifts by a gas lift system. The former change allowed better use of the heat generated during catalyst regeneration and the latter change permitted more rapid catalyst circulation that allowed a higher catalyst/oil ratio to be used in the reactor.

An early design of the TCC unit is shown in Figure 11 (*32*). The bead catalyst is continuously added to the top of the reactor from a catalyst hopper. The bead moves at a constant rate down the reactor and catalyst is continuously removed at the bottom. As removed, the spent catalyst is transported by bucket elevator, and later a gas flow, to the top of the regenerator. Catalyst is fed from the kiln feed hopper to the Thermofor kiln where it passes through a series of semi-independent burning zones. Combustion gas is independently added to, and the flue gas removed from, each burning zone. Between zones, heat is removed from the catalyst by steam generation in cooling coils. Regenerated catalyst is removed from the bottom of the kiln and transported by a second bucket elevator to the reactor feed hopper at the top of the reactor. The TCC reactor is a cylindrical steel vessel that is 11 to 16 feet in diameter with catalyst beds ranging from 5 to 35 feet deep. The amount of catalyst in the reactor is adequate to give a catalyst:oil ratio of about one. The catalyst flows into the reactor through a long vertical pipe and this is sufficient to overcome the pressure in the reactor and allow catalyst to flow; the pipe also serves as a seal for the top of the reactor. Catalyst flows into the reactor through a series of pipes, one for about each 10 ft^2 of reactor cross section; the bottom of these pipes control the height of catalyst in the reactor. If a bucket-lift elevator stops operating, valves automatically close feed lines to maintain catalyst bed height. Before the catalyst enters the bottom of the reactor a steam purge is used to strip adsorbed products from the catalyst; this increases the product yield and decreases the amount of carbon burned in the regeneration step. The steam-stripped catalyst flows from the reactor through an elaborate distribution system that consists of a series of perforated plates, as illustrated in Figure 12 (*33*).

The regenerator of a typical 10,000 barrels per day (BPD) unit had an overall height of 120 feet and an internal cross-sectional area of 100 ft^2. A kiln usually had 7-10 burning zones (Figure 13) (*33*).

One of the first improvements in the TCC process was to switch to a concurrent flow of catalyst and feed. This change improved the thermal efficiency and allowed the use of heavier feedstocks. This change necessitated several changes in the hardware (*34*).

However, the process was a success only after overcoming major operational and hardware problems. A method had to be designed to allow removal of catalyst fines that were formed during catalyst transport. Means of adding catalyst and feed uniformly to the reactor required solutions to hardware

134

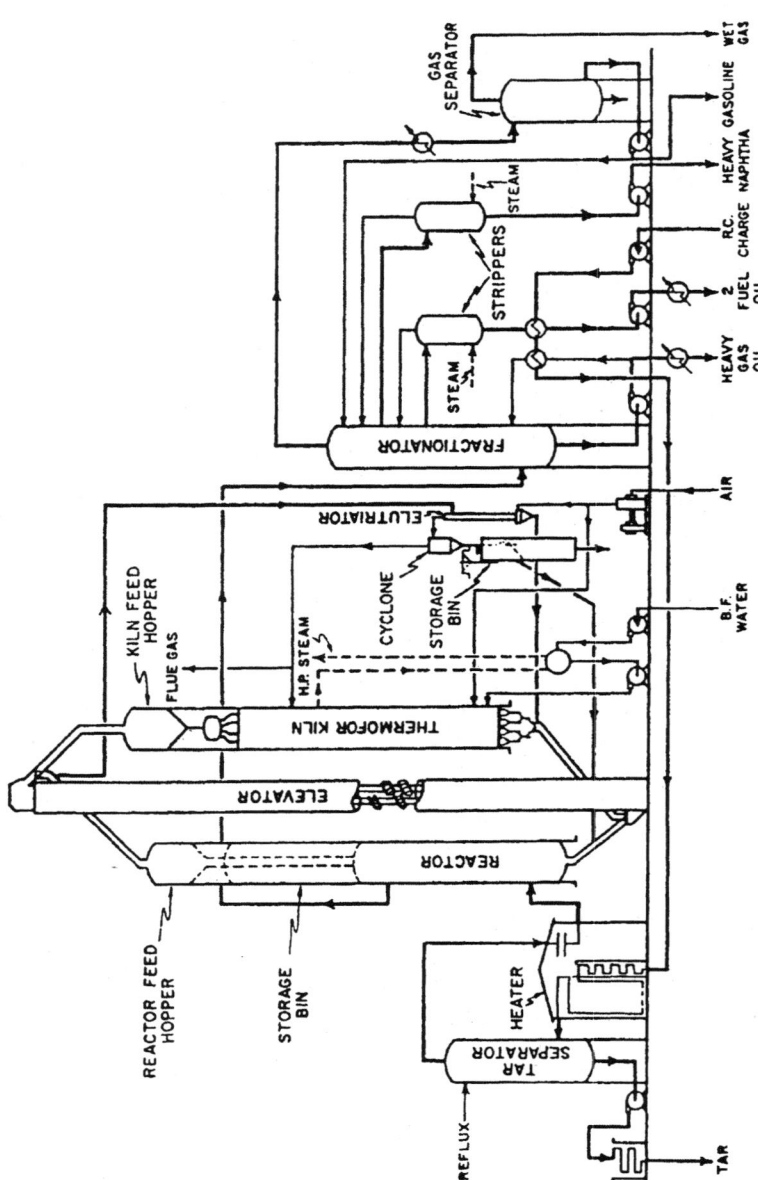

Figure 11. Flow diagram of TCC unit (early design). (Reproduced with permission from reference 32. Copyright 1954 Elsevier.)

135

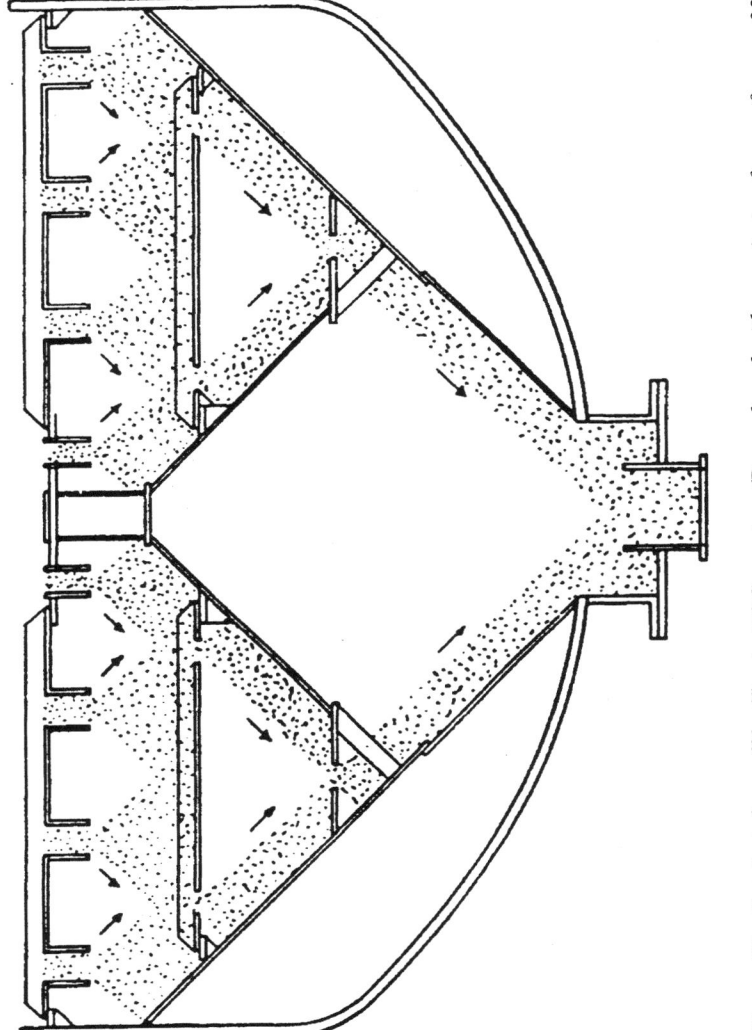

Figure 12. Catalyst-drawoff baffles in TCC reactor. (Reproduced with permission from reference 32. Copyright 1954 Elsevier.)

136

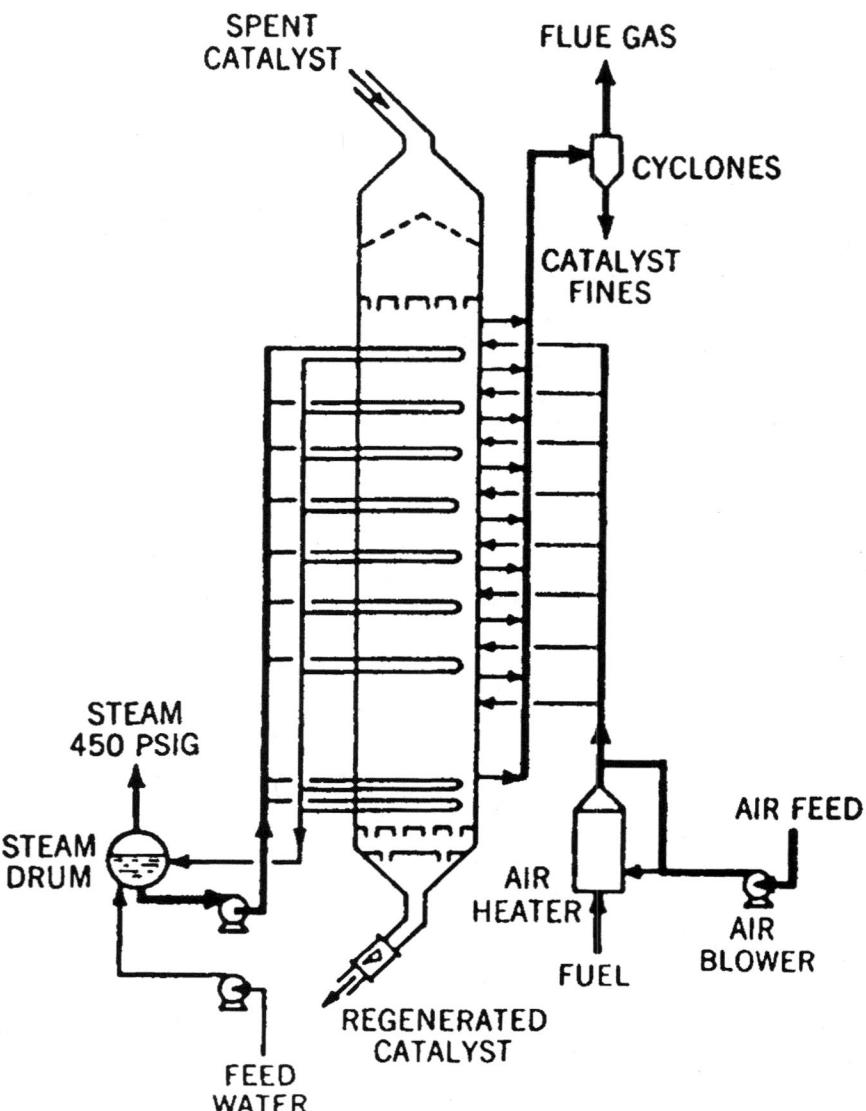

Figure 13. TCC multizone regenerator (kiln). (Reproduced with permission from reference 33. Copyright 1947 American Society for Mechanical Engineers.)

problems. The ability to continuously add catalyst and to maintain pressure had to be solved. Overall, the moving bed system provided a marvelous example of good engineering to overcome numerous problems. The introduction of the moving bed reactor provided a solution to the problem of intimate contact between solid catalyst and the gas in the reactor, stripper and regenerator; it suffered from the utilization of large catalyst beads with the problems associated with diffusion limitations on reaction rate and increased contributions of secondary reactions. In the 1960s, a visitor entering the lobby of the main research building was confronted by a large working model of a later version of the TCC unit.

Houdriflow Process

The Houdry process using fixed bed reactors encountered difficulties during WWII. First, much of the hardware was constructed of special alloys to permit operation at the severe conditions needed for the process and the turbo compressors, which used high-grade alloys and were imported from Switzerland, were no longer available (4). Thus, the Houdry Process Company turned to the moving bed technology since the materials requirements were much simpler and were much more readily available under the wartime restrictions. To accomplish this, the Houdry Process Company reimbursed Socony-Vacuum for some of the development costs associated with the TCC process (about two million dollars), and began licensing the TCC process (28). By 1947, the Houdry Process Company had developed the Houdriflow process (Figure 14) (34) and, during the same time period, Socony-Vacuum developed their air-lift TCC process (Figure 15) (35). While there were some differences in hardware, the major difference between the two processes was that the Houdriflow used flue gas to lift the catalyst whereas Socony-Vacuum used air for this purpose. Both were successful, so that by 1956, Houdriflow units with a capacity of 280,000 BPD were licensed and, following the first air-lift TCC unit that Socony-Vacuum brought on stream at Beaumont in October 1950, there were 54 Socony-Vacuum and licensed air-lift TCC units in operation by 1956 (28). During this period, talented engineers made improvements in the process and hardware so that eventually both the TCC and Houdriflow processes stacked the reactor and regenerator on top of each other in a single vessel as shown for the improved Houdriflow unit (Figure 16) (36).

Early in the 1950s, the TCC and other moving bed processes were competitive with FCC. Both processes produced similar product yields with the light feeds and the relatively low severity cracking and regeneration conditions that were applicable at that time. As the demand for cracking heavier feeds and the installation of larger units increased, it became apparent that TCC could not continue to compete successfully with FCC. It was cheaper to build the lower height FCC units and they were much less complex mechanically. Thus, the

138

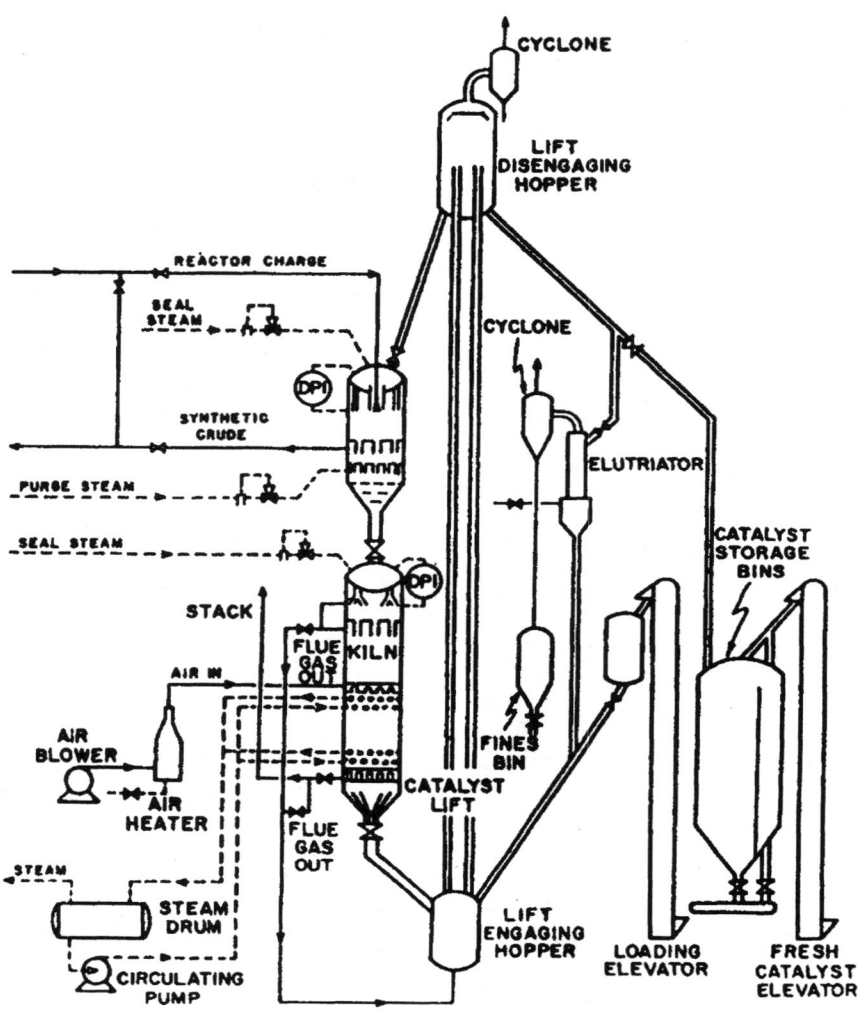

*Figure 14. Flow diagram of Houdriflow unit (original design). (Reproduced
with permission from reference 32. Copyright 1954 Elsevier.)*

139

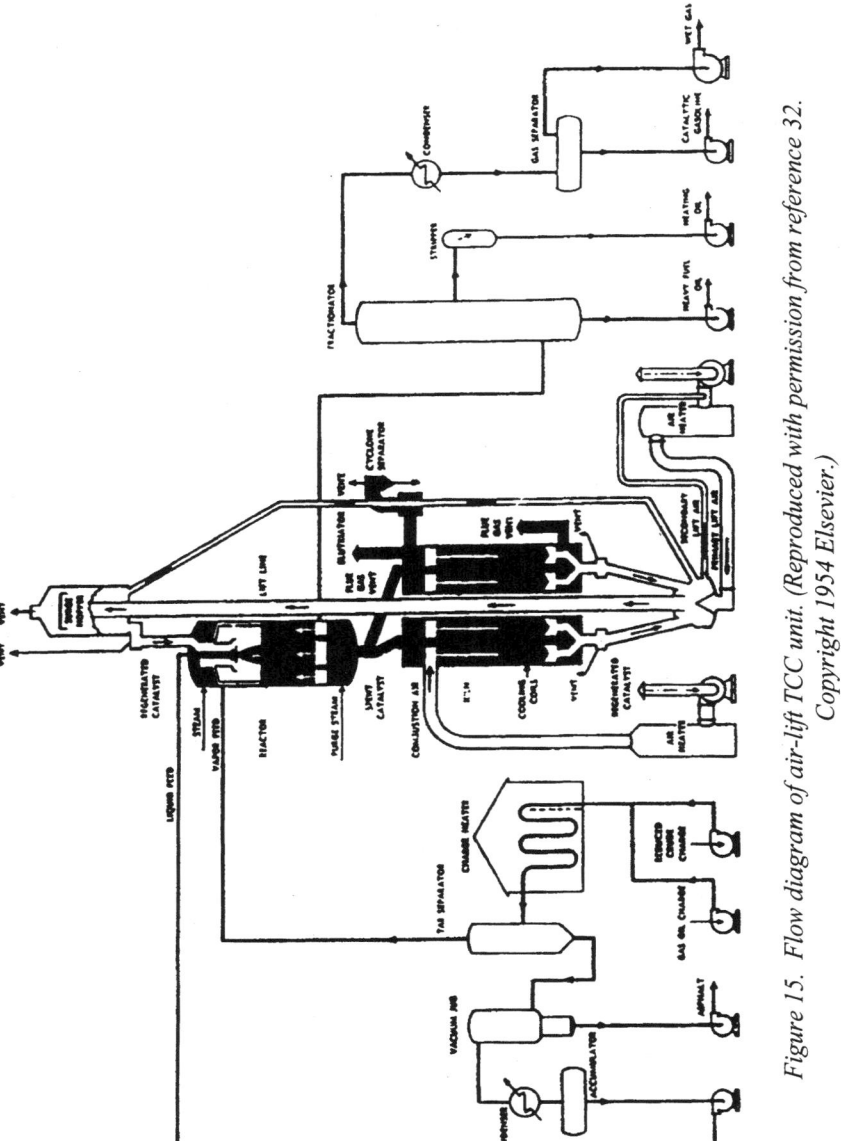

Figure 15. Flow diagram of air-lift TCC unit. (Reproduced with permission from reference 32. Copyright 1954 Elsevier.)

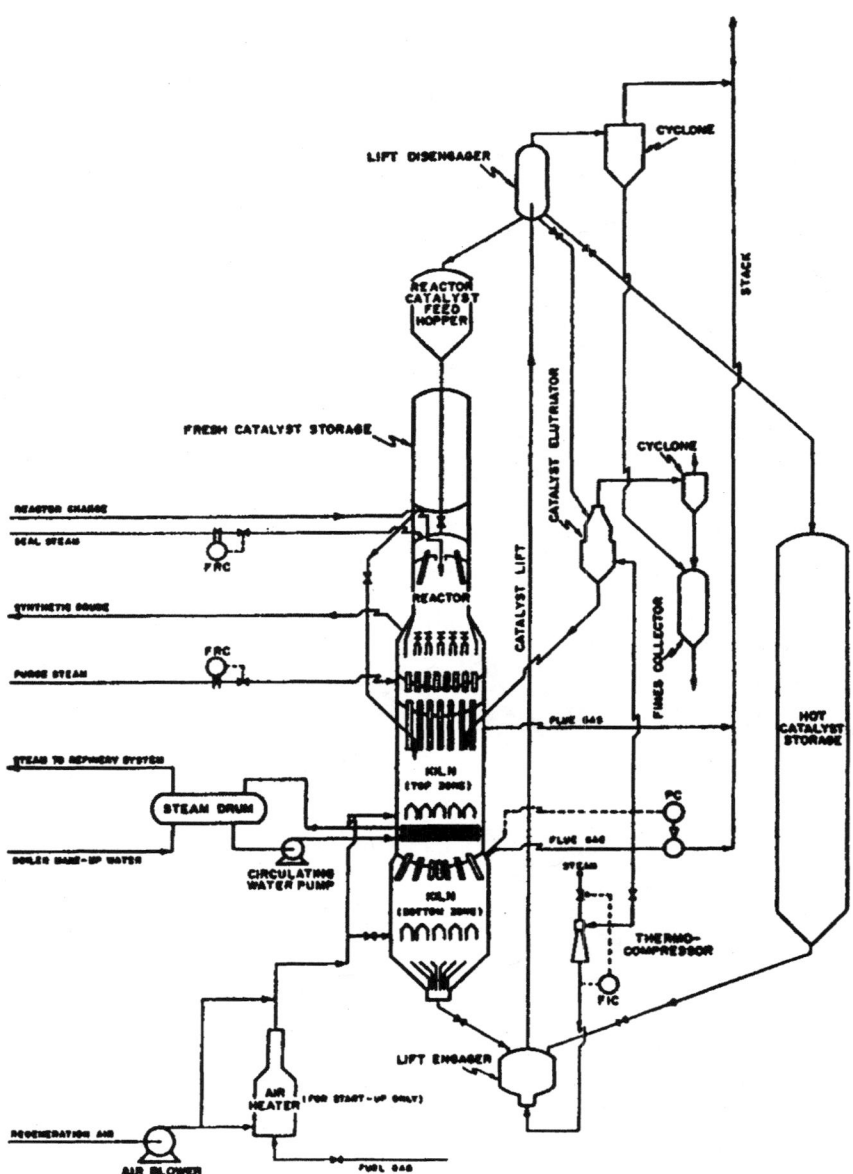

*Figure 16. Flow diagram of Houdriflow unit (improved design).
(Reproduced with permission from reference 32. Copyright 1954 Elsevier.)*

number of licensed TCC and Houdriflow units began declining in the mid-1950s.

There had been a continuing "debate" between the Houdry Process Corp. (HPC) and Socony-Vacuum over the moving bed agreement. In 1952, the HPC brought suit against Socony-Vacuum for $50 million; however, the rapid decline in the popularity of the moving bed process led to the suit being settled out-of-court for a lower amount (4).

Diffusion and Ahlborn Wheeler

Ahlborn Wheeler was employed at Houdry during the period of development of the Houdriflow process (37). Trained at Princeton University, the U.S. center of catalysis research during the 1930-1950 period, Wheeler joined the Shell Research and Development Company in Emeryville, California. Here, he was associated with Otto Beeck and participated in the classic adsorption studies that led to the relationship between metal properties and catalytic activity, such as the relationship of hydrogenation activity to the lattice spacing and the d-band holes of the metal. Houdry was a brilliant scientist with a bent for developing theoretical frameworks to describe complex operations, so Bert Wheeler fit in very well. In two long, classic papers (38,39), Wheeler extended the concepts introduced by Thiele and provided the theoretical concepts still utilized to relate catalytic activity and selectivity to the physical properties of the catalyst. Emmett described Wheeler during his class on catalysis as a very brilliant person who spent too much time fishing (today he may have said, "He would need to spent too much time at the Betty Ford Institute.") Unfortunately, Wheeler found it necessary to be employed for short times with several companies, and during the 1970s he "dropped out."

Fluid Catalytic Cracking – FCC

As frequently happens with a radical new invention, the originators overestimate the value of the advance and ask for more royalties than "the market will bear." This appears to be the case with the Houdry Process Co. Thus, Standard Oil Company (New Jersey) began to look at other options rather than pay the royalties. For example, in 1938 eight companies (Standard Oil (New Jersey), Standard Oil (Indiana), Texas Co., Shell, Anglo-Iranian, M. W. Kellogg, UOP and I. G. Farben) organized a consortium, Catalytic Research Associations (CRA), with the purpose to develop a process for cracking oil that would not infringe on the Houdry patents. At the first meeting, held on November 30, 1938, the eight member companies agreed to develop a process using catalyst in form of a powder (40).

The catalyst particles in the Houdry-type cracking processes, both fixed and moving bed, were initially granules, then pellets or beads of about 3 mm diameter. Later, fluid catalytic cracking used beds of catalyst in the form of fine powder, initially made by grinding and later by spray drying of microspheres, with 50 to 80 wt% of the particles in the size range of about 50 to 150 microns, with the remainder down to 20 or even 10 microns (That part was quickly lost from the unit, as fines.) (*41*),

The moving bed-type process that eventually "won" was fluid catalytic cracking (FCC). The early developments for this process were accomplished by Standard (New Jersey). Work with fixed-bed reactors during the late 1930s convinced E. V. Murphree, vice-president in charge of development, to conclude that the only viable approach was to use circulating catalyst processing that would allow steady-state operations (*4*). He also made the decision to utilize a powdered catalyst (*4*).

It is amazing that Murphree found time to be able to participate in this development. The Advisory Committee on Uranium was formed by President Roosevelt in 1939 (*42*). On June 15, 1940 the committee was put under the National Defense Research Committee (NDRC). About this time the committee was enlarged and reassigned to report directly to Vannevar Bush, who reported directly to the President. Contracts relating to the diffusion and centrifuge processes were to be recommended to Bush by a group of eminent chemical engineers, called the Planning Board. E. V. Murphree was the chairman with W. K. Lewis, L. W. Chubb, G. O. Curme, Jr. and P. C. Keith as the other members. Murphree was a member of the Atomic Committee, S-1, of the Office of Scientific Research and Development (OSRD), that was headed by Dr. James B. Conant, president of Harvard University. The S-1 Executive Committee consisted of H. C. Urey, E. O. Lawrence, J. B. Conant, L. J. Briggs, E. V. Murphree and A. H. Compton (*4*). To be involved to this extent in the administrative efforts to organize what became the Manhattan Project and to lead the research effort at Standard (New Jersey) required a person with exceptional ability.

Suspensoid Process

In 1934, R. K. Stratford at the Standard (New Jersey) Canadian affiliate (Imperial Oil) discovered that fine clay discarded from lube oil treating had catalytic activity (*23*). Thermal crackers at their Sarnia refinery were revamped by 1940 to "Suspensoid Cracking" by adding small amounts of catalyst (2-10 pounds/barrel). The catalyst was utilized on a once-through basis. Subsequently, a 5,000 BPD unit was constructed for the Suspensoid Cracking process. Standard was examining a number of other approaches to improve contact between catalyst and oil. In 1936, Standard (New Jersey) purchased the rights to a patent (*43*) by an independent inventor, W. W. Odell, to a process that utilized a jiggling bed (catalyst particles in a semi-stationary state or moving concurrently with the gas) of fluidized solids (*4*). Odell's patent used the term "fluid bed" and F. A. Howard introduced the name fluidized bed cracking later in the development of the process (*4*).

Professor Lewis and the Dense Catalyst Bed

Johnig indicates that the pneumatic transport design that Standard (New Jersey) intended to use required tubes so long that they had to be folded into upflow and down flow sections (*44*). There was concern that the catalyst concentration would be higher in the upflow section due to catalyst settling. Professor Lewis was asked to define slippage over a wide range of gas flows. In another version, work with a conical reactor operated so that particles moved upward more slowly than the gas (a hindered settler) (*4*). The Standard engineers considered this a complicating factor in the design and a disadvantage. Professor Lewis was consulted and he pointed out that some slippage was an advantage; on this basis he was authorized to conduct fundamental studies to learn how to take advantage of the slippage.

In any event, Professor Lewis, working with Professor E. R. Gilliland, made the surprising discovery that a stable dense bed could be maintained at velocities far exceeding the Stokes' Law free fall velocities of the individual particles (*37*). This work showed that the prior concepts, indicating that a uniform particle size was required to maintain a stable catalyst bed and that gas velocities had to be kept lower than the free falling velocity of the finest particle in the bed, were incorrect. This discovery led to the basic patent for the FCC process (*45*). Thus, the degree of slippage between a rising gas stream and fine solids suspended in the gas was great enough, under proper conditions, to establish a dense phase of solids. The high concentration of solids provided the major economic advantage of small reactor volume and low pressure drop through the catalyst bed.

Standard Oil (New Jersey - Now ExxonMobil)

At about the same time that Professor Lewis made his discovery, the concept of a standpipe to build up pressure was conceived by Standard (New Jersey) workers. Without the standpipe, catalyst could not be circulated at the high rates required to transfer all of the heat released in the regenerator over to the reactor. This advance provided a simple means of circulating the catalyst, together with eliminating the need for complicated mechanical devices such as buckets, pumps or lock hoppers.

These two discoveries offered the potential for a radically new design for a catalytic cracking process. Jersey Standard applied an intensive effort, both funding and personnel, and within three years had commercialized the process (46). In this instance, Standard (New Jersey) was taking a high risk within an area that was a proven commercial success. In contrast, Houdry had to sell the concept of catalytic cracking as well as his process for doing so, and it took him about three times as long to get his concept developed to the commercial stage.

The initial FCC reactors operated in the upflow mode; a simplified flow diagram is illustrated in Figure 17 (47). The process consists of regenerator, regenerated catalyst hopper, catalyst recovery, reactor and product fractionator. Regenerated catalyst from a hopper flowed by gravity through a standpipe, which provided the needed pressure head, and then through a slide valve which controlled the rate of catalyst addition to the oil-feed line. The oil, vaporized in the tube furnace, diluted the catalyst and then flowed into the reactor where the oil was cracked to products and coke accumulated on the catalyst. The reactor size was chosen to provide a vapor velocity sufficient to maintain fluidization and to accomplish the desired cracking rate.

Spent catalyst was separated from the product by cyclones and then held in the spent catalyst hopper. Slide valves controlled the rate of spent catalyst transfer from the storage hopper to an air stream where the dilute suspension was transported to the regenerator. The regenerator was sized so that the air needed to combust the coke was adequate to maintain fluidization and with a height to maintain catalyst holdup sufficient to remove the desired amount of coke. Leaving the regenerator, the catalyst was separated from the flue gas and was held in the regenerated catalyst hopper. Catalyst fines in the gas stream were recovered in a Cottrell or other type of precipitator. Make-up catalyst was added to compensate for loss, which was in the range of 0.12 to 0.4 lb./barrel of feed in the early units (47,48). Conversion was controlled by the catalyst/oil ratio and the reactor temperature. Reactor temperature was very uniform (constant within 5°F throughout) and was controlled by oil-preheat temperature and by the temperature and circulation rate of catalyst from the regenerator. Likewise, temperature throughout the regenerator was uniform and was controlled by catalyst flow and the coke content. The pressure at the bottom of the reactor was near atmospheric and was only 10 psi higher at the top of the reactor.

The demonstration of the above concept (49) was accomplished in 1940 using a 100 BPD pilot plant. The first of three 12,000 BPD plants was on-stream in May 1942, and the other two shortly thereafter. Johnig indicates that this was the largest construction effort carried out in the petroleum industry up to that time. Johnig (44) concludes that the early developments taught that "innovations need to be developed and applied fast in order to generate favorable economics." Jersey Standard certainly continued to innovate and to apply the innovations, although the gap between commercialization became longer as the process matured (Table 2).

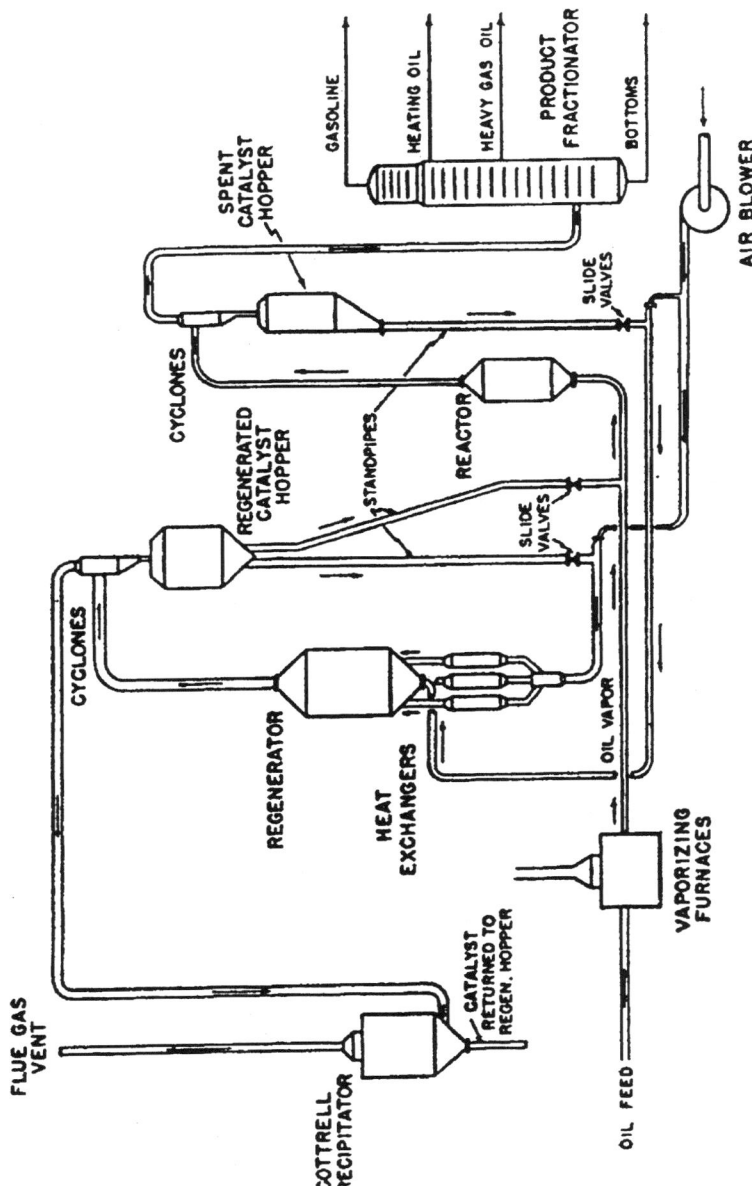

Figure 17. Flow diagram of fluid cracking unit-upflow design. (Reproduced from reference 48. Copyright 1943 American Chemical Society.)

Table 2. Time scale for development of Fluid Catalytic Cracking by Standard Oil Co. (New Jersey; now Exxon)

PROCESS	TYPE	DEMO	COMMERCIAL	YEAR
Model I	Upflow	100 BPD (1940)	12,000	1942
Model II	Downflow	100 BPD (1941)	12,000	1943
Model III[a]		--		1949
Model IV	Downflow; lower height, no slide valves; J-shaped transfer lines	--	11,000	1953
Flexicracking		--	18,000	1979

a. Much of the development work was done by M W Kellogg; the process is often referred to as Kellogg side by side units.

From an analysis of the early operational data from the 100 BPD pilot plant, it was suggested that it was not necessary to take all of the circulating catalyst overhead and through the cyclones. By operating with a greater height of dilute phase it was possible to withdraw catalyst from the dense phase rather that by use of the cyclone. It therefore became possible to control the inventory in the reactor and regenerator vessels independent of the catalyst circulation rate of gas velocity. Recognition of these factors led to the concept of a downflow operation as illustrated in Figure 18 (*50*) and to an improved unit (Model II) (Figure 19) (*43*). Even before the first Model I reactor was in operation the design for the Model II reactors was underway. The regenerator was still located higher than the reactor and very long (100-150 foot) standpipes were used. The long standpipe associated with the regenerator caused large pressure drops across the regenerated catalyst slide valves and these eroded very rapidly.

The management of U.S. refineries was far ahead of the U.S. government in foreseeing the need for 100-octane aviation fuel. Heron (*51*) indicates that J. H. Doolittle (later Lt. General, USAAF and destined to subsequently gain fame as the leader of the first bombing raid on Japan in WWII), then Aviation Manager of Shell, risked his future by persuading Shell management to heavily expand their production of 100-octane gasoline. The Army resisted expanding their use of 100-octane gasoline and the new plants operated far below their design

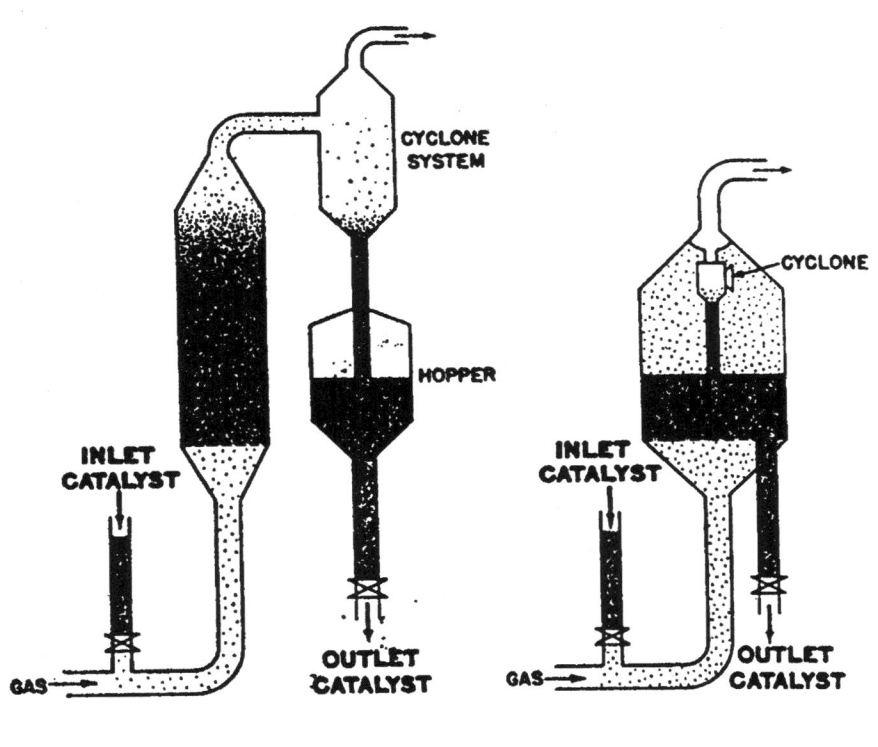

UPFLOW   **DOWNFLOW**

*Figure 18. Original and modern methods of fluid-catalyst circulation.
(Reproduced with permission from reference 50. Copyright 1953 American
Institute of Chemical Engineers.)*

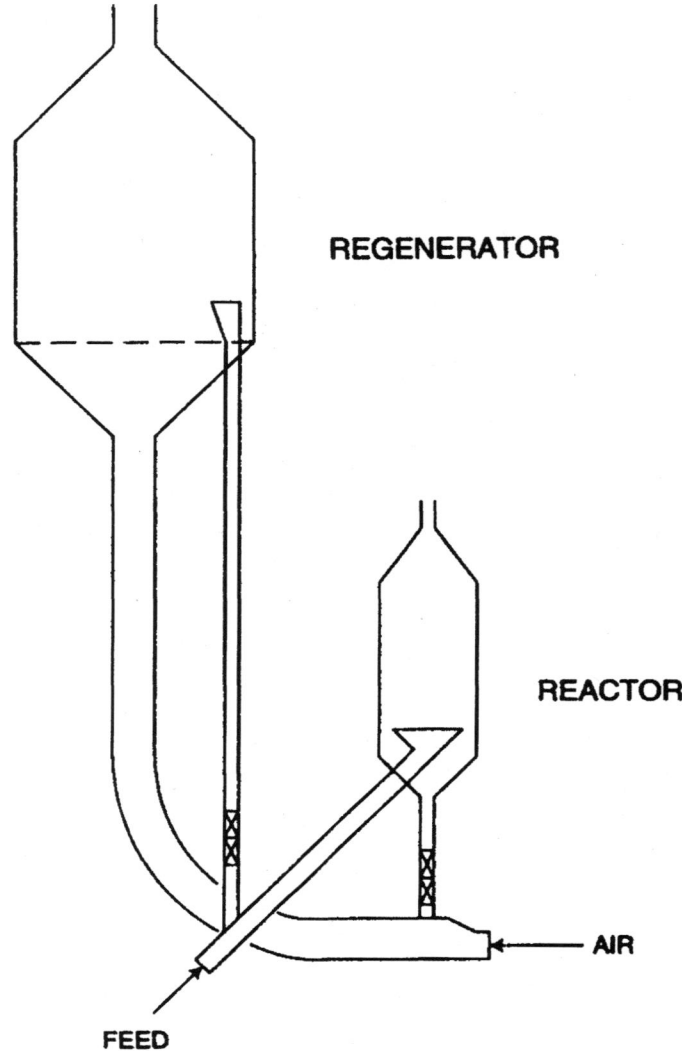

REGENERATOR

REACTOR

AIR

FEED

Figure 19. Esso Model II. (Reproduced with permission from reference 89. Copyright 1997 PennWell Books.)

capacity (52). At that time the military, presumably with total faith in the advances that could be accomplished by refinery scientists and engineers, wanted one fuel to serve the needs for all types of military engines, from motorcycles to high-performance aircraft. To overcome this view of the War Department, the Air Corps at Wright Field in Dayton, Ohio published in 1935 their results showing a 15-30% power increase in switching from 75-octane fuel to 100-octane fuel. Engine manufacturers were impressed by this paper and designed higher compression engines that would only use 100-octane fuels. Eventually the War Department relented, and ordered all future aircraft to have engines that operated on 100-octane fuel.

How could the management of these oil companies justify this long-term gamble? Heron (51) wrote that it "...was really competition for technical prestige." He contended that it was really more than prestige since a supplier of aviation fuel would gradually lose sales unless they had prestige as one of their selling points.

Jersey Standard had many officials, from the president down, serving on various committees devoted to preparation for war in advance of the U.S. entry into WWII. In the case of fluid catalytic cracking, Jersey Standard was driven not by perceived war demand but by the need to find a way to produce gasoline for autos that could compete with the fuel produced by Houdry catalytic cracking units. However, without the wartime conditions, it is likely that Jersey Standard would have taken longer than three years to go from research to an operating commercial unit. The first commercial fluid units were tremendously expensive because of their overpowering size and complexity (53).

Patents covering fluid catalytic cracking were pooled in response to the Petroleum Administration for the War (P.A.W.) recommendation. This arrangement was similar to that of alkylation processes. The need for and the degree of cooperation during the war is illustrated by the fact that Shell installed their new isomerization process units not in their refineries but in those of its competitors (53). The P.A.W. set royalties for fluid catalytic cracking at $0.05 per barrel for all products. Jersey Standard, because of its larger effort, received one-half of the royalties and the other half was shared by Anglo-Iranian, Texaco, Shell, M.W. Kellogg and UOP (53).

The Catalytic Research Associates generated a tremendous research effort (53). In terms of scientific manpower, the effort was surpassed only by the Manhattan Project (4). At Jersey Standard about 400 men worked on the project. "The leaders were E. V. Murphree, D. L. Campbel, H. Z. Martin and C. W. Tyson - known to their associates as 'The Four Horsemen' (54)." During this period Notre Dame football was a "religion" in the U.S. The backfield of one of their most famous undefeated teams was dubbed "The Four Horsemen" by Grantland Rice, the dean of U.S. sports writers. To be compared to the fabled Four Horsemen was high praise.

During 1935 to 1945 inclusive, the fluid catalytic cracking research totaled about $34 million and was borne entirely by Jersey Standard (54). In contrast,

HPC asked Jersey Standard to pay about $25 million in cash and a total payment of about $50 million as royalty fees - all of this without the opportunity to participate in ownership of the Houdry process (44). Without question, Jersey Standard was correct in its decision to develop its own new process. In hindsight, HPC erred by overpricing their process and later had to make dramatic adjustments to overcome the effects of this decision.

The first commercial 12,000 BPD fluid catalytic cracking plant was designed using limited operating experience with a 100 BPD pilot plant; thus, it is not surprising that operating problems were encountered following startup. In spite of the scaleup factor, seven additional plants were being built by Jersey Standard's affiliates and several others were being constructed by other companies. In proceeding at this fast pace, Jersey Standard was taking a large risk even though the $90 million construction at the Baton Rouge plant was shared about equally by Jersey Standard and the government through the Defense Plant Corporation. Imposed upon the technical uncertainties was the fact that they were scaling up by a factor of 120 and that this was based on the recently discovered and poorly understood area of fluidization. Furthermore, the war effort resulted in a shortage of trained personnel. In addition, union organizers were especially active during the wartime period and labor problems required significant amounts of management's time.

Petroleum industry consultant Sy Shulman used to tell the story of the start-up of the first FCC unit, and it sounded a little like the Trinity atomic bomb test conducted in the New Mexico desert a few years later. Large gaggles of bigwigs were gathered on a knoll, a safe distance away from the unit, as the engineers started up the flows in the various parts of the system. Cheers went up, and there were broad smiles and handshakes as the unit came on-stream and hydrocarbon vapors started flowing through the bed. In a very few minutes the smiles turned to puzzlement and then concern, as the oval-shaped metal shell visibly puffed in and out in a rhythmic cadence accompanied by an "oom-pah" sound resembling a giant tuba. Fearing that the unit would shake itself apart and burst open, spewing hot vapors and catalyst everywhere, the panicked engineers quickly shut the unit down to figure out what had gone wrong. You can bet that a lot of people had some really anxious moments until a fix was worked out, and the bursting bubbles were diagnosed and cured. Thus was born the fluidized bed reactor baffle system, without which no self-respecting fluidized unit of any size could operate (55).

Another problem encountered as soon as the first plant started up was that it was impossible to raise the catalyst temperature high enough in the regenerator to enable sufficient heat to be transferred to the reactor to cause the degree of cracking that was required to take place. Two engineers in charge, Vice-President M. W. Boyer and H. J. Vorhees, analyzed the problem and developed an explanation that accounted for the low temperature. The first plant was very tall, 250 feet, and was designed to circulate 40 tons of catalyst per hour. Three heat exchangers to control the catalyst temperature were connected in parallel and these were located below the regenerator. Catalyst temperature could be controlled by the flow rate of catalyst through the heat exchangers. It was deduced that in the commercial plant the catalyst did not circulate through the heat exchangers as planned but instead recirculated so that more heat was removed than predicted based on the catalyst flow rate through the regenerator. The two engineers were faced with two choices: (1) go back to a pilot plant which in all likelihood would have to be built, since only a 100 BPD plant was available or (2) find a solution by making changes in the commercial plant. The second option was followed and, in effect, the first 12,000 BPD commercial plant then became a pilot plant. They decided to extend the pipes from the heat exchangers up far enough into the regenerator to end close to the catalyst distribution grid. The two engineers decided that the pipes should be capped and that restricting holes would be provided around the top of the pipes. Their solution - actually a large gamble – worked, and plant operation became satisfactory.

Operating experience showed that the regenerator standpipes could be much shorter, and this was accomplished in the Model III design (Figure 20). For this model, W. K. Kellogg did much of the development work. However, catalyst was still distributed into the fluid beds through perforated grids and erosion was still a problem. The design for Model IV attempted to eliminate the corrosion problem by eliminating catalyst control valves. Catalyst flow in this model is controlled by changing the pressure differential between the reactor and regenerator, or through adjustment of the control air flow rate. The valves shown in Figure 21 are used only during start up and shut down. In this design, catalyst flow is essentially horizontal at the bottom of the U-bend and there is a tendency for the catalyst to defluidize. To overcome this, the U-tube is fitted with an extensive fluidization system, and this, as well as reactor temperature control, required more supervision that the earlier models employing slide valves.

The latest model is known as Exxon Flexicracker (Figure 22) (56). In this model, the U-tubes are replaced by a standpipe that is followed by upwardly sloped laterals, referred to as J-bends. This model takes advantage of riser cracking and spent catalyst in the J-tube is controlled by a slide valve. Reactor temperature is controlled by pressure differential between the regenerator and reactor.

It is revealing to consider the advances made in the size of the reactors as the FCC units were utilized during the years. First, enormous gains in throughput were obtained from experience gained through operating the units and from the application of improvements developed by talented engineering. Thus, over the years, several Model II plants were gradually revamped, and at very low cost relative to a new unit. For example, a plant with a design capacity

152

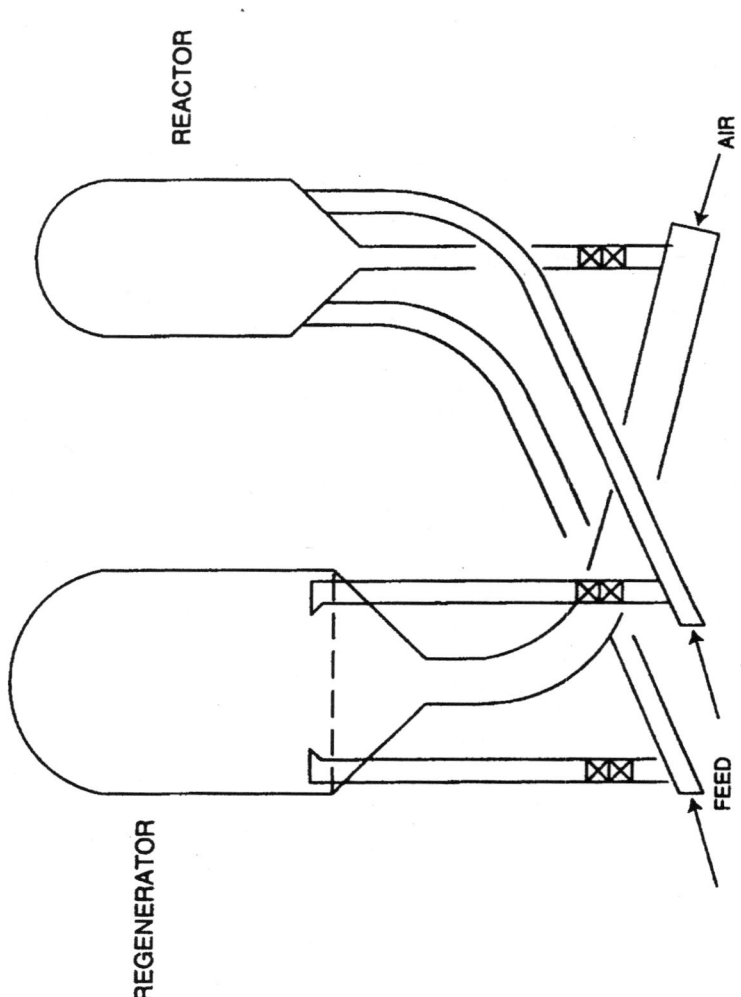

Figure 20. Esso Model III. (Reproduced with permission from reference 89. Copyright 1997 PennWell Books.)

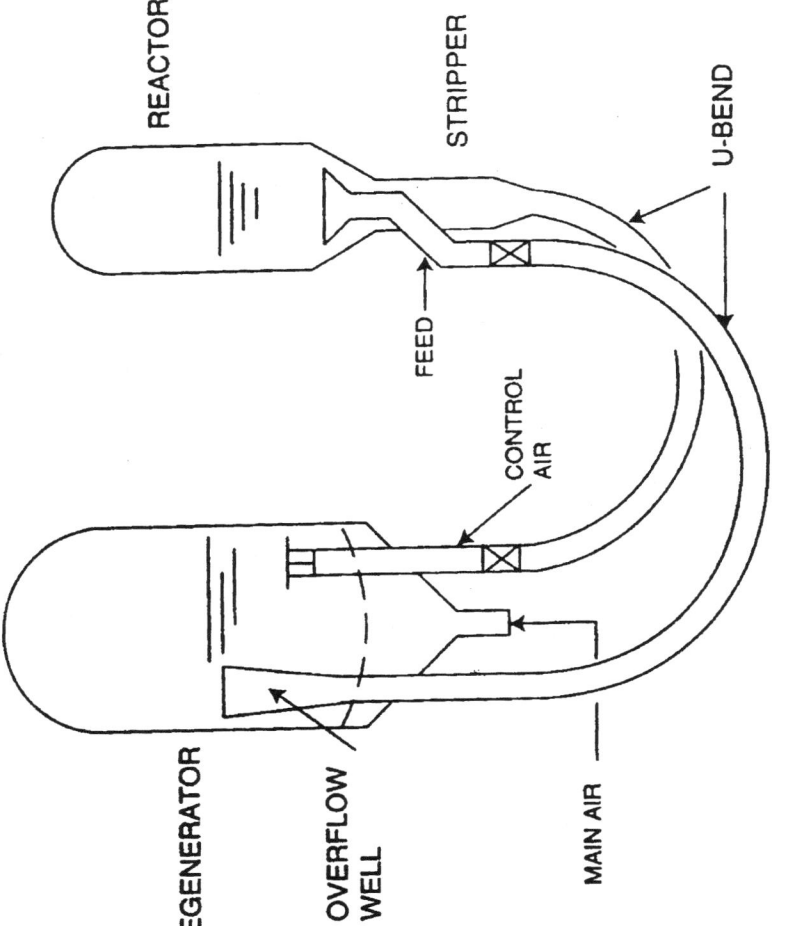

Figure 21. Esso Model IV. (Reproduced with permission from reference 27. Copyright 1990 Freund Publishing House Ltd.)

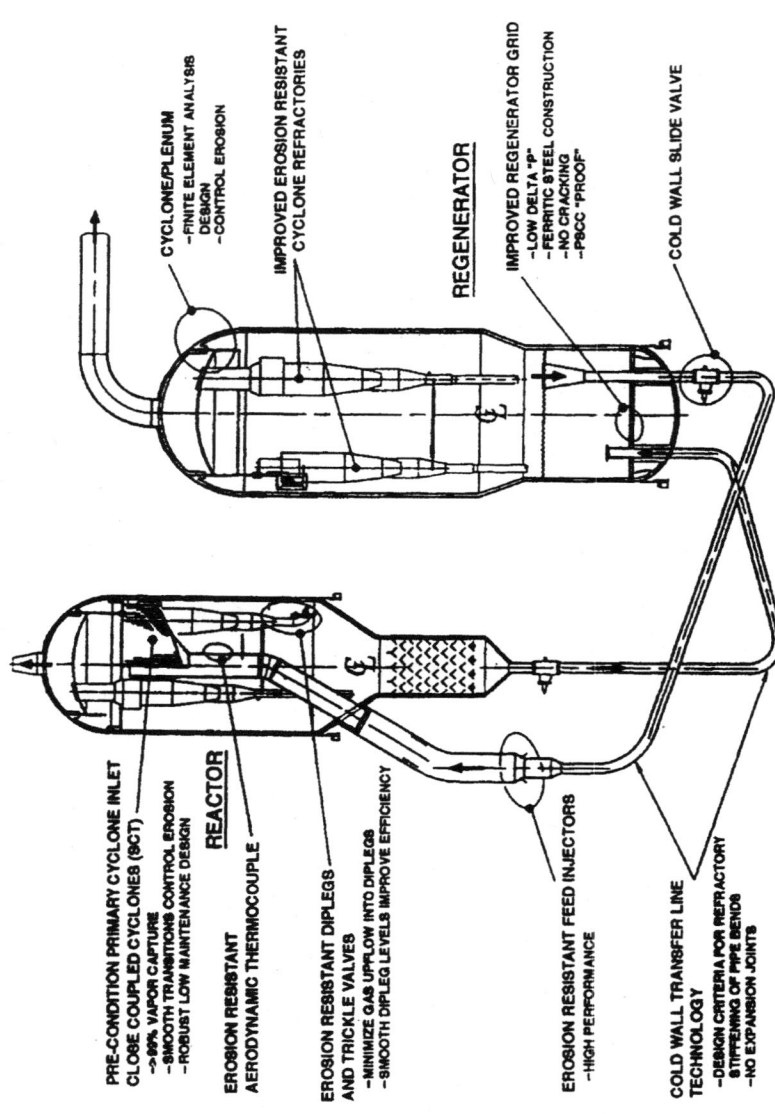

Figure 22. Exxon Flexicracker. (Reproduced with permission from reference 56. Copyright 1996.)

of 15,000 BPD was gradually increased to 50,000 BPD by, among other changes, modifying operating conditions to permit substantial increases in regeneration capacity, by improving carbon-conversion relationships through spent catalyst stripper development, and by making various other process and mechanical improvements. By making these evolutionary improvements, Standard (New Jersey) was able to obtain a throughput that was equivalent to the cost of three of the initial FCC units but having to provide the funds for the construction of only one unit.

The first 12,000 BPD FCC Model I plant was 250 feet tall. Modifying the design to a downflow mode in Model II allowed for the height of the unit to be lowered to about 230 feet for a larger 20,000 BPD unit. This size reduction continued, as is illustrated in Figure 23, showing that, for plants of similar capacity, the Model IV unit is about 30% shorter than the Model III unit (37). The reduction in size for 20,000 BPD plants is shown to scale in Figure 24 where three units of later design are compared to Jersey Standard's Model II unit (27).

Other companies did not abandon the FCC field to Exxon. As noted above, in 1939 UOP switched its funding for development in cracking from fixed-bed to fluid-bed processing. Fluid catalytic cracking grew steadily over the years, helped by improved process designs as well as by improvements in such critical catalyst properties as porosity and attrition resistance. Larger and larger FCC units were built, while the practical size limit for a Houdriflow Moving Bed unit was only about 40,000 BPD, so refiners tended to go for the bigger units and the focus of catalyst development shifted toward FCC catalyst (41).

Installed catalytic cracking capacity in the U.S. in 1955 was over 3.3 million barrels of feed per stream day, distributed as follows:

Fluidized bed plants	2,300,000
TCC units	570,000
Houdry fixed bed (not all operating)	250,000
Houdriflow units	200,000

By about 1968, there were 134 fluid catalytic cracking units operating in the U.S., using 350 to 370 tons per day of FCC catalyst supplied principally by Davison, American Cyanamid, Nalco and Filtrol. There were over 360 FCC units operating worldwide at the end of the 20th century, and gasoline yields had gone from 13% in 1910 by distillation, to the low 20 percents via thermal cracking, into the 30 percents with Houdry's catalytic cracking process, up into the 40's with the advent of crystalline zeolite catalysts in the early 1960's, to about 48% and higher by the late 1980's, and higher still into the 21st century (41).

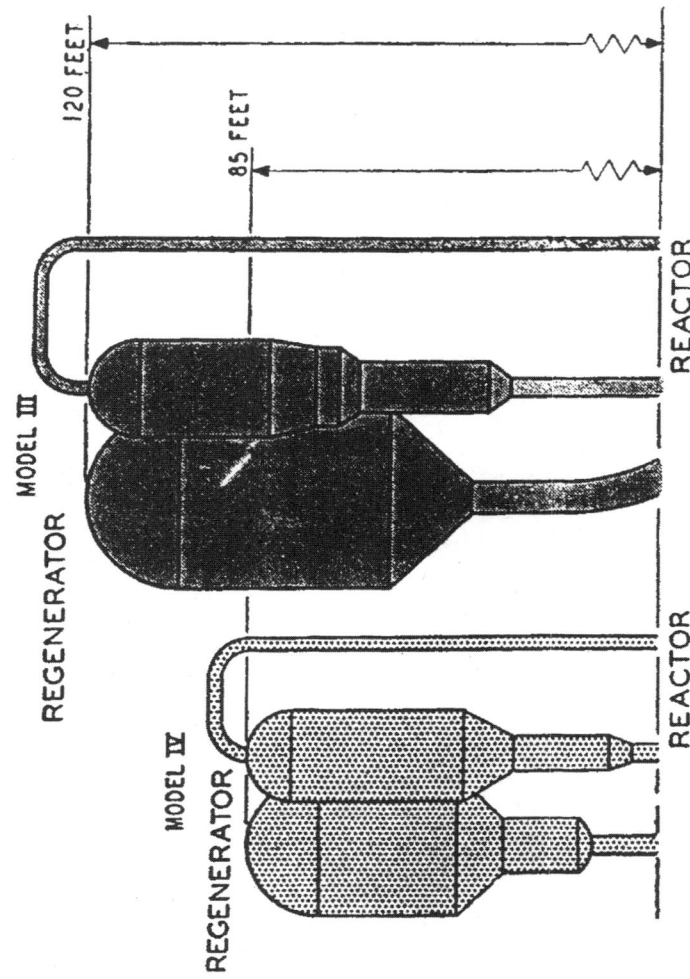

Figure 23. Schematic comparison of Model III and IV fluid units. (Reproduced from reference 46. Copyright 1951 American Chemical Society.)

157

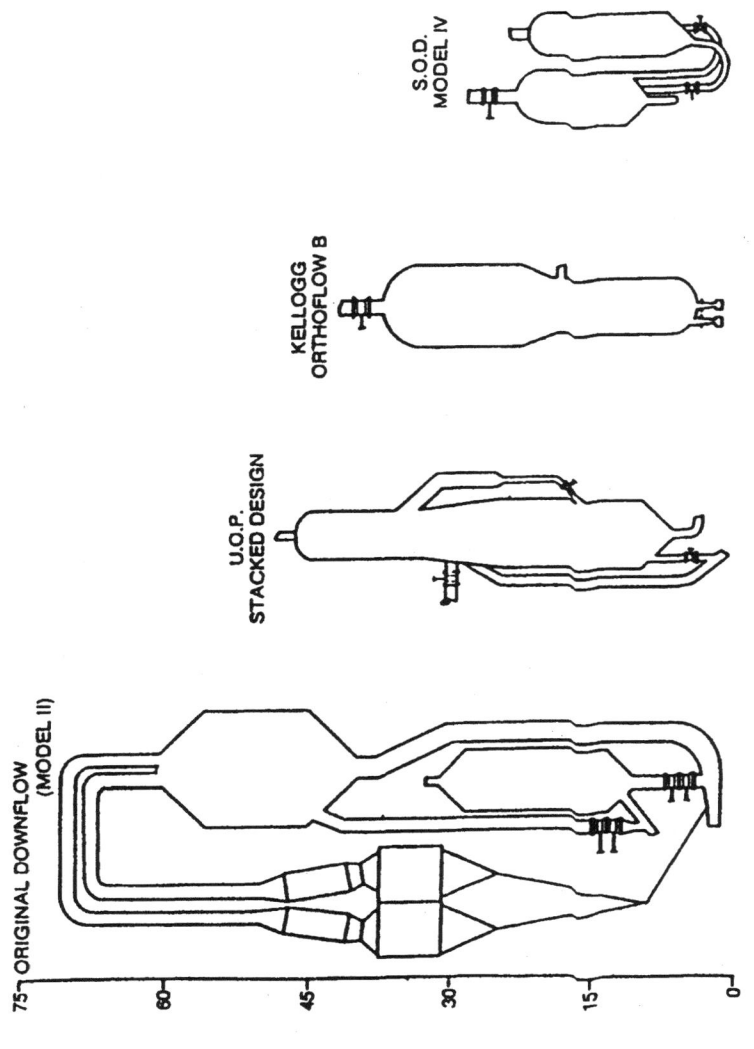

Figure 24. Evolution of a 20,000 BPD FCC from 1943 to 1953. (Reproduced with permission from reference 27. Copyright 1990 Freund Publishing House Ltd.)

Senator Kilgore's Committee Report

Senator Harley M. Kilgore headed a committee to investigate technological mobilization for WWII (*57*). Kilgore utilized his position to keep before the American public the ills of cartels and to expose the dangers in the business relationships between Japanese and American companies which occurred in the period prior to WWII. The Japanese utilized cartel-like arrangements, such as patent licensing, pooling arrangements, and exchange of technology. The Japanese companies purchased technical information and observed American technical achievements. The industries most involved included oil, aircraft, machine tools and electronics. Kilgore pointed out that this occurred while American citizens were not even permitted to land on islands mandated to Japan by the League of Nations to determine whether or not they were being fortified.

The Japanese were able to acquire technical information and some processes needed for the production of high-octane aviation gasoline even before firms in the U.S. could obtain the information (*58*). Much of the knowledge came from the Universal Oil Products Company. In 1928, Osaka financial interests incorporated Japan Gasoline Company as the mechanism to acquire technology for the Dubbs process, and paid UOP one million dollars for this. The Dubbs technology proved unprofitable, and only one plant was built. Thus the Japanese turned their interest to the alkylation process that Ipatieff developed at UOP to produce isooctane. The Japanese also were interested in catalytic cracking to produce a higher yield of gasoline even though this process was not in general use until about 1938. By August 1938, UOP and Japan Gasoline had worked out a three-pronged agreement. With the first, Japan Gasoline obtained the process for making isooctane. The second permitted the Japanese "to acquire the rights under all of Universal's processes in the entire petroleum field" that would be developed through December 31, 1946. Universal began to design isooctane units to be used by Mitsubishi Oil Company and Nippon Oil. The designs were delivered to Japan in 1939 and Universal engineers were dispatched to Japan to help build these plants. Prior to signing the agreements, UOP informed the U.S. War and State departments, and the War Department expressed no objection. This is strange in view of the holding in secret from about 1936 to 1939 the UOP patent covering the production of aromatics using dehydrocyclization catalysts (*59*). Thus, it appears that while the U.S. War (now Defense) Department was holding up a patent that could lead to the production of toluene, a high octane fuel for high performance aircraft, the U.S. State Department had no problem with UOP allowing the Japanese to learn of the latest developments in catalytic cracking. This is ironic since dehydrocyclization, as a commercial process, was not practiced during WWII but catalytic cracking, and the associated alkylation process, primarily developed by UOP (*60*), produced much of the aviation fuel used by the Allied Forces.

UOP carried out a very extensive program of research in catalytic cracking and hydrotransforming in order to arrive at a type of process which would satisfactorily meet the Japanese requirements (59). Universal provided the Japanese with research and written test results from pilot plant runs and all information concerning other Universal catalysts. Universal gave lectures on a variety of topics and even reviewed the notes taken by the Japanese to be sure that they comprehended the presentations. On December 20, 1939 the State Department declared that "there should be no future delivery to certain countries of plans, plants, manufacturing rights, or technical information," and since this involved the production of aviation gasoline, Universal terminated the lectures to the Japanese. Universal personnel were satisfied that the information given the Japanese would enable them to proceed on their own. In June, 1940, Japan Gasoline filed suit in the U.S. against Universal, seeking "Performance of the contracts, or the payment of $10,000,000.00." The case was postponed due to the war. Unfortunately for the U.S., the Universal officials' opinion was correct and the Japanese were able to plan and construct five catalytic cracking plants with a total daily capacity of fifteen thousand barrels. Each plant included units to manufacture isooctane as well.

Ironically, Shell Oil Company, which owned 50% of UOP voting stock and was a licensee for all UOP patents up through 1947, complained that the information UOP provided to them in 1938 concerning cracking and catalytic reforming was incomplete. Standard Oil Company of Indiana (then Amoco and now a part of BP), a paid-up licensee through December 1947 and a minor holder of UOP stock, requested fifteen reports and other information similar to that obtained by the Japanese (58). UOP refused, since under the agreement the American company would have to purchase the catalyst but the Japanese would not. However, the Japanese did purchase a years' supply of the catalyst. The Ocon Petroleum Process Corp., in September 1940, indicated that discussions between its technical people and the technical people of Okura Company, resulted in the conclusion that the Japanese company needed the "catalytic naphtha reforming process." A unit with a 3,000-barrel-per-day capacity was capable of producing 1,650 barrels of aviation gasoline. The company was willing to demonstrate this process for the Japanese in Mexico, indicating that the company was afraid to demonstrate it in the U.S. Even though the "moral embargo" had been declared three months earlier, Ocon was willing to demonstrate the process outside the U.S. (58).

Cracking Catalysts

We can best examine cracking catalyst history by dividing it into two eras: the pre-zeolitic era from 1941 to 1964, and the crystalline zeolite era. Catalysts for the various processes had fairly similar properties. They were for the most part either synthetic silica-alumina or acid-treated clays. Fresh condition surface areas ranged as high as 600 m²/g. The catalysts were quite thermally stable, surviving temperatures up to 600°C, unless high partial pressures of steam were present (41).

Amorphous Silica-alumina - Clay Catalysts

Houdry had found acid treated clay to be a superior cracking catalyst. This was fortunate since acid treated clays had been used in the U.S. for decolorizing petroleum and other products. For example, Filtrol was founded in 1922 by "three typical Californian style west coast entrepreneurs" (61). The bentonite or montmorillonite clays were leached with sulfuric acid at its boiling temperature in a process called acid denning, washed, filtered, dried and then sized to meet the needs of its application. The early clays that Filtrol produced were used almost exclusively to treat lubricating oils. Later on, similar treatments were used with kaolinite and halloysite clays

Filtrol initially supplied Houdry with catalysts, and the first was "Super Filtrol". Soon after this, Filtrol made and supplied catalysts to Standard (New Jersey) for their FCC process and to Socony-Vacuum for the TCC process. Filtrol continued to supply catalyst for fixed-bed, FCC and TCC operations following WWII.

The Houdry Process Corp. was interested in identifying the location of clay deposits that could be used for the preparation of their catalysts in their own plant. Ted Cornelius was assigned to survey deposits in the southeastern states (62) in the U.S. To accomplish this, Cornelius traveled with a geologist through many of the states and collected samples for testing at each site. In making his report to the company after he returned from surveying sites in Georgia, he showed maps of several regions in Georgia where he had obtained samples. A meeting attendee noted that there was a significant area in one of the maps where samples had not been obtained and asked, "Why did you fail to get samples from that area?" Cornelius' answer was quick, "Too many rattlesnakes!"

For the synthetics, which Houdry introduced in 1940, one processing method was to hydrolyze an aluminum salt in the presence of freshly prepared silica hydrogel. The mixed hydrogels were then filtered, washed, dried and calcined. Forming into pellets, beads or, later, microspheres, could be accomplished either before or during the drying operation (41). Another approach was to blend an alumina hydrogel with a silica hydrogel. In all cases, thorough washing was necessary to remove soluble compounds, especially those of sodium, to prevent neutralization of the catalyst acidity, and therefore activity, and to enhance thermal and hydrothermal stability. Alkali metals are known to be pretty good fluxing agents, and ammonium ion exchange was employed to help reduce the residual sodium (41).

In 1942, Grace introduced a low aluminum amorphous silica-alumina catalyst (*63*). The research efforts to prepare synthetic amorphous silica-alumina catalysts that were superior to clay based catalysts were widespread throughout the industry. In 1942, American Cyanamid, based upon cooperative work with UOP, began manufacturing a synthetic silica-alumina catalyst that contained 12-14 % alumina (*64,65*). The early catalysts were manufactured following UOP's specifications (*66*). The initial catalysts were made in granular form but microspheroidal catalysts manufacture was started in 1946, initially by Cyanamid, and Davison followed suit in 1949. Microspheres were claimed to improve fluidization properties and to reduce catalyst attrition losses. However, refiners, who had historically tended to resist change, did not immediately accept the microsphere catalyst and it was not until 1955, nine years after introduction, that they completely replaced ground catalyst (*67*).

The early amorphous silica-alumina catalysts were rather dense materials. Operating on the basis that if a little is good more must be better, higher aluminum containing materials were made and tested. However, with the early formulations it was observed that when the aluminum content was increased the activity increased only marginally but the coke content increased substantially (*68*).

Continued research led to improvements so that the first commercial high alumina catalyst was produced by American Cyanamid Co. in 1954, this time in cooperation with Shell Development Co. (*53*). These catalysts contained about 25% alumina and had significantly higher pore volumes than earlier formulations. These new high-alumina catalysts had higher activity with essentially the same product distribution, exhibited a slower deterioration of activity and had better attrition properties. In later years, with zeolitic catalysts, high alumina contents in the catalyst matrix were employed for stability enhancement, and then for a host of other improvements in overall performance.

Initially synthetic catalysts commonly contained about 10 to 15% alumina by weight, and it turns out that the proportion of tetrahedrally coordinated aluminum is at a maximum at about 13% alumina, corresponding to a molar SiO_2/Al_2O_3 ratio of about 6 (which is interestingly similar to modern zeolites). The significance of this, as discussed below, is that acidic sites are associated with tetrahedral Al, and cracking is an acid-catalyzed set of reactions. Eugene Houdry had recognized very early that acidity was a key factor in cracking catalysts, and had attempted to maximize this in his preparations (*69*).

One of, if not the, first detailed descriptions of the acidity and activity of cracking catalysts was provided by C. L. Thomas (*70*). Following Pauling's formulation of solid state chemistry, Thomas proposed that alumina substitutes into the tetrahedral positions normally occupied by silicon. Since the valence of aluminum is +3 and silicon is +4, the substitution of aluminum so that it is surrounded by four oxygen anions leaves a charge deficit of -1; Thomas offered the view that a proton is present to balance the charge. Thomas offered a graph showing that the catalytic activity paralleled the variation in acidity of the solid as the Al/Si ratio varied from 0.01 to 100 (Figure 25). Thomas was at UOP during the war and conducted extensive work on synthetic cracking catalysts; however, the paper describing catalysis was published after he left UOP.

Much research was devoted to the question of poisoning. Alex Mills and co-workers at the Houdry Labs did an extensive series of studies to characterize the nature of the cracking catalyst surface and poisoning mechanisms. Among other things, it was shown that *less* basic heterocyclic nitrogen compounds do not inhibit activity as strongly as their *more* basic relatives, and that, at equal basicity, inhibition *increases* with molecular weight. Therefore, base strength, and thus acid strength, was important, as well as active site geometry. Using quinoline, they established that slowly reversible chemisorption was occurring (*71*).

Titration in benzene with *n*-butylamine gave values of about 0.5 meq of acidity per gram of catalyst. Mills and co-workers showed, however, that not all those acid sites were equally active. Only 0.02 millimoles of quinoline per gram were chemisorbed on a cracking catalyst with 273 m^2/g surface area. Since the quinoline only covered about 2% of the surface and resulted in an order of magnitude decrease in activity, this clearly showed the existence of only a limited number of strongly active sites. Paul Emmett showed, with deuterium exchange, that each aluminum atom does *not* produce an independently active site, and suggested that small amounts of water "activated" Lewis acid sites, presumably forming protonic acid sites. He estimated that a site occupied about 10 nM^2 of catalyst surface.

A turnover number for cracking of a model compound, like isopropylbenzene, was measured by Mills et al. (*72*) in 1950 at 340 molecules per active site, quite an achievement at that time. It was concluded that activity inhibition could be minimized by (a) removal of the nitrogen bases, (b) use of short process cycles or high catalyst to oil ratios, and (c) conducting the cracking reactions at a higher temperature.

As indicated above, Socony-Vacuum had developed a bead catalyst for their TCC process. Sodium silicate and acidified aluminum sulfate solutions were rapidly mixed in the proper proportions to form a rapid setting silica-alumina sol. Just before setting, the sol was passed over a fluted cone to form droplets which descended through hot mineral oil where they formed a droplet that set to produce a rigid hydrogel about 8 mm diameter. The beads then pass to a saline solution where temperature and time are adjusted to harden the bead and control the density of the ultimate catalyst. After a complex protocol involving washing, drying and calcination, the bead has become even harder and has shrunk to a diameter of about 3 mm. The early bead catalyst contained about 10% alumina, a surface area of 420 m^2/g, a particle density of 1.10 and average pore diameter of 4.7 nm. Doing studies to understand the chemistry involved in the bead catalyst fabrication provided Plank (*73*) with an early exposure to catalyst synthesis and the relationship of their physical properties to catalytic activity - good training for his subsequent work with zeolites.

In 1957, the competition was primarily among Nalco, Filtrol and American Cyanamid for the cracking catalyst business. At that time catalysts based on natural clay material were selling for $270/ton but the synthetic silica-alumina was priced at about $400/ton; this made it difficult for Nalco to compete (*74*).

Kaolin or other clays were mixed with amorphous silica-alumina and then formed into microspheres. One variation of this is the "house-of-cards" effect. By including the clays, the synthetics could compete with Halloysite - a natural clay that, when dried, forms tubes and these, after acid treating to develop acidity, are an excellent cracking catalyst. There is not a large amount of Halloysite in the world, so that it is not readily available. Kaolin gave good structure to the amorphous material. Nalco was able to get Chevron interested and they put in the kaolin-containing catalyst; this followed its initial use in a small plant in Michigan. Davison then jumped into this market, and they were followed by everyone. Nalco then got involved with ICI at the request of Exxon.

Other synthetic formulations besides silica-alumina were studied over the years but never widely used, such as boria-alumina, silica-zirconia-alumina, and especially silica-magnesia. In the early 1950's Cyanamid introduced a silica-magnesia catalyst which was tried by several refiners. It gave substantially higher gasoline yields, but lower product octane than silica-alumina. After several months of commercial trial, the roof fell in, so to speak. This particular formulation did not regenerate completely and built up high levels of residual carbon, causing it to lose its activity. The entire catalyst inventory in the commercial units had to be discarded and replaced with silica-alumina (*41*).

It was this kind of disastrous experience that led refiners to adopt a highly cautious attitude regarding catalyst changes. While both Cyanamid and Davison later developed improved silica-magnesia and silica-alumina-magnesia catalysts, which were tried in several commercial units in the early 1960's, acceptance was slow. The regeneration problem was overcome by adjustment of pore size distribution, and the steam stability problems were tamed with surface fluoride treatment. If it hadn't been for the advent of zeolite catalysts, silica-magnesia and/or silica-alumina-magnesia might have eventually made the grade (*67*).

Introduction of Zeolite Catalysts

Zeolites (molecular sieves) have been known since Cronstedt recognized stilbite in 1756 (*75*). They found limited application in areas such water softening by ion exchange, but they did not attract widescale attention until synthetic methods were developed, particularly by Professor Richard Barrer beginning in the late 1930s. Eventually, Robert Milton, Donald Breck and others at Union Carbide became interested in these materials for separation of air into nitrogen and oxygen, and they worked out synthesis procedures that

164

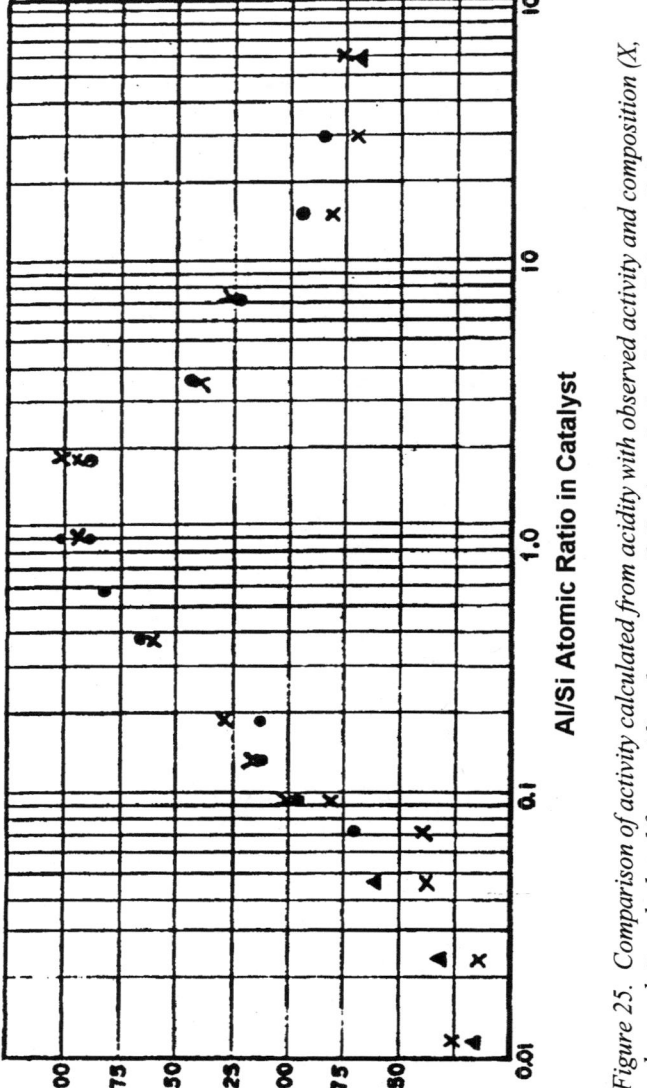

Figure 25. *Comparison of activity calculated from acidity with observed activity and composition (X, observed; ●, calculated from acidity plus excess Al₂O₃; ▲, calculated from SiO₂ or Al₂O₃ content). (Reproduced from reference 70. Copyright 1949 American Chemical Society.)*

were applicable at the commercial scale (*76*). Jule Rabo, when he began working at Union Carbide, recognized that they did not have the expertise to judge the potential of zeolites as catalysts. Thus, Rabo made arrangements to visit people in the research laboratories of major oil companies. At Texaco, he met with Dr. Robert Eischens (*77*). After listening to Rabo, Eischens asked about the pore size of the zeolite catalyst. When Rabo responded that it was about eight Angstroms, Eischens indicated that no refiner would be interested in a catalyst with pores that were that small. Eischens was an innovator who introduced infrared (IR) spectroscopy to catalysis scientists and pioneered in the use of IR to characterize supported metal and acid catalysts. At the time he first talked to Rabo about the zeolites, he was responsible for refinery research, and his outlook reflected that of industry.

It should be noted that the most radical innovation in cracking catalysts since Eugene Houdry's introduction of aluminosilicates about thirty years earlier was the incorporation of crystalline aluminosilicate zeolites, or molecular sieves. Mobil Oil introduced Durabead 5 for the moving bed market in 1962, mostly for their own units, and this was followed in a short time by HZ-1, produced by the Houdry Division of Air Products and Chemicals in partnership with Engelhard-Minerals & Chemicals. Mobil later came out with an improved Durabead 9 product, but one of the present authors (Flank) remembers someone from Mobil saying, off the record, of course, that if their units weren't obligated to buy the in-house product, they would put HZ-1 into their moving bed units.

In quick succession, fluid catalyst products were announced and introduced to the petroleum industry in the Spring of 1964 by Davison and Esso (which eventually became ExxonMobil), and also by Filtrol, and the world witnessed a small revolution. In two years, 60% of the fluid bed units were using zeolitic catalysts, and in two more years about 85 % of those units were using them. Nalco and finally Cyanamid were forced to offer zeolite catalysts to stay in the business (*67*). In about 10 years, over 90% of a very cautious industry had embraced zeolites.

Ed Rosinski and Charles Plank

Edward J. Rosinski was employed in 1956 at Socony-Vacuum after having completed his bachelor's degree in chemistry and chemical engineering at Temple by attending night classes after working full-time during the day (*78,79*). Also, in the few hours that he was not attending classes or working, Rosinski built a house for his family. Rosinski reasoned that activity and/or selectivity might be improved if he could prepare silica-alumina with a uniform pore size rather that the broad size distribution of catalysts then in use (*79,80*). He felt that non-uniform and uniform pores of his catalyst might interact and work in concert to produce a new and enhanced cracking effect. Rosinski was an assertive individual with a "damn the torpedoes, full speed ahead" outlook in life and in research. His initial efforts were to form uniform pores in synthetic silica-alumina by incorporating an organic substance during the preparation and then generate the porosity later by burning out the organic. The selectivity results of his catalyst were favorable, but they were lost during the regeneration step because of the steam that was generated. At this point, Rosinski began collaborating with Charles J. Plank, and joined Plank's large group. One reason that Plank may have been attracted to Rosinski's work was his earlier studies on the preparation of silica with selective adsorption properties for the compound that was templated during preparation. Plank's studies were based upon work by Dickey and Pauling (*81,82*) who reported that silica gel prepared in the presence of the methyl orange dye adsorbed preferentially methyl orange from a mixture with ethyl orange, and vice versa. Plank and Rosinski's thought processes were very similar and their discussions stimulated each other. Even after many hours of discussion with Rosinski about his thought process leading to using zeolites, one author (BHD) was as much in the dark about this process as when the discussions started. Plank, a Ph.D. chemist from Purdue, was much lower-keyed than Rosinski and had a more aristocratic outlook. Plank was very successful in the stock market, and during the 1960s he would spend the morning with the Wall Street Journal to keep up with his investments, and the afternoon in discussions with George Doughtry and/or Rosinski about the data generated during catalyst testing the previous day. Plank saw humor in most situations.

One of Plank's favorite scientists was George Halsey. Halsey was present at the first spring symposium of the Philadelphia Catalyst Club. He was seated far in the back of an auditorium style meeting room on the University of Pennsylvania campus. During the discussion, Halsey got so engrossed in the discussion that he did not take time to go to one of the aisles; rather, he proceeded to the stage and the chalkboard by the most direct route: stepping over the backs of the seats as he progressed to the lower stage level. On another occasion, the Gordon Conference on Catalysis was dominated by discussions of silica-alumina and catalytic cracking. Tiring of all the talk about this topic, late in the week Halsey arose during the discussion period to give a ten-minute discourse on why the snow does not fall under a bridge. Halsey was soon to be banned from further Gordon Conferences; among the reasons for this was that he waited until about 3:00 AM to knock on the door of Sir Hugh S. Taylor, the Dean of U.S. catalysis, to inform him that he had a phone call. Taylor went to the phone, located in an enclosure, and Halsey proceeded to affix the lock on the door so that Taylor could not get out. Needless to say, Taylor was not happy when he was finally released from the phone booth hours later. Plank delighted in telling these Halsey stories, as well as many others. With regard to the Gordon Conference, Plank had priorities - first among these was an afternoon of golf, then poker and drinking following the evening presentations, and finally attendance of those papers of interest to him.

The Plank group was large and emphasized preparation and activity testing of an enormous number of zeolite catalysts. They soon incorporated a zeolite in an amorphous silica-alumina matrix and found that the initial activity of this catalyst was vastly superior to any amorphous silica-alumina catalyst that was utilized at that time. With this encouraging result, they quickly improved upon their catalyst formulation and a number of patents were obtained. They obtained a synergetic effect with their catalysts since they had superior activity and selectivity over that expected from the sum of the two when tested separately.

The initial work by Rosinski and Plank was done using zeolite X which has a high Al/Si ratio. During repeated regenerations, dealumination occurred so that the zeolite was not sufficiently stable. Fortunately, the conversion of fresh and steamed catalysts were tested. Thus, it was learned that catalysts exchanged with rare earth ions were exceptionally stable toward the steaming that would be encountered during catalyst regenerations. The group continued to make improvement in the rare-earth X-zeolite (REX) catalysts. This is shown (Figure 26) by a comparison of gasoline yields at 65% crude conversion: the standard catalyst gives 45% gasoline, the early zeolite 51%, an improved zeolite 54% and an improved zeolite containing a carbon oxidizing promoter gives 57% gasoline. At the same time, 4% carbon on catalyst (● on each curve in Figure 26) is attained at increasing conversion levels. Thus, with the standard catalyst (curve ----- in Figure 26) the 4% carbon content is attained at about 55% crude conversion but with the improved zeolite catalyst the 4% carbon content is not attained until the conversion is about 85%.

It was crucial that some metal ion be incorporated into the X-zeolite-containing catalyst, because as noted above, it did not have adequate steam stability. Plank and Rosinski found that rare earths were most effective and used them in the mixed ratio that could be purchased on the open market.

Mobil immediately incorporated zeolites into their moving-bed bead catalysts and in their fluid-bed cracking catalyst formulations. The superiority of these catalysts assured their quick acceptance, and a large increase in profitability. This success attracted imitators and lawsuits. In 1973, the Court indicated that these catalyst had resulted in enormous savings in crude oil, reduced capital plant investment and refining operating costs and had saved the industry more than two billion dollars in the U.S. alone (*80*).

One measure of the value of a catalyst is the volume% change that occurs during cracking. While predicting the theoretical volume requires many assumptions, in general the feed is more dense than the transportation fuels product so that there is a potential volume increase. The data in Figure 27 show that the improvements introduced with the clay and/or amorphous silica-alumina were approaching a limiting improvement of recovering about 90% of the volume. With the introduction of the zeolite cracking catalyst, there was a significant improvement and the volumetric expansion is now nearing another plateau that approaches 110%.

Typical Catalyst Structure

A bird's eye view of zeolite structure would view these materials as crystalline inorganic oxide open frameworks with three-dimensional pore or channel networks. The framework can be considered as a giant anionic crystal comprising AlO_4 and SiO_4 tetrahedra linked by shared oxygen atoms. A cubo-octahedron or sodalite cage is characteristic of the faujasite structure present in Type X and Type Y zeolite, and this can be considered a secondary building block, along with double 6-rings which connect the sodalite units in a tetrahedral array to form supercages with interconnected 0.8 nanometer windows. Type X has a SiO_2/Al_2O_3 molar ratio of ~ 2.5, while Type Y is about double that, and is more stable (*41*).

To charge-balance the anionic framework, there are several structurally distinct cation locations, originally filled by sodium ions after synthesis, and then ion-exchanged to introduce acidic species into the product. Hydrothermal treatment after ion exchange results in at least partial framework dealumination, leading to hydroxo-aluminum cations as well as the functionally similar rare earth cations, which provide acid sites throughout the structure as well as increased thermal stability.

Acidity has been ascribed to various combinations of Bronsted, Lewis or defect-type sites. Some people emphasize ordering as a critical property, but on the other hand, a good case can also be made for the importance of lattice defects in providing acidic hydroxyl groups. It can be argued that silica-alumina was active because of partial *ordering* of a disordered structure, while zeolite catalysts were active because of partial *disordering* of an ordered structure (*41*).

Catalyst Characterization

Activity testing has generally been conducted on steam-deactivated catalyst, with the trend over the years being toward higher severities, and the testing unit has gotten a lot smaller as well. There were CAT A, CAT D, CAT D-1, D+L, and CAT C tests, employing several hundred grams, that were used for many years, and finally the Davison microactivity (or MAT) test was developed, using only a few grams, and it became universally employed for just about everything. It has become a highly valuable workhorse in the lab. While it's difficult to make comparisons because product distributions have changed, it is nonetheless interesting to note that, over a 20-year period, the mean microactivity for equilibrium FCC catalyst from most of the operating units, as collected by Davison, went from 68 in 1974 to 67 in 1994, staying in a narrow range from 65 to 70.

Other frequently used characterization parameters include surface area and pore size distribution, X-ray crystallinity, unit cell size, and chemical composition (especially rare earth loadings). Surface areas can range from 100

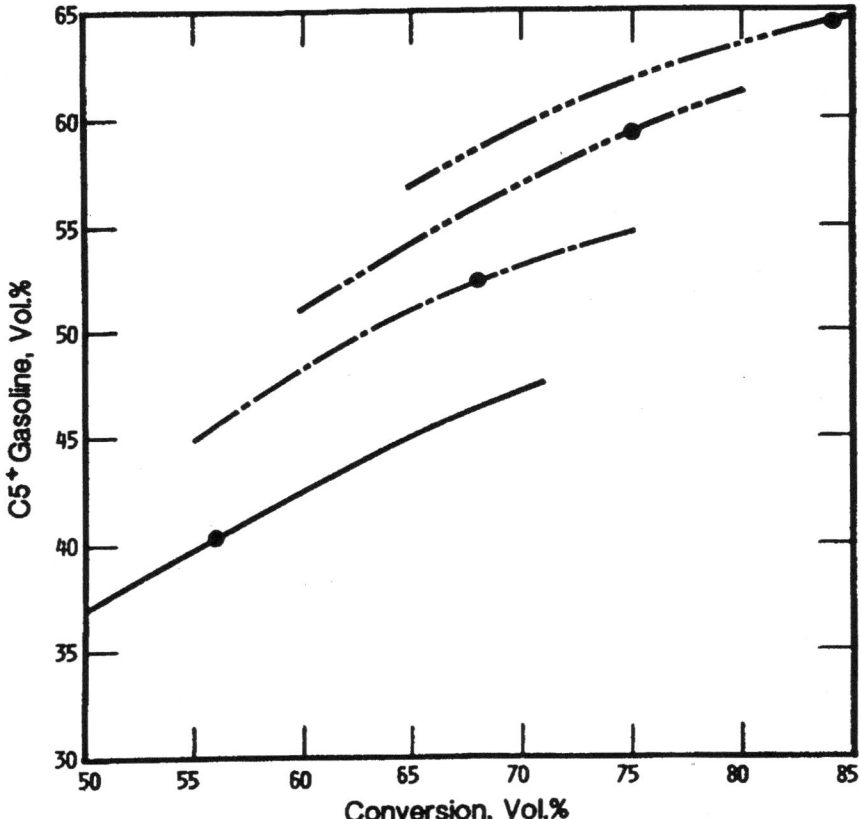

Figure 26. Improvements in zeolite cracking catalyst selectivity (----, standard silica-alumina gel; --- - ---, early zeolite catalysts (REHX); — - - ---, improved zeolite catalyst (REHY and copromoter) and; ●, point where 4% coke (on charge) occurs. (Reproduced from reference 78. Copyright 1983 American Chemical Society.)

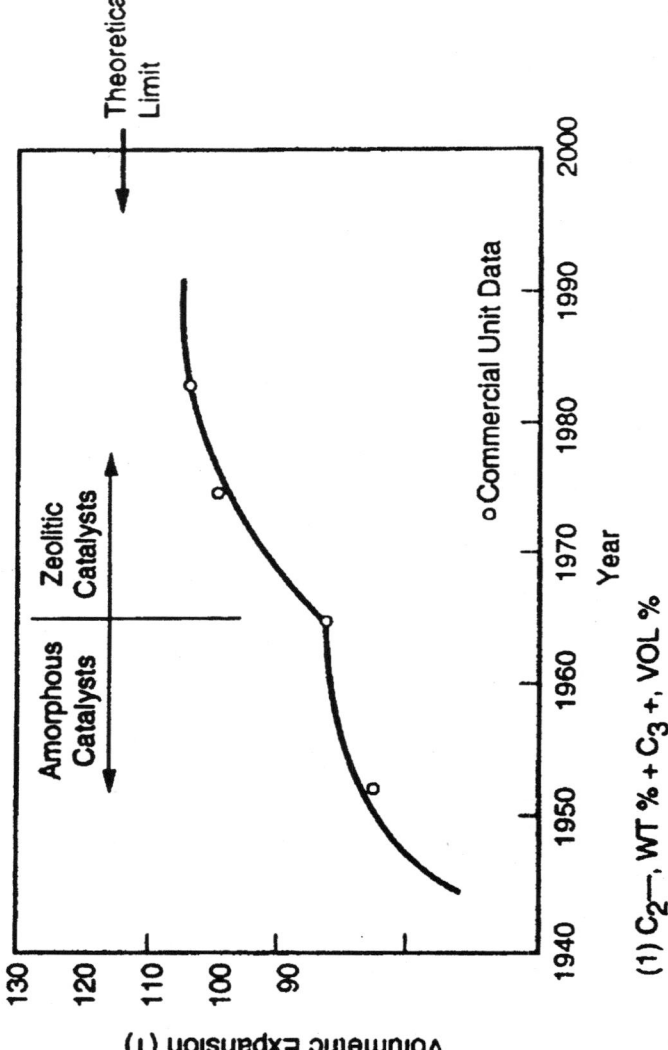

Figure 27. FCC volume expansion, historical and future.

to 400 m^2/g, zeolite content can run from 15 to 40%, and rare earth can range from zero to 3% in fresh catalyst.

An interesting survey was conducted by Engelhard a few years back of 15 companies and their testing philosophies. The conclusion drawn was that the choice of steaming procedure and MAT procedure, and what combination of them is chosen, will affect the observed ranking of catalyst performance. Furthermore, if you want to realistically assess catalysts, the steaming conditions as well as the MAT conditions should be related to commercial experience.

It can be added that curves are more useful than fixed-point comparisons, and that it is possible, by poor choices, to make low-activity catalysts seem artificially attractive, or high-activity catalysts appear relatively non-selective, or to make real differences get small enough to be lost in the data scatter, so proper testing control is essential for meaningful comparisons (*84*).

Competitive Catalyst Distinctions

Operationally, the basic difference between the Type X and Type Y structures was largely a matter of thermal stability, although acid strength-related differences in selectivity were also found. While the Type Y zeolite was much more stable, it was also much more expensive. Davison initially marketed their XZ-15 catalyst, containing steam-stabilized Y zeolite admixed with low-alumina silica-alumina, at a price of $800 per ton. This was quite a high price at the time, but despite that, about 15 refiners tried it. At the same time, Filtrol introduced their Grade 800 catalyst, and although they didn't officially claim that it contained molecular sieves, the product distributions clearly pointed to their presence (*67*).

It was rumored that Filtrol produced a Type X zeolite concentrate from a treated clay and mixed it with halloysite clay to make the Grade 800 product. They sold it for $399.50 per ton, and it gained very rapid acceptance by refiners, who knew a bargain when they saw one. Within a year or so of its introduction, it was being used in about 45 fluid cracking units, representing about a third of the industry. Although Davison's XZ-15 was clearly a better catalyst, Grade 800 garnered the bulk of the fluid zeolitic catalyst business on the basis of price (*67*).

In response to this competitive pressure, Davison introduced a new molecular sieve catalyst in June of 1965. It was called XZ-25 and it was priced at $450 per ton. This was based on rare earth-exchanged Type X zeolite in a high-alumina semi-synthetic matrix, and had a lower zeolite content and a little less hydrothermal stability than the XZ-15. Acceptance was very rapid, and within half a year it was being used in 36 units, in eight of which it displaced the Filtrol Grade 800, reputedly on the basis of advantages in activity, selectivity and stability. Ultimately, XZ-25 was used in more than 116 units (*67*).

In less than two years, 60% of the fluid units were using zeolitic catalysts, with Davison and Filtrol sharing the business. In four years, that number was up

172

close to 85%, and both Cyanamid and Nalco had been forced to make licensing arrangements to stay in business. Of the roughly 134 fluid cracking units in operation, 112 used zeolitic catalysts exclusively and two used a mixture of zeolitic and synthetic silica-alumina types. No longer were the three synthetic catalyst manufacturers marketing exactly the same catalysts. The market split was approximately as follows in 1968 (*67*):

Producer	Percent of Units	Catalyst Price Per Ton
Davison	50%	$450 (XZ-25); $490 (XZ-36)
Filtrol	30%	$399.50 (Grade 800)
Nalco	10%	$525 - $550
Cyanamid	10%	$550 (TS-150)

While Filtrol's catalyst was generally conceded not to be quite as good as the others, they killed everybody on price. Only Davison came close to meeting them, and got the lion's share of the business on the basis of their price-quality relationship and their service. The producers all claimed that the only ones profiting from molecular sieve catalysts were the refiners and Union Carbide, the inventor, and at that time also the supplier, of synthetic molecular sieves.

Union Carbide had assembled some formidable research talent, and for several decades produced a great variety of molecular sieve inventions of enormous value to the petroleum and other industries, but they chose to commercialize only selected ones, relying on partnerships and royalties to exploit the rest. Bob Milton's zeolite X and Don Breck's zeolite Y, along with Jule Rabo's catalytic modification, were licensed to the catalyst market and generated very substantial royalties from the cracking catalyst manufacturers. Filtrol, however, adopted a policy of flaunting all adversely held patents, and became embroiled in a sea of lawsuits as a result. They eventually lost in court after a lengthy battle, and had to pay out over $28 million to Union Carbide and also to Mobil Oil, the owner of a series of infringed zeolite catalyst inventions by Charlie Plank and Ed Rosinski.

About 1968, Davison introduced one of the most popular FCC catalysts ever, called CBZ-1, a semi-synthetic formulation based on stabilized Type Y zeolite. It held sway as a high-gasoline yield product of choice until about 1975 and beyond, when their AGZ-50 and Super-D series came in, offering better attrition resistance to refiners starting to worry about high catalyst inventory losses and white plumes from their units that were getting the attention of local pollution control officials. These stable catalysts managed to maintain that all-important high gasoline yield, and held prominence for close to ten years (*84*).

By 1978 fluidized cracking catalysts were a $130 million per year market, and the three major producers in the U. S. employed somewhat different approaches, with a zeolite content rising from about 10% to about 15 to 20%.

Nalco and American Cyanamid had dropped out of the business. The Davison Division of W. R. Grace dominated the market with about a 55% share of total tonnage. They started with a dilute sodium silicate containing 5% silica, to which they added metakaolin and then sulfuric acid, forming silica hydrogel which was aged to establish the appropriate pore size. This was co-precipitated with aluminum sulfate, forming an aluminosilicate gel. Separately, reactive silica and alumina was heated with NaOH for several hours near the boiling point to bring about a reactive crystallization reaction. The zeolitic product was treated with a base-exchange solution containing rare earth ions to remove sodium, and then spray dried with the aluminosilicate matrix. The sodium content is further lowered by hot ammonium sulfate exchange, and flash drying decomposes the ammonium ion (85).

Filtrol, which had an estimated capacity of 100 to 150 tons per day and a bit over 25% of the market, used halloysite or bentonite treated with sulfuric acid to leach out some of the alumina. This was treated with ammonium hydroxide to produce a hydrated alumina, to which was added some of the leached clay, forming a gelatinous mixture of clay and hydrous alumina. The remainder of the leached clay reacts with sodium silicate and is heated to effect crystallization. After base exchange, the resulting zeolite is mixed with the clay-alumina matrix and spray dried to form microspheres (85).

It had recently been shown that thermal disruption of the kaolinite mineral layer structure produced a highly reactive disordered aluminosilicate with a silica to alumina molar ratio of two. Higher calcination temperatures produced what is known as mullitized kaolin, which also contained some reactive free silica (86). This served to supply the additional silica needed for the synthesis of zeolite Y, which has a silica to alumina molar ratio in the 4.5 to 5.0 range.

Engelhard's Minerals and Chemicals Division, which now had close to 20% of the market and a capacity of 100 to 150 tons per day with an average price of about $1000 a ton, marketed their product through the Houdry Division of Air Products and Chemicals, continuing the marketing relationship they had with pelleted moving bed products in the Western Hemisphere. They started with spray dried kaolin clay with an average particle size of 65 microns and a controlled size distribution, which were then calcined to produce reactive metakaolin. This, along with some more highly calcined kaolin, was treated with a caustic solution in an agitated tank under controlled time and temperature conditions to produce *in situ* zeolite and sodium silicate. After filtration and base exchange to remove sodium and incorporate some rare earth ion, flash drying produced a flowable product for shipment to refineries (85).

The thermal stabilities of the various forms of zeolites X and Y can be compared to show some of the options available for catalyst formulation. Ultrastable Y had the highest thermal and hydrothermal stability, but not the highest activity. Rare earth-exchanged Y was active, almost as stable, but not as selective and not particularly cheap. A lower level of ion exchange sacrificed some stability, but changed the product distribution. Zeolite X in its various forms was cheaper but not as stable. Trade-offs had to be made between stability, activity, product distribution and cost, for a variety of special market niches, leading to a proliferation of targeted catalyst products. Since the matrix makes up the majority of the catalyst, most of the physical properties of the catalyst, such as attrition resistance, density, specific heat and pore volume, are controlled by the composition of the overall binder system, or matrix (*41*).

Research and development efforts were focused on addressing several of the problems of the day. Sulfur oxides (SO_x) were becoming more stringently controlled, and stack gas scrubbers or use of hydrodesulfurized gas oil feed were expensive options, so catalyst makers developed new products that worked in conjunction with a Claus unit to first reduce the sulfur to hydrogen sulfide and then recover it as elemental sulfur. Carbon monoxide, formed at the higher regenerator temperatures resulting from improved zeolite catalysts, needed to be oxidized to improve heat recovery, and additives were developed to accomplish this. Metals tolerance and octane improvement were needed as well, and producers vied with each other to address the needs of the market (*85*).

By 1983 the domestic picture showed Davison with half the business, and Engelhard coming up to a quarter, while Filtrol dropped to a quarter under Kaiser ownership. Davison's high-stability DA series was supplanting the Super D and Octacat octane catalysts and the GX and DZ resid catalysts. Filtrol's leading FCC product was ROC-1, while Engelhard's were the Ultrasieve and HEZ series. Some refiners made a habit of swinging back and forth between suppliers, telling the local pollution control people that the particulate emissions problem they had recently had was fixed, since they had just switched to a new catalyst supplier with a more attrition-resistant product. It's surprising how many times they were able to recycle that story (*67*).

In the mid-1980's, catalysts were being supplied around the globe not only by Davison, Filtrol, and Engelhard, but also by Ketjen, Katalistiks, Crosfield, and Catalysts & Chemicals, in, according to Oil & Gas Journal, 34 series of products with a total of 160 varieties. (That has shrunk some since then.) In addition to activity levels and octane enhancement, this proliferation of catalyst products addressed such specialized needs as bottoms cracking, special feedstocks, low coking and low regenerator severity, metals resistance, nitrogen tolerance, enhanced thermal and hydrothermal stability, special product distribution requirements, SOx reduction, low density adjustment for circulation-sensitive units, CO combustion promotion, enhanced attrition resistance, and a number of other sales niches. And catering to the small remaining moving bed market, Engelhard introduced its EMCAT series of non-fluid cracking catalysts.

Davison introduced its XP line of catalysts in 1986 to combine low coke and gas makes with bottoms cracking and attrition resistance, and in 1991 they built a new plant and came out with a rare earth-free product based on an improved ultrastable Zeolite Y.

Paul Weisz of Mobil Oil, in a CHEMTECH article (*87*), showed a product distribution comparison between a zeolite catalyst and an amorphous silica-alumina catalyst, making clear the dramatic difference between them:

	Amorphous Silica-Alumina Catalyst	Zeolite Catalyst
Paraffins	13%	24%
Olefins	17%	3%
Naphthenes	41%	23%
Aromatics	29%	50%

The advantageous characteristics of zeolite cracking catalysts were generally summarized by Paul Venuto and Tom Habib in 1979 as follows, and they're still true (*88*):

1. High conversion activity, allowing more throughput
2. Stability of activity, especially with respect to coke deactivation
3. Very high selectivity to gasoline
4. Low coke and dry gas makes
5. Thermal and hydrothermal stability in the regenerator
6. High porosity and diffusion accessibility
7. Attrition resistance (not always realized, but then you need to withdraw catalyst from the unit *somehow!*)
8. Resistance to poisoning by heavy metals, nitrogen, etc.
9. Low cost, especially considering the benefits

Zeolite catalysts have probably saved refiners at least a billion dollars a year, and saved our economy over 5 billion dollars a year in imported oil we didn't need because of the enhanced selectivity and gasoline yields.

Zeolite Catalyst - Mobil Versus Esso (Exxon)

Kimberlin applied for a patent in February 5, 1957 which was issued on February 14, 1961 and was assigned to Esso Research & Engineering Co. (*80*). This patent taught the reduction of sodium content by base exchange, a process similar to that used with amorphous synthetic silica-alumina catalysts. This patent included in the body, but not the claims, the use of base exchange with rare earths. The court held that Kimberlin's patent indicated that (*80*): (1) the zeolite should be used as the sole catalyst in the cracking zone, (2) that crystalline materials adversely affect the catalytic performance of conventional amorphous catalysts ["By whatever means prepared, the final catalyst was amorphous and, indeed, if any crystalline material were present in the final product, hydrocarbon conversion was seriously adversely affected and byproduct formation increased." (*80*)], and (3) that the random size pore openings of the conventional amorphous catalysts are undesirable. Furthermore, Kimberlin did not teach a technique for better gas oil cracking performance than the conventional silica-alumina; and it taught that as much as one-third of the original sodium should be retained in the zeolite for best results. After the Rosinski and Plank patent issued, Exxon applied for a reissue of its Kimberlin patent, seeking to eliminate the word "sole" from its claims. By early 1965, the examiner had allowed the Kimberlin Reissue, subject to the pending interference litigation between Mobil and Exxon. A blocking patent situation existed; thus, two adverse parties held patents so that neither party could practice the advance without infringing the other's patent (*80*). In order to resolve this impasse, Mobil and Exxon negotiated a non-exclusive cross-licensing agreement about July 1, 1966. Most Mobil research employees believed that in this high-stakes game, Mobil blinked first and gave in to the Exxon threat. In any event, Exxon now was in the game. After filing a suit against Mobil, W. R. Grace procured a license under the Kimberlin-Exxon patent (*80*). This occurred only after Exxon had used economic pressure by removing about three million dollars of its annual business from Grace and then threatening to send all of the remainder of its catalytic business elsewhere (*80*).

Zeolite Catalyst - Riser Cracking

It was soon realized that in the FCC process some cracking occurred in the riser transport tube following catalyst addition. The introduction of the much more active zeolite catalyst gradually made the industry aware that a significant fraction of the cracking was occurring in the riser and that the "reactor" was predominantly a region where catalyst-product separation and unwanted secondary reactions were occurring. Furthermore, because of less secondary, gas producing reactions, the yield selectivity of the cracking that occurred in the riser was superior to that in the larger reactor section. The first riser cracker was constructed by Shell in their Anacortes refinery (*89*). The improvement of the yield structure dictated that most, or all, new cracking units be of the type that employed riser cracking. In addition, some of the other reactors were converted to riser cracking through internal modifications.

When the advantages of the zeolite catalyst were recognized, Mobil came to Nalco and wanted enough royalty to make the catalyst cost $800/ton (*75*). At that time Nalco only got $300/ton for their catalyst that was based on synthetic silica-alumina and clay, and they told Mobil that nobody would think of paying $800/ton. However, by 1965, Nalco, Filtrol and Davison were into the business with zeolite containing catalysts, and the competition among them went on for years.

In summary, there have been many advances in catalysts since the initial use of homogeneous aluminum chloride. Management desires revolutionary advances as contrasted to evolutionary advances. This viewpoint had become practically the only driving force behind research during recent years and has resulted in dramatic cuts of 50% or more in research effort in nearly all refining companies. The philosophy of revolutionary advances is great in theory. Unfortunately, the results over many years indicate that it is a theory at variance with facts. The advances in FCC catalyst performance are illustrated in Figure 28 (*90*). There is only one revolutionary advance shown in this figure: the introduction of the X-zeolite catalyst. Within five years, this revolutionary advance - X-zeolite - had been made obsolete by the evolutionary advance of the superiority of the Y-zeolite catalyst. In this, as in the case of the introduction of FCC, the revolutionary advance became obsolete within a short period. Furthermore, revolutionary advances "fly in the face" of conventional wisdom and, because of this, cannot be predicted so that they can be included in the research management strategic plan. On the other hand, evolutionary advances can frequently be foreseen and their development can be planned by management. This conundrum continues to confound the industry.

Summary

The initial thermal cracking process was developed only after many earlier alternatives had been tried and found to be unsatisfactory. The development of this process can be traced to the employment of highly trained chemists who combined a sound scientific training with Edisonian approaches to research. The resulting thermal cracking process was a revolutionary advance. As important as the scientific advances, and perhaps even more important, was the gradual development of hardware to permit the refiner to operate at much more severe conditions than they had been accustomed to doing. Burton was successful because of his scientific ability and for his management skill that allowed him to rise to offices high enough in the company for him to strongly influence the ultimate decision. His process was a revolutionary advance and provided his company phenomenal returns on investment. However, the company failed to remain competitive in the area and the Burton process was replaced during a few years by processes developed by others who continued to work and innovate in this area.

Attempts to introduce catalytic cracking met with failure on numerous occasions. Only with the recognition of the need for a process involving regeneration was its commercialization possible. Only the strong personality and the entrepreneurship of Eugene Houdry allowed this revolutionary process to attain commercialization. Hardware advances were as, or more, significant than the scientific advances. The revolutionary advance gained rapid acceptance and captured a significant fraction of the cracking refinery capacity. The Houdry process likewise provided phenomenal returns on the investment. Soon

178

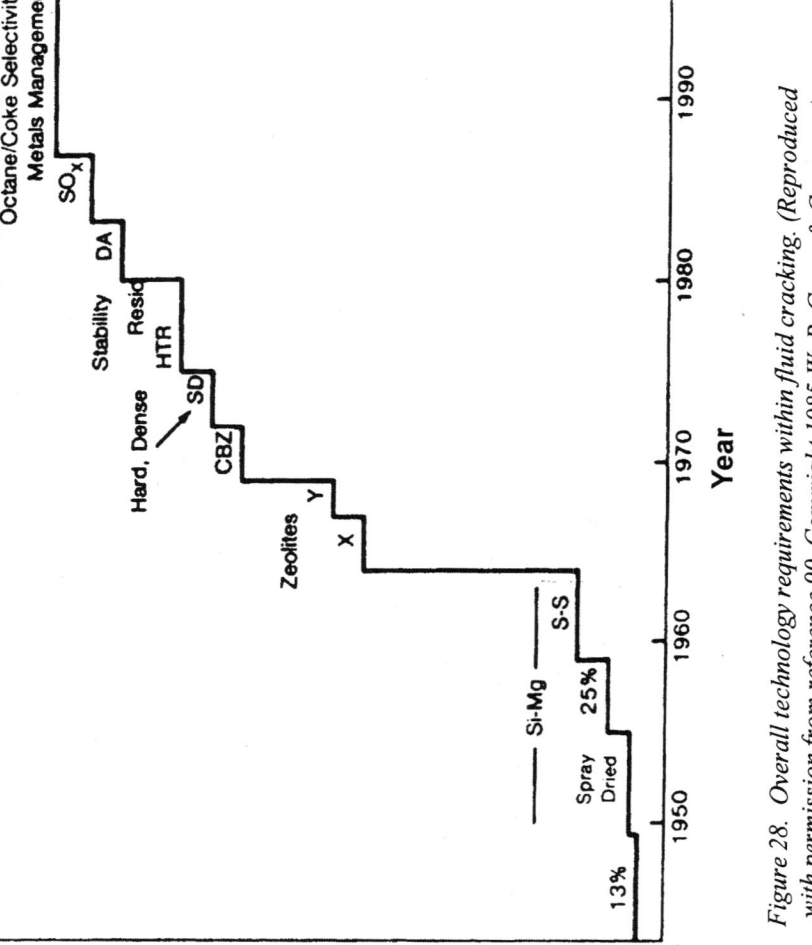

Figure 28. Overall technology requirements within fluid cracking. (Reproduced with permission from reference 90. Copyright 1985 W. R. Grace & Company.)

after the introduction of the revolutionary advance, evolutionary improvements led to the initial advance becoming obsolete.

Fluid catalytic cracking was another revolutionary advance; however, the developers of this process had the advantage of making it in a "proven area." Unlike the above two examples, the developer, Exxon, continued to made evolutionary advances in the process and has therefore remained competitive with others who entered the field. The evolutionary advances have continued but the pace of their introduction has slowed so that longer time intervals have been required between each advance. The return on investment has been outstanding even though the initial process was introduced during WWII when price controls were in effect and prices, including royalties, were regulated.

There have been three revolutionary advances in cracking processes: the initial regeneration process in 1932, the moving bed processes in the 1940's and riser cracking in the 1970s. An optimistic person would anticipate another revolutionary advance within the next few years.

The rise and fall of the various cracking processes during the period between 1913 and 1957 is shown in Figure 29. The dominance and the replacement of various processes are evident, as described above. Today, if we ignore hydrocracking, FCC is totally dominant in processes involving catalytic cracking.

In catalysis, there have been two revolutionary advances: the initial introduction of the acid-washed clay catalyst and the introduction of the rare earth X-zeolite catalyst. There have been many evolutionary advances that result in small, but significant, improvements in catalysis. The initial discovery was made about 1930 and the zeolite catalyst was introduced in 1963. On the basis of time intervals between revolutionary advances, one would anticipate a revolutionary advance in catalysis within the next few years.

The introduction of a successful process has been exploited very rapidly by petroleum refiners. Thus, in a few years a new process goes from its introduction to its maximum utilization (Figure 30). In each instance, the introduction of these processes has proven to be financially rewarding. However, it appears that the evolutionary process provides about the same return as the revolutionary process (Table 2). Perhaps this is why refiners are so reluctant to be the first one who makes the revolutionary advance.

The data in Figure 31 show that there was a phenomenal growth in the use of catalytic cracking during 1913-1930. Growth in catalytic cracking in the U.S. has increased since 1930 but at a slower rate of growth. Today catalytic cracking may be considered a mature process on the basis of the refinery capacity. Even so, it is noted that the capacity growth over many years has been at an exponential rate. The return on the research investment has been exceptionally high (Table 3).

To a scientist, the most surprising aspect of catalytic cracking has been the extensive litigation. It would appear that refiners have been willing to spend more money on the "protection of advances" than in making the advance. Thus,

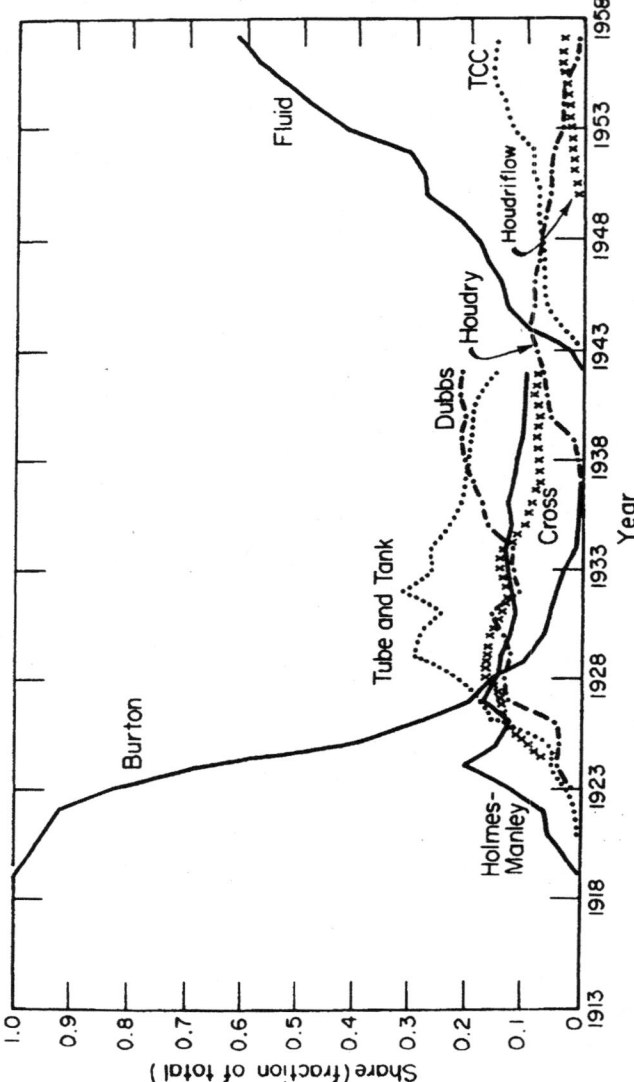

Figure 29. Share of cracking capacity by process, 1913-1957. (Reproduced with permission from reference 8. Copyright 1962 Massachusetts Institute of Technology.)

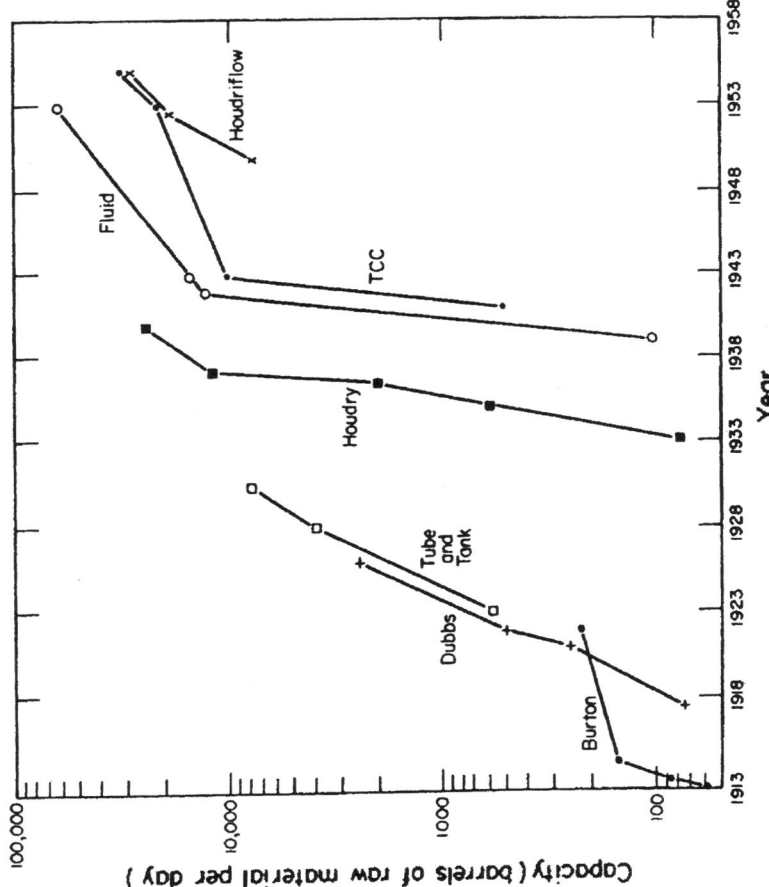

Figure 30. Growth in size of single cracking units, 1913-1957. (Reproduced with permission from reference 8. Copyright 1962 Massachusetts Institute of Technology.)

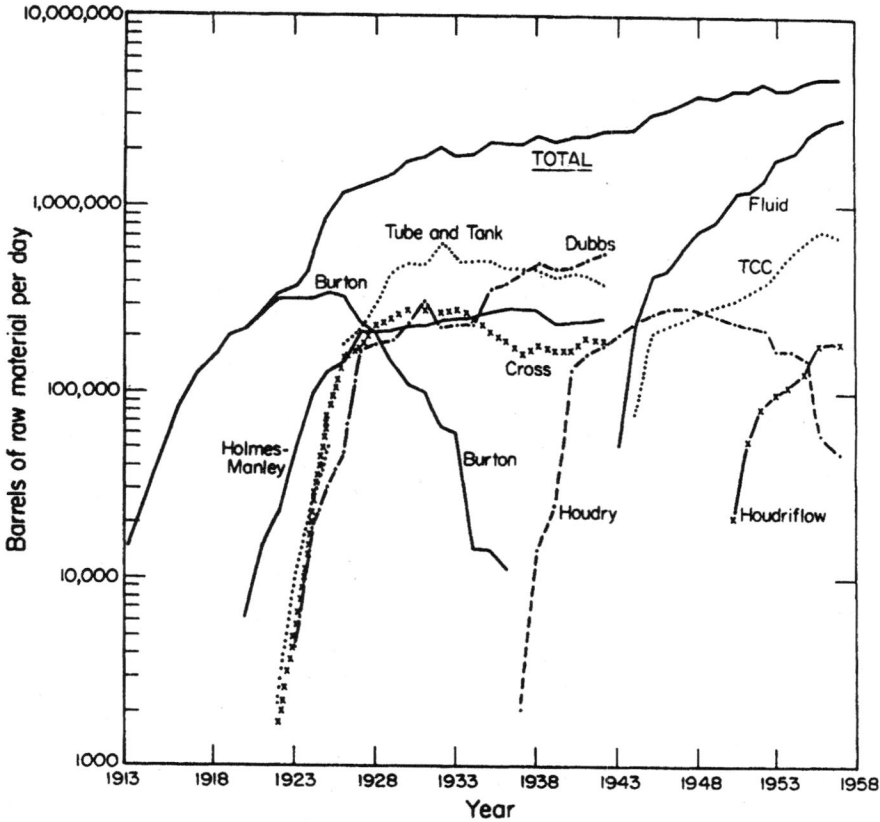

Figure 31. Growth in U.S. cracking capacity, total and by process, 1913-1957. (Reproduced with permission from reference 8. Copyright 1962 Massachusetts Institute of Technology.)

Table 3. Cost of and returns from cracking process innovations, 1913-1957 (from ref. *4,8*)

Process	Cost of Innovation		Returns from Innovation		Approximate ratio of returns to cost ($ per $)
	Period over which expenses incurred	Estimated amount	Period over which returns calculated	Estimated Amount	
Burton	1909-1917	$236,000	1913-1924	$150,000,000+	600+
Dubbs	1909-1931	7,000,000+	1922-1942	135,000,000+	20
Tube and Tank	1913-1931	3,487,000	1921-1942	284,000,000+	80+
Houdry	1923-1942	11,000,000+	1936-1944	39,000,000	3.5
Fluid	1928-1952	30,000,000+	1942-1957	265,000,000+	9
TCC	1935-1950		1943-1957	71,000,000+	
Houdriflow	1935-1950	5,000,000+	1950-1957	12,000,000	16

184

the history of catalytic cracking involves at least four components: (1) recognition of scientific and engineering advances, (2) development of the complex process and hardware needed to implement the advance, (3) introduction of similar variations on the initial advance, and (4) large-scale legal battles to protect one's position. While the data are not readily available, it is certain that the legal budget devoted to protecting a refinery innovation is much greater than the research and engineering budget that permitted the development of the advance in the first place.

References

1. Williamson, H.F.; Andreano, R.L.; Daum, A.R.; Klose, G.C. *"The American Petroleum Industry: The Age of Energy,"* Northwestern University Press, Evanston, IL, 1963.
2. Reese, K., *Today's Chemist at Work,* American Chemical Society: Washington, DC, Nov/Dec., 1993, p. 59.
3. Burton, W.M. *Ind. Eng. Chem.* **1918**, *10*, 484.
4. Enos, J.L. *"The History of Cracking in the Petroleum Refining Industry: The Economics of a Changing Technology,"* Ph.D. Thesis, MIT, 1958.
5. Dubbs, J. U.S. Patent 1,123,502, January 5, 1915.
6. Dubbs, J. U.S. Patent 1,135,506, April 13, 1915.
7. Private communication, George Tobiasson to B. Davis.
8. Enos, J.L. *"Petroleum Progress and Profits. A History of Process Innovation,"* The M.I.T. Press, Cambridge, MA, 1962.
9. Schmerling, L. "Gustov Egloff, `Mr. Petroleum'," presentation 147th National Meeting, Philadelphia, PA, ACS, April 5-10, 1964.
10. Anonymous, Science Illustrated, Sept., 1947, pp 80.
11. Reprint from "Chemical Bulletin" and based on information provided by Dr. Louis Schmerling, courtesy UOP library.
12. C & E News, October 13, 1997, pg. 7.
13. Pines, H. in *"Heterogeneous Catalysis. Selected American Histories";* Editors, B. H. Davis and W. P. Hettinger, Jr.; ACS Symp. Series, 222 (1983) pp 23-32.
14. Trumble, M.J. U.S. Patent 1,281,884, October 15, 1918.
15. Dubbs, J. U.S. Patent 1,392,629, October 4, 1921.
16. James, M. *"The Texaco Story: The First Fifty Years, 1902-1952,"* The Texaco Co., New York, 1953.
17. Cross, W.M. U.S. Patents 1,203,312, October 31, 1916 and 1,326,851, December 30, 1919; Cross, R. U.S. patent 1,255,138, February 5, 1918.
18. Brooks, B.T. in *The Science of Petroleum* (A. E. Dunstan, A. W. Nash, B. T. Brooks and H. Tizard, Eds) Oxford University Press, London, Volume III, pp. 2078-2087 (1938).

19. McKnight, Jr., D. *"A Study of Patents on Petroleum Cracking,"* U. Texas Publication No. 3831, August 15, 1938.
20. Nat. Petr. News, 29, No. 49, December 8, 1937, p. 9.
21. McAfee, A.M. *Ind. Eng. Chem.*, **1913**, *7*, 737.
22. Friedel, C.; Crafts, J.M. English Patent, 4,769, December 15, 1877.
23. McAfee, A.M. U.S. patents 1,099,096 June 2, 1914 and 1,144,304, June 22, 1915.
24. Oblad, G. in *"Heterogeneous Catalysis. Selected American Histories,"*; Editors, B. H. Davis and W. P. Hettinger, Jr.; ACS Symp. Series, 222 (1983) pp 61-76.
25. Newton. R.H.; Shimp, H.G. *Trans. Am. Inst. Chem. Engrs.* **1945**, *41*, 197.
26. Lassiat, RC.; Thayer, R.H. *Oil Gas J.*, **1946**, *45* [13], 84.
27. Avidan, A.A.; Edwards, M.; *Rev. in Chem. Eng.* **1990**, *6*, 1-71.
28. Berkman, S.; Morrell, J.C.; Egloff, G. *"Catalysis. Inorganic and Organic,"*: Reinhold Pub. Corp., New York, 1940.
29. Egloff, G.; Morrell, J.C.; Thomas, C.; Block, H. *J. Am. Chem. Soc.* **1939**, 61, 3571.
30. Drake, L.C. in *"Heterogeneous Catalysis. Selected American Histories;"* Editors, B. H. Davis and W. P. Hettinger, Jr.,; ACS Symp. Series, 222, (1983) 451-462.
31. Noll, H.D.; Holdom, K.G.; Bergstrom, E.V. *Petroleum Refiner* **1946**, *25* [5], 237.
32. Shankland, R.V. *Advan. Catal.* **1954**, *6*, 271.
33. Hagerbaumer, W.A.; Lee, R. *Trans. Am. Soc. Engrs.* **1947**, *69*, 779.
34. Farrer, G.L. *Oil Gas J.* **1951**, *51* [32], 120.
35. Anonymous, *Petroleum Refiner* **1951**, *30* [9], 164.
36. Faragher, W.F.; Noll, H.D.; Bland, R.E. *Proc. 3rd World Petroleum Congr.*, Hague, Section IV, 138 (1951).
37. Mills, G.A. in *"Heterogeneous Catalysis. Selected American Histories;"* Editors, B. H. Davis and W. P. Hettinger, Jr.,; ACS Symp. Series, 222 (1983) pp 179-182.
38. Wheeler, A. *Advan. Catal.* **1951**, *3*, 249.
39. Wheeler, A. in *"Catalysis;"* Editor, P. H. Emmett; Reinhold, New York, Vol. II, 105-158, 1955.
40. Squires, M. *Adv. Chem. Eng.* **1994**, *20*, 1.
41. Flank, W.H. Symposium on the Catalytic Inventions of Eugene Houdry, ACS National Meeting, New Orleans, March, 1996.
42. Baxter, Jr., J.P. *"Scientists Against Time;"* Little, Brown & Co., Boston, 1946.
43. Odell, W.W. U.S. patent 1,984,380, December 18, 1934.
44. Johning, E.; Martin, H.Z.; Campbell, D.L. in *"Heterogeneous Catalysis. Selected American Histories;"* Editors, B. H. Davis and W. P. Hettinger, Jr.; ACS Symp. Series, pp 273-292.

45. Lewis, W.K.; Gilliland, E.R. U.S. Patent 2,603,608, February 21, 1950.
46. Murphree, E.V. *Advances in Chem. Series* **1951**, *5*, 30.
47. Murphree, E.V.; Fischer, H.G.M.; Gohr, E.J.; Sweeny, W.J.; Brown, C.L. *Am. Petroleum Inst.* **1943**, *24*, III, 91.
48. Murphree, E.V.; Brown, C.L.; Fischer, H.G.M.; Gohr, E.J.; Sweeny, W.J. *Ind. Eng. Chem.* **1943**, *35*, 768.
49. Campbell, L.; Martin, H.Z.; Murphree, E.V.; Tyson, C.T. U.S. Patent 2,451,804, October 19, 1948; filed December 27, 1940.
50. Gunness, R.C. *Chem. Eng. Progr.* **1953**, *49*, 113.
51. Heron, S.C. *"Development of Aviation Fuels,"* Graduate School of Business Administration, Harvard University, p. 607.
52. Beaton, K. *"Enterprise in Oil. A History of Shell in the United States,"* Appleton-Century-Crofts, Inc., New York, 1957, p. 562.
53. Larson, H.M.; Knowlton, E.J.; Popple, C.S. *"New Horizons. History of Standard Oil Company (New Jersey) 1927-1959,"* Harper and Row, New York, 1971, 167-168.
54. Popple, S. *"Standard Oil Company (New Jersey) in World War II,"* Standard Oil Company (New Jersey), New York, 1952, pg. 15.
55. Personal communication, S. Shulman to W. H. Flank.
56. Shaw, D.F.; Walter, R.E.; Zaczepinski, S.; "FCC Reliability Mechanical Integrity," NPRA Annual Meeting paper AM96-24 (1996).
57. Maddox, R.F. *West Virginia History* **1997**, *55*, 127.
58. U.S. Congress, Senate, Committee on Military Affairs, "Cartel Practices and National Security, Hearings before a Subcommittee of the Committee on Military Affairs, 78th Congress, 2nd Session, 1944, Vol. 16, 2011.
59. Grosse, A.; Morrell, J.; Mattox, W. *Ind. Eng. Chem.* **1940**, *32*, 528.
60. Mullen, W. *"Unlikely Hero. A Polish Immigrant's High-Octane Role in Winning the Battle of Britain,"* Chicago Tribune Magazine, July 15, 1990, Section 10.
61. Yanik, S.J., An historical look at Filtrol's FCC activities.
62. Personal communication, T. Cornelius to B. H. Davis.
63. Montgomery, J.A. "Guide to Fluid Catalytic Cracking, Part One," Grace Davison, 1993.
64. L. B. Ryland, M. W. Tamele and J. N. Wilson in "Catalysis," (P. H. Emmett, Ed.), Reinhold Pub. Corp., Vol VII, pp 1-91.
65. Anon., Chem. Eng., 58[12] (1951) 224.
66. L. Thomas, N. K. Anderson, H. A. Becker and J. McAfee, Proc. Am. Petroleum Inst. (24th Annual Meeting), 24 (1943) 75.
67. History of Fluid Cracking Catalysts, internal memo, Houdry Division, Air Products & Chemicals Corp., 1968.
68. L. B. Ryland, Oil Gas J., 53[51] (1955) 115.
69. Milliken, Jr., T.H.; Mills, G.A.; Oblad, A.G., "The Chemical Characteristics and Structure of Cracking Catalysts," *Discussions Faraday Soc.* **1950**, *8*, 279-290.

70. C.L. Thomas, Ind. Eng. Chem., 41 (1949) 2564.

71. Mills, G.A.; Boedeker, E.R.; Oblad, G.A. *J. Am. Chem. Soc.* **1950**, *72*, 1554.

72. Heinemann, H.; Mills, G.A.; Hattman, J.B.; Kirsch, F.W. *J. Ind. End. Chem.* **1953**, *45*, 130.

73. C.J. Plank and L. C. Drake, J. Colloid Sci., 2 (1947) 399.

74. Personal communication, W. P. Hettinger, Jr. to B. H. Davis.

75. A.F. Cronstedt, Akad. Handl. Stockholm, 17 (1756) 120.

76. W. Breck, Zeolite Molecular Sieves: Structures, Chemistry and Use, John Wiley, New York, 1974.

77. Personal communication, R. P. Eischens to B. H. Davis.

78. C.J. Plank in "Heterogeneous Catalysis. Selected American Histories," (B. H. Davis and W. P. Hettinger, Jr., eds), ACS Symp. Series, 222 (1983) pp 253-272.

79. Personal communication, E. J. Rosinski to B. H. Davis.

80. Mobil Oil Corp. vs. W. R. Grace & Co., Civil No. 14589, U.S. District Court, District of Connecticut, Nov. 2, 1973.

81. H. Dickey, Proc. Nat. Acad. Sci. U.S., 35 (1949) 227.

82. L. Pauling, Chem. Eng. News, 27, (1949) 913.

83. Flank, W.H. "A Philosophy for Testing," ACS Symposium Series 411, Washington, DC, 1989, p. 92.

84. Tongue, T. "Davison and the Development of Fluid Cracking Catalyst," *Davison Catalgram*, **1992**, *83*, 1.

85. "Refinery Catalysts are a Fluid Business," *Chemical Week, July 26, 1978, pp 41-44.*

86. Flank, W.H. *Clays and Clay Materials* **1979**, *27*, 11-18.

87. Weisz, P.B. *Chemtech* **1987**, *17*, 368.

88. Venuto, P.B.; Habib, E.T. *Catalysis Reviews – Science & Engineering* **1978**, *18*, 1-150.

89. J. W. Wilson, Fluid Catalytic Cracking, PennWell Books, Tulsa, OK, 1997.

90. J.B. Hattman, Davison CATALAGRAM, No. 72, 1985, p. 3, published by W. R. Grace & Co.

Chapter 6

The History of Fluidized Catalytic Cracking: A History of Innovation: 1942–2008

R. P. Fletcher[1,2]

[1]Albemarle Catalysts, 2625 Bay Area Boulevard Suite 250,
Houston, TX 77058
[2]Current address: Intercat Inc., P.O. Box 412,
Sea Girt, NJ 08750

The Fluidized Catalytic Cracking Unit (FCC) has passed through a remarkable period of innovation which has resulted in its commanding the leading position as the primary cracking asset in most high conversion refineries. The history of this process began in one man's laboratory (Houdry) and was catapulted by the emergence of World War II into one of the primary means of producing aviation jet fuel. The prominence of the FCC unit continued to increase following the war within the refining world. The FCC unit has passed through a number of fundamental shifts in innovation including the advent of zeolitic catalysis, residue cracking, and most recently, as an engine for producing chemical feedstocks (propylene). Additionally, world fuel markets are now shifting towards diesel which is once again providing a driving force for further innovation. While the FCC unit is a mature technology, it is far from stagnant. The coming several decades are sure to continue this trend.

Prehistory

From Horse Drawn Carriages to Automobiles (1859-1913)

The first commercial production of petroleum began in Titusville, Pennsylvania in 1859*. The "horseless carriage" (Figure 1) made its debut in North American cities shortly after this well head came on-line. By the turn-of-the-century it is estimated there were approximately 8,000 horseless carriages being driven in North America. However, by 1910 this number had grown to approximately 500,000 automobiles (1).

Initially, motor gasolines were derived largely from straight run products distilled from crude. A typical gasoline yield was approximately 20 vol% with an octane number of about 50-55. The first gasoline shortages in the history of the automobile then began to be felt by these initial car owners.

The refiners of the day responded by increasing refinery charge rates to meet this new demand. However, as is the case today, a driving force existed to maximize high valued product yields, and thereby, profitability. It was at this point that the innovators of the day began to search for a means to increase the amount of gasoline available for this new market.

The Advent of Thermal Cracking (1913-1942)

This demand led William Burton, of Standard Oil Co. of Indiana, to commercialize the first thermal cracking process in 1913. This discovery launched the rapid development of several competitive thermal cracking designs.

The most popular design of the day was the Dubbs process (2). The Dubbs process (Figure 2) was licensed by the Universal Oil Products Co. (now UOP), which was founded in 1913 by T. Ogden Armour by purchasing the Standard Asphalt Co. and the Jesse Dubbs patents. UOP guaranteed a 23% yield of gasoline and a 24 hour cleanout period for the 250 bpd Dubbs unit!

---

* The author wishes to acknowledge that many have written histories of the FCC in past years. (Several exceptional accounts may be found in references 1-31.) The intention in this history has been to combine key points from several of these accounts plus FCC's most recent advances to produce a more up to date, and hopefully, complete history. Where appropriate, the author has quoted previous authors and provided credit via the footnotes. As a trumpet of warning, many of our industry's more interesting "side bars" of history are being lost as our most experienced engineers retire. May we all redouble our efforts to ensure that this "voice tradition" not be lost!

Figure 1. America's love affair with the automobile begins.

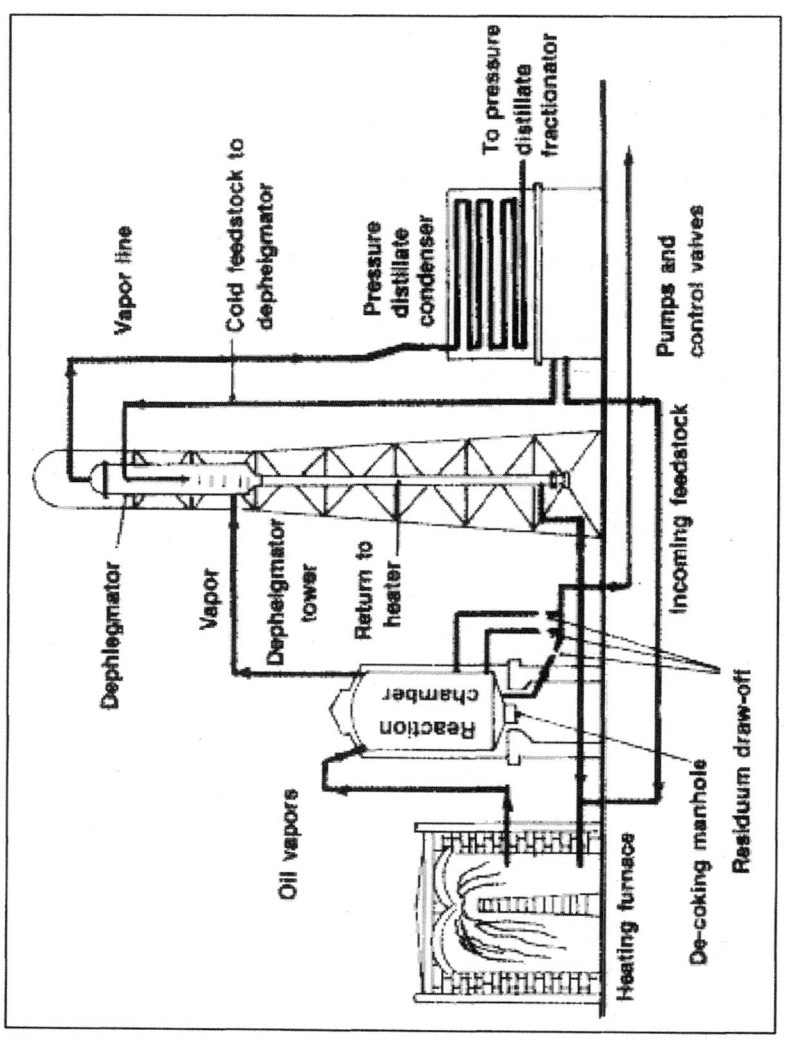

Figure 2. The Dubbs thermal cracking process (Adapted with permission from Oil and Gas Journal, January 8, 1990. Copyright 1990 Oil and Gas Journal.)

The Dubbs process (*3*) was a continuous thermal cracking process in which a relatively clean feedstock was cracked in furnace tubes with the continuous removal of heavy cracked residue from the system. This process was patented (1,049,667) in 1913.

The use of thermal cracking in these early refineries increased such that by 1943 it was estimated that nearly 50% of the nation's gasoline demand was being met by this process. The inclusion of tetraethyl lead in 1925 together with the improved product yields supplied by thermal cracking led to gasoline yields approaching 60 vol% with premium octane levels as high as 79. However, the high concentrations of olefins and diolefins led to gasoline stability issues (gum forming compounds) resulting in yet another driving force among innovators for improved gasoline quality.

In the 1920's a direct connection between engine performance and gasoline composition was discovered through the invention of the octane engine. The identification of engine knock as a function of premature ignition, the invention and use of anti-knock compounds, combined with the newly defined octane scale led to an advanced understanding of gasoline quality. It was evident that thermally cracked gasoline was superior to straight run gasoline and that the soon to be invented, catalytically cracked gasoline was even better.

The Birth of FCC – Fixed Bed to Fluidized Cracking

A major drawback of all the thermal processes of the day was the production of low valued coke. Alternative process routes were being developed to address this issue. One development pursued was hydrocracking which suppresses coke formation by the circulation of high pressure hydrogen. However, the route successfully pioneered by Houdry led to an eventual refinery configuration with the FCC centered as the primary means of conversion in today's modern refinery. Houdry's route appeared to preempt hydrocracking. Table I provides an overview of the major milestones of FCC (4, 5, 6, 7).

Fixed Bed Design: The Houdry Process

Eugene Houdry (*8, 9*) is credited with the discovery of activated clay as a catalyst for converting petroleum into high-octane gasoline. In the 1920s, Houdry (Figure 3) discovered while experimenting with the removal of sulfur from hydrocarbon vapors, that catalytic activity could be restored to catalysts via combustion of cumulative carbon. This single observation was a major breakthrough for the refining industry which led to the development of the Houdry process and eventually the FCC process itself.

194

Table I. Evolution of FCC

Year	Development
1915	McAfee of Gulf Refining Co. discovered that a Friedel-Crafts aluminum chloride catalyst could catalytically crack heavy oil
1928	Acid activated clay (Houdry)
1936	Use of natural clays as catalyst greatly improved cracking efficiency and was used in the Houdry process
1938	Catalytic Research Associates (CRA) was formed. This was a result of a new requirement by the United States government which decreed a general patent pooling covering the catalytic cracking know-how of all companies under the famous "Recommendation 41" agreement which lasted for the duration of World War II plus 7 years. The original CRA members were: Standard of New Jersey (Exxon), Standard of Indiana (Amoco), Anglo Iranian Oil Co. (BP Oil), the Texas Company (Texaco), Royal Dutch Shell, Universal Oil products (UOP), the M. W. Kellogg Co., and IG Farben (dropped in 1940).
1940	Synthetic catalyst plant (Houdry SV)
1942	First commercial FCC unit (Model I up flow design) started up at Standard of New Jersey's Baton Rouge, Louisiana Refinery. Fluid catalyst developed (Exxon)
1943	First down flow designed FCC unit was brought on line. This was the Exxon Model II unit. TCC beaded catalyst (SV).
1947	The first UOP stacked FCC unit was built. Kellogg introduced the Model III FCC unit.
1948	Microspheriodal catalyst (Davison Division of W.R. Grace)
1950s	Evolution of bed cracking process design continues
1951	Kellogg introduced the Orthoflow A design
1952	Exxon introduced the Model IV
1953	High alumina catalysts were introduced
1961	Kellogg and Philips developed and put the first resid cracker on stream at Borger, Texas
1964	Spray dried zeolite bearing catalyst (Mobil). USY & REY catalysts (Davison Division of W.R. Grace).
1971	Riser cracking
1973	First complete combustion regenerator
1974	Pt CO combustion promoter (UOP with assistance from Mobil), elevated feed distributors
1975	Antimony nickel passivation (Phillips)
1981	Two-stage regeneration for processing residue feedstocks (S&W, Total)

Table I. *Continued.*

Year	Development
1982	High efficiency feed injection (S&W, Total)
1983	First two-stage regenerator with external dense phase cooling for highly contaminated residue feeds (UOP & Ashland)
1984	SOx reducing additive (Arco)
1985	Mobil Oil started introducing closed cyclone systems in its FCC units
1986	ZSM-5 octane additive (Mobil)
1990	UOP VSS introduced
1993	Ramshorn riser termination (S&W), Direct coupled cyclones (Texaco), Post-riser vapor quench (Amoco)
1994	MSCC unit commercialized (Coastal Corp.), structured packing in strippers (Shaw S&W)
1995	NOx and gasoline sulfur reducing additives (Davison)
1997	DCC unit commercialized for ultra high propylene (Shaw S&W)
2004	Riser Separation System (RSS) commercialized (Shaw S&W)
2005	RxCAT technology commercialized by UOP

Houdry's long interest in racing cars led him to recognize the importance of gasoline quality (10). Hundreds of catalyst variations were tried at random and simple tests were used for screening. A tail gas burned with a blue or invisible flame indicated uninteresting low molecular weight gas. Color stability and gum formation were tested by hanging bottles of motor fuel on a clothes line at the back of the laboratory. Motor performance was determined in Houdry's Bugatti racing car, by driving up a calibrated hill. In the end, Houdry settled upon acid-activated clay from Filtrol and established air regeneration to burn coke off the catalyst.

In June 1933, Houdry signed a joint development agreement with the Sun Oil Company and Socony-Vacuum which resulted in him leading a team of engineers to commercialize his fixed bed catalytic cracking process. Houdry and his team developed a process that consisted of cracking, stripping, and regeneration cycles. This process consisted of large motor operated control valves which switched between vessels undergoing cracking, followed by stripping, and regeneration phases. The cracking cycle lasted for just 10 minutes before the catalyst activity dropped to the point that demanded regeneration.

In March 1937, a 15,000 barrel per day Houdry unit shown in Figure 4 was started up in Paulsboro, New Jersey using Houdry's semi-regenerative process. This process incorporated innovations new for his day including a molten salt heat control technique as well as motor control operated valves with timers. By 1943, 24 "Houdry" units were on stream or under construction charging a total of 330,000 barrels per stream day (11).

Moving Bed Design: the Thermofor Catalytic Cracking Unit

A serious disadvantage of Houdry's process was the semi-continuous, cyclic nature of the operation. The complexity and limitations of the Houdry process were apparent to engineers of the day and a race was on for a simpler, more elegant solution. The next step in this evolutionary process was to move the catalyst between vessels in place of the complex piping and control schemes to used to alternate between oil contact, steam stripping, and combustion air regeneration phases.

The first step in the direction of a continuous process utilized buckets and conveyers to transfer spent catalyst from the reactor to a Thermofor kiln. The Thermofor kiln was in use at that time for burning coke off the Fullers earth used in the filtration of lube oils. The idea of transferring catalyst between a reaction and regeneration zone led to the eventual development of the early bucket elevator TCC, the Houdriflow, the airlift TCC, and eventually the Fluid Catalytic Cracking unit.

Figure 3. Eugene Houdry

The catalytic cracking unit at Marcus Hook, 1938.

Figure 4. Houdry unit at Marcus Hook, 1938

The first 500 bpd semi-commercial bucket elevator Thermofor Catalytic Cracker went on-line at the Paulsboro refinery in 1941 with the first 10,000 bpd TCC unit being built by the Magnolia Oil Company at its Beaumont, Texas refinery in 1943. By the end of World War II, TCC capacity was at nearly 300,000 bpd.

Further developments in this process included an air lift design that pneumatically transported catalyst between vessels thus relieving a maximum cat-to-oil constraint of approximately 1.5 present in the "bucket lift" design (Figures 5, 6). This technology development was launched in October 1950, and by 1956 there were 54 Socony-Vacuum licensed units in operation. The development of synthetic catalyst beads was an additional step forward for the TCC process. However, there remained several disadvantages of the TCC process when compared to the Fluidized Catalytic Cracker. The inherent technical advantages present with the FCC led to the eventual demise of the TCC process.

Catalytic Cracking and World War II

The most dramatic aspect of Houdry's discovery was observed during the initial two years of World War II. The high-octane gasoline being produced via this process provided very high quality aviation fuel resulting in increased performance of aviation engines which led to a substantial advantage in the war effort.

"The increased performance meant that Allied planes were better than Axis planes by a factor of 15 percent to 30 percent in engine power for take-off and climbing; 25 percent in payload; 10 percent in maximum speed; and 12 percent in operational altitude. In the first six months of 1940, at the time of the Battle of Britain, 1.1 million barrels per month of 100-octane aviation gasoline was shipped to the Allies. Houdry plants produced 90 percent of this catalytically cracked gasoline during the first two years of the war." (12)

The Houdry Process, TCC and the FCC units were highly valued during World War II as a source of LPG olefins. These olefins were converted into 100 plus octane aviation jet fuel in sulfuric acid based alkylation units. The FCC naphtha which was produced in these processes was blended with the aviation gas (avgas) produced in the alkylation plant.

After World War II most of these alkylation units were subsequently shut down and the process received little attention until the phase out of lead. The alkylation plant then experienced a revival as a source of high-quality, high-octane gasoline. In many of today's high conversion refineries, the FCC unit and the alkylation unit are usually operated hand-in-hand.

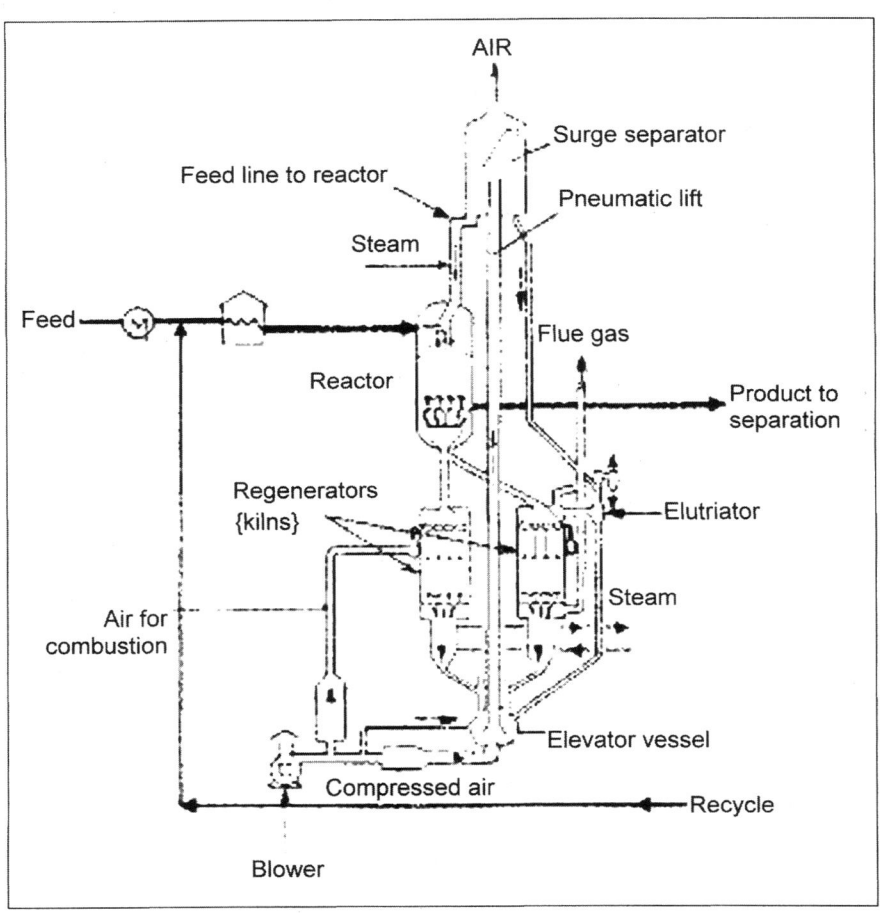

Figure 5. TCC elevation diagram (Adapted with permission from Oil and Gas Journal, *January 8, 1990. Copyright 1990 Oil and Gas Journal.)*

200

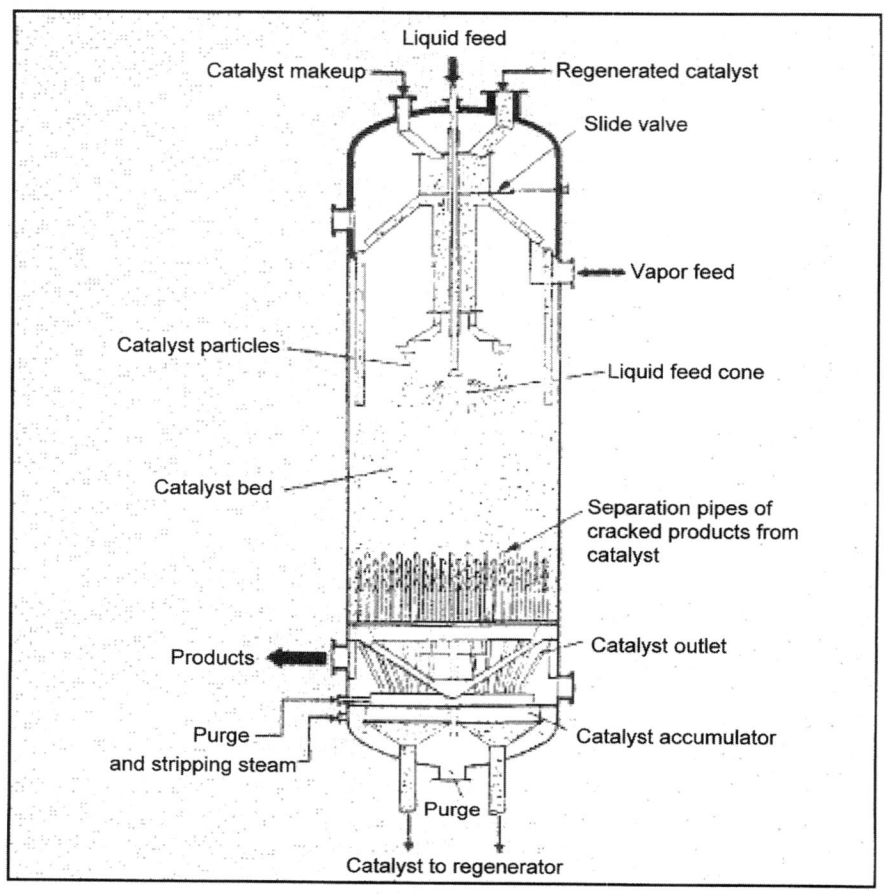

Figure 6. TCC reactor system (Adapted with permission from Oil and Gas Journal, *January 8, 1990. Copyright 1990 Oil and Gas Journal.)*

Fluidized Catalytic Cracking: the FCC Unit

In the 1930's, Standard Oil of New Jersey (now Exxon) attempted to license the Houdry technology but were discouraged by the high license fee set by Houdry of $50,000,000 (*13, 14*). This fee, adjusted via the Consumer Price Index, is over $750,000,000 in today's currency[*]! This led the Standard Oil Co. of New Jersey (Jersey) to develop new catalytic cracking technology. Their initial work was based upon the fixed bed concept but was quickly refocused upon the more efficient fluid bed design to avoid the inefficiencies and complexities of the cyclic fixed beds (*15*).

In 1938, the Jersey formed a consortium of eight companies, which was named the Catalytic Research Associates, or CRA, with a charter to develop a catalytic cracking process which would operate outside of Houdry's patents. These initial eight companies included: Jersey, M.W. Kellogg Co., Royal Dutch Shell, the Standard Oil Co. of Indiana, Anglo-Iranian Oil Co. (today's BP), Universal Oil Products Co. (today's UOP), the Texas Corp. (which would become Texaco) and IG Farben (which was eventually dropped in 1940). Over 1000 professionals were employed to develop the FCC process during the course of time in which the CRA consortium was in existence.

In 1934, R. K. Stratford at Jersey's Canadian affiliate (Imperial Oil Co.) discovered that the spent clay used in lube oil treating had catalytic effects. Four thermal crackers were eventually revamped to "Suspensoid Cracking" by adding 2 to 10 pounds of powder/barrel feed in 1940. This catalyst was used in a once through mode and improved the yield selectivities from the thermal cracking process (*16*).

Jersey chose not to develop this process further but was likely instrumental in influencing the designers of the first FCC unit. The cracking of hydrocarbon vapors with these clays gave interesting yield selectivities. Furthermore, it was observed that the best yield selectivities were seen in the initial moments of the cracking reaction. This may have led researchers in the direction of a continuous operation. In addition, the transport of fine powders was clearly easier than conveying pellets or beads.

Several key observations made by Jersey were instrumental in overcoming the many large engineering obstacles in design of the first few FCC units. These include: the concept of the standpipe, the dense bubbling bed, and liquid feed injection.

Breakthrough work was carried out at the Massachusetts Institute of Technology (MIT) which demonstrated that a dense bed of fluidized particles could be maintained at gas velocities (1-3 ft/sec) which far exceeded the Stoke's Law settling velocity of the particles (0.1 ft/sec or less) (*17,18*).

---

[*] The Consumer Price Index indicates $1 in 1935 is worth $15 in 2007.

A second critical discovery by Jersey was that a column of fluidized particles could be used to re-pressure solids. This enabled solids operating at a lower pressure to be transferred into a higher pressure zone. This afforded the design team a mechanism to transfer the large amount of surplus energy from the regenerator to the reactor thus establishing the FCC "heat balance". Furthermore, the development of the standpipe eliminated the use of compression screw conveyors that were a source of attrition and particle agglomeration.

The concept of the standpipe was only accepted within the halls of Jersey after a 100 ft standpipe was erected complete with pressure gauges. This standpipe was filled with catalyst and aerated. The pressure gauges confirmed the pressure build-up. Furthermore, the catalyst flowed out as if it were water when the bottom valve was opened (*19*).

Another landmark development in these early days was the injection of oil as a liquid rather than as a gas. This made effective use of the large amount of heat released in the regenerator during the combustion of carbon. However, there was concern that mud would be formed resulting in a blockage of catalyst circulation. A successful test run to confirm this operating mode was carried out after the first upflow unit in Baton Rouge (PCLA #1) began operations.

An additional observation made by Lewis at MIT was that the density of solids in upflow piping exceeded that of solids in downflow piping. Upflow allowed catalyst particles to settle or "slip" which could not be predicted or estimated from available theory as Stoke's law applied only to individual particles and not to clusters of particles. Catalyst slip is a well known concept used in riser reactors today.

The Jersey pilot plant was then modified by Kellogg with standpipes and a slide valve to demonstrate the new technology. This pilot plant was circulated with a Super Filtrol clay catalyst. Oil was introduced into this unit on July 17, 1941 which provided the process and engineering data for the design of the world's first FCC unit, PCLA #1 (Powdered Catalyst Louisiana, #1).

The construction of PCLA #1 was started on September 16, 1940 and was completed on May 1, 1942. By mid-1941, even before the first cat cracker had started up, design engineers had recognized that a dense bed could be maintained thus enabling a bottoms draw off reactor/regenerator in which the catalyst exited from the bottom of the vessel while the cracked products/gases flowed out the top. This design would have significant advantages over the then current upflow designed reactor system and was included in the Model II unit as was liquid feed injection.

PCLA #1 - The World's First Operating FCC Unit

PCLA #1 (Figures 7, 8) was a genuine engineering marvel for the day (*20*). The unit was 19 stories tall, contained 6,000 tons of steel, 85 miles of pipe,

Figure 7. PCLA No. 1 unit, Baton Rouge, Louisiana, 1942

3,500 yards of concrete, 209 instruments, and 63 electric motors. The cold oil feed (28 API South Louisiana reduced crude), after being preheated in various exchangers, including catalyst coolers and a waste heat gas cooler, was passed with steam through the coils of the vaporizer furnace. It was then flashed in a vaporizer tower where 85% was taken overhead. The overhead vapors and steam were passed through the superheater furnace and injected at 800°F and 17 PSI into a stream of catalyst at 1050°F from regenerated catalyst standpipe at an approximate catalyst to oil weight ratio of four. Cracking took place at 900-925°F in the reactor, a cylindrical vessel 15 feet in diameter and 28 feet high. The catalyst and cracked vapors flowed into cyclones for separation. The cracked vapors were passed to the fractionator for separation into finished products.

About 99.9% of the catalyst was separated from the cracked gas oil vapors by use of three cyclone separators in series which dropped the catalyst through aerated dip legs into the spent catalyst hopper. Flow of spent catalyst to the transfer line was regulated by a slide valve. The spent catalyst was picked up by a stream of air and carried into the regenerator.

The regenerator was a cylindrical vessel 20 feet in diameter and 37 feet high, having a catalyst inventory of about 68 tons. Regeneration was carried out at 1050°F. The regenerator bed temperature was controlled by recycling catalyst

204

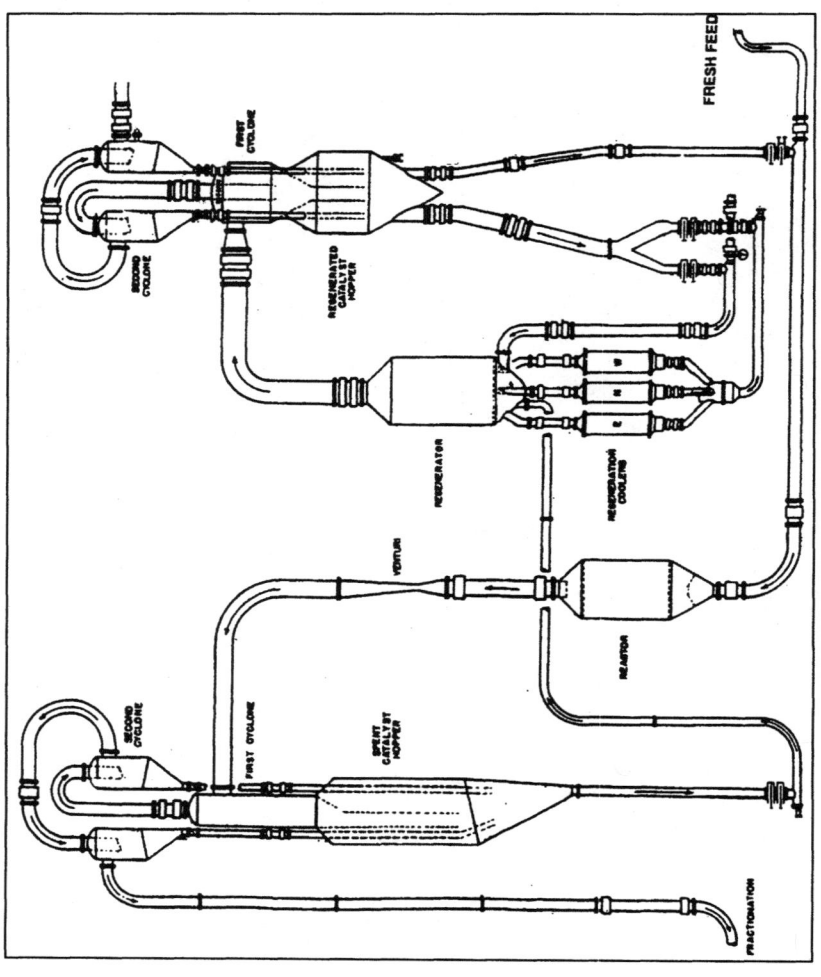

Figure 8. Esso Model I unit design (Courtesy of KBR)

from the regenerated catalyst hopper through three catalyst coolers in series back to the regenerator. A similar recycling system which bypassed the coolers was used to control catalyst density in the regenerator. These recycle rates were controlled by slide valves and recycling was accomplished by air injection.

Regenerated catalyst and gas flowed out the top of the regenerator to three cyclone separators connected in series, where, approximately 99.9% of the catalyst was separated from the flue gases and deposited in the regenerated catalyst hopper. The stack gas was then passed through a waste gas fresh feed heat exchanger to a Cottrell electrostatic precipitator where 96-98% of the fine catalyst not separated by the cyclones was recovered and returned by air injection to the inlet of the tertiary regenerated catalyst cyclone.

Regenerated catalyst to the reactor was drawn from the regenerated catalyst hopper through a standpipe in which slide a valve controlled the flow rates. The 113 foot regenerated catalyst standpipe operated with a pressure head of 31 PSI.

The total cost for PCLA #1 was $3.8 million ($57 million in 2007).

Startup of this new unit began on April 22, 1942. Successful operation was achieved on May 25, 1942. By June 3, 1942, the fresh feed rate had been increased to 16,600 barrels per day (128% of design), limited by availability of reduced crude and Cottrell precipitator gas velocities.

Only a very brief announcement of this major milestone in refining history was made three days later. It was noted that a 10 day run at 120% of design capacity had been achieved and "...that the first large-scale units of such intricate design and revolutionary principle were usually subject to 'children's diseases,' but in this case only the mildest forms occurred and have already been remedied." (*21*)

The first products resulting from this unit were in line with expectations based on the 100 b/d pilot plant operations. The reactor feed had a 31.3 API gravity South Louisiana reduced crude including a slurry oil recycle (17 API) rate of 4% on feed. The reactor temperature was 910° F. with the cat to oil ratio of 3.5. A 53.5 vol% conversion was obtained with "Super Filtrol" natural clay catalyst. The regenerator operated at 1052° F., the spent catalyst containing 1.6 wt% carbon while the regenerated catalyst was reduced to 0.5 wt%. Flue gas excess oxygen was 4.4 wt%. The observed yields for PCLA #1 are recorded in Table II.

By August 1942, the unit had been converted to avgas production. At this point the unit was operated with a synthetic DA-1 catalyst (Davison Division of W. R. Grace) and a light Coastal gas oil feed (31 API). The unit was operated with the 975°F riser outlet temperature and a 10.5 cat-to-oil ratio. The conversion of the feed was 65 vol% with dry gas and coke yields of 11.0 and 4.7 wt% respectively. The avgas yield was 25.7 vol% along with 14 vol% heavy naphtha and 35 vol% gas oil.

Two additional Model 1 units were being constructed, one for Standard Oil of New Jersey at their Bayway New Jersey refinery and one for its Humble Oil &

Table II. PCLA No. 1 yield selectivities

Product	Yield (vol%)
Dry gas (C3-, wt%)	6.7
Butanes	8.9
Naphtha	43.6
Ron	93
MON	79
Cycle oil	39.9

Refining Co. affiliate at Baytown, Texas. The three Model 1 units operated continuously throughout the war on avgas operations, after which they switched to motor gasoline type operation. Prior to its final shutdown in October 1963, PCLA #1 reach the maximum throughput of 41,000 barrels per day.

The successful startup of the PCLA #1 resulted in an explosion of new units being built. By the end of World War II, 34 FCC units were operating, with capacity exceeding 500,000 barrels per day in the United States for 20 different companies.

The Baton Rouge Model II unit, PCLA #3, is the world's oldest operating FCC unit. This unit together with its sister unit, PCLA #2, continue to operate after more than 63 years of operation.

Engineering Creativity: 60 Years of Innovation

Synthetic Catalyst Era

After the successful startup and operation of the PCLA #1, there began a long period of explosive development of the FCCU unit designs (22, 23, 24, 25). This period represents an era of advancement in engineering concept and design rarely seen in the industrial world. A summary of FCC unit designs and when they were first launched is included in Table III.

It is worth noting that after the dissolution of the CRA collaboration, the solutions to process and reliability issues which emerged and are highlighted below were rarely the result of collaboration between companies. Rather, this has been primarily a series of step-change improvements by independent engineers resulting in a process that currently meets and surpasses the needs of society today. The challenge for the engineers of today and tomorrow will be to ensure that this legacy of innovation continues. History, which has yet to be written, will be the judge determining how well we as a society of engineers meet this challenge.

Table III. Global overview of FCC unit designs

Date	Type	Key Features
1942	Model I	• Side-by-side configuration • Fast fluidized upflow riser & regenerator • Low pressure • High elevation external cyclones • Catalyst coolers • Full feed pre-vaporization
1942	Model II	• Side-by-side configuration • Bed cracker • Internal multiple cyclones • Down flow catalyst design • Increased coke burning capability • Higher cat-to-oil rates • Liquid feed injection
1945	Model III	• Side-by-side configuration • Higher pressure operation • Lower regenerator elevation • Improved spent catalyst regeneration • Elimination of electrostatic precipitator and waste heat boiler
1945	UOP SXS	• Side-by-side configuration • Internal cyclones • Simplified design
1947	UOP Stack	• Stacked configuration • Improved air compression for smaller vessels • Gravity flow of spent catalyst to regenerator • Catalyst stripping introduced
1951	Orthoflow A	• Stacked configuration • Elevated reactor with a low elevation regenerator • Internal stripper • In-line plug flow valves • Vertical tubes for catalyst transport
1952	Model IV	• Side-by-side configuration • Higher pressure & velocity, low elevation vessels

Continued on next page.

Table III. *Continued.*

Date	Type	Key Features
1952	Model IV	• Elimination of slide valves • U bend standpipes • Catalyst flow controlled by differential vessel pressures
1956	Shell Berre (& other subsequent designs)	• Side-by-side configuration • Third vessel for catalyst stripping • Two stage fluid cracking – riser cracking followed by conventional bed cracking • Expander turbine for power recovery • Catalyst fines recycle for improved particle size distribution control • FCC advanced control • Stripper cyclones
1953	Orthoflow B	• Stacked configuration • Positions of reactor/regenerator reversed for reduced utility consumption • Bubbling bed cracking
1960	Orthoflow C	• Stacked configuration • Return to Orthoflow A vessel configuration • Implementation of the riser concept for zeolitic cracking • Two risers, fresh feed & recycle • Bed cracking of recycle streams • Flexibility for bed cracking • Steam coils in regenerator enabling residue cracking
1960	UOP Riser Cracking	• Side-by-side configuration • Vertical riser • Baffled annular stripper surrounding riser
1967	Texaco	• Side-by-side configuration • Dual bent risers, gas oil & recycle cracking • Flexibility for bed cracking
1971	Gulf	• Side-by-side configuration • Vertical riser with feed injection at two levels • Counter-current regenerator

Table III. *Continued.*

Date	Type	Key Features
1973	Orthoflow F	• Stacked configuration • Two staged regeneration in same vessel • Spent catalyst distributor in the first regenerator • Straight vertical external riser folded for side entry into reactor vessel • Multiple nozzle feed injection system • Inertial separation at riser discharge into reactor vessel
1978	UOP High Efficiency	• Side-by-side configuration • High superficial velocities for small catalyst inventory • Fast dilute phase regeneration in small vessel combustor discharging into a disengaging chamber
1979	Exxon Flexicracker	• Side-by-side configuration • High elevation disengaging-stripper vessel with low elevation regenerator • Single straight external vertical riser ending in a proprietary separation system
1979	Ultra Orthoflow	• Stacked configuration • Vertical riser terminating in roughcut cyclones • Internal baffled stripper
1981	Total Resid Cracker	• Side-by-side configuration • External straight vertical riser with two sets of multi-directional feed nozzles • Two stage regeneration with external cyclones in the second stage
1981	Kellogg HOC	• Stacked Orthoflow configuration • External straight vertical riser with lateral cross-over entrance with roughcut cyclone discharging into disengaging vessel • Heat removal via bed steam coils and external catalyst coolers
1983	UOP RCC	• Side-by-side configuration • Stacked two-stage regenerator with stage one in the top position

Continued on next page.

Table III. *Continued.*

Date	Type	Key Features
1983	UOP RCC	• Flue gas from stage one passes into the bed of stage two thereby enabling one flue gas system • Equipped with external catalyst coolers
1994	MSCC	• Stacked unit configuration with the regenerator above the reactor • Ultra low contact time achieved via feed injection perpendicular to a shaped falling curtain of catalyst • Reaction products pass to external cyclones while the spent catalyst falls into a stripper and air lifted into a bubbling bed regenerator
2004	DCC	• Shaw S&W side-by-side configuration for maximum propylene production • Straight vertical riser terminates into a reactor bubbling bed • Low operating pressure unit • Additional steam injection into riser for very low hydrocarbon partial pressure
2004	RxCat	• UOP Side-by-side configuration for delta coke limited operations • Spent catalyst recycle line from reactor stripper to mixing vessel at base of riser with slide valve control • Regenerated catalyst and spent catalyst mixed prior to acceleration and contacting with fresh feed
2004	Superflex	• KBR Stacked configuration for maximum propylene production • Naphtha cracking unit

The lack of centrifugal fans during World War II had a large impact on early FCC unit design as seen in the Model I and Model II configurations. These designs were characterized by lower operating pressure regenerators at high elevation. The Model I unit contained three external cyclones in series with discharge into a catalyst hopper in order to obtain a fluidized catalyst downflow to the reactor vessel.

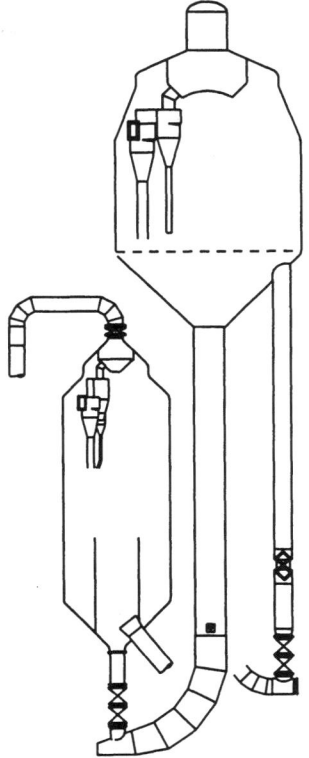

Figure 9. Esso Model II unit design (Courtesy of KBR)

The Model II unit (1945, Figure 9) presented substantial improvements over the Model I unit. This unit included larger vessels with internal cyclones. These units included catalyst downflow and increased carbon burning capacity. This allowed increased conversion at higher cat-to-oil ratios. Liquid feed injection was first introduced in the Model II unit.

The conclusion of World War II brought improvements through the availability and utilization of air compression equipment. This led to the Model III unit which operated at higher pressures and lower elevation. This unit also improved spent catalyst regeneration.

Also in 1945, UOP introduced its first FCC unit design to the marketplace. This side-by-side configuration included internal cyclones with a simplified flow pattern. In 1947, UOP introduced its very successful stacked configuration as shown in Figure 10 in which air compression was used to obtain smaller and better regenerators. This was the first FCC design which included a spent catalyst stripper.

UOP Stacked Fluid Cracking Unit

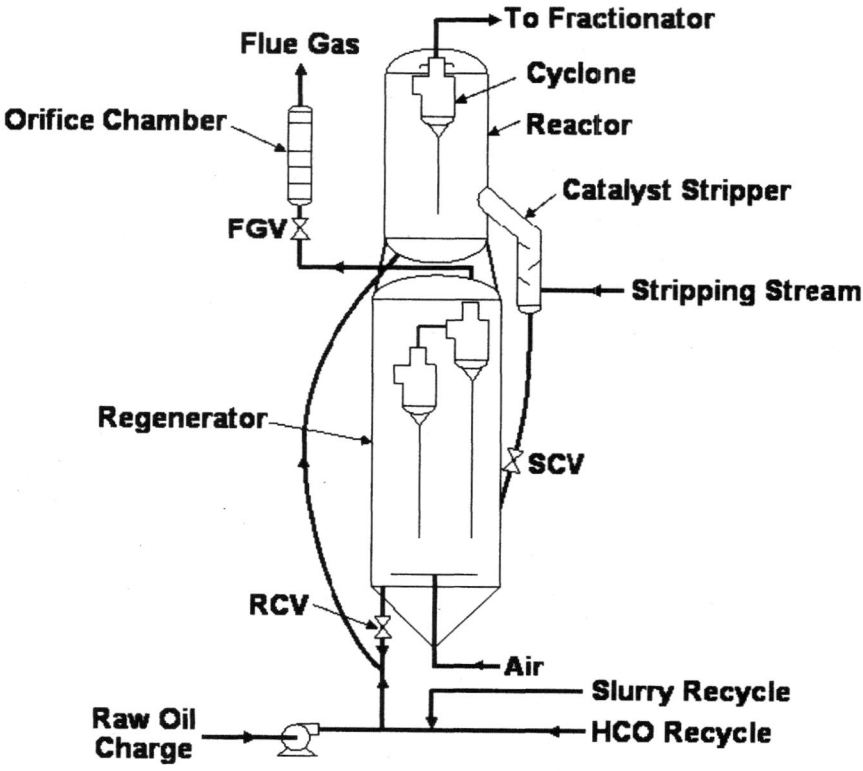

Figure 10. UOP Stack design (Courtesy of UOP)

In 1951, the M.W. Kellogg Co. introduced its Orthoflow A concept (Figure 11) in bed cracking design. This cat cracker was designed with a low elevation regenerator topped by a higher elevation reactor with an internal stripper. Kellogg introduced in this design its plug valve which was placed in the direct line of catalyst flow rather than at right angles, as were slide valves.

The Model IV unit (Figure 12) was introduced to the refining world in 1952 which included many technological advances in the FCC operation. This unit consisted of a side-by-side configuration with much smaller vessels operating at higher pressure and higher internal velocities. The catalyst circulation was simplified by the elimination of the slide valve control flow together with the introduction of the "U bend" concept. Catalyst flow was manipulated via changes in the differential pressure between the reactor and regenerator.

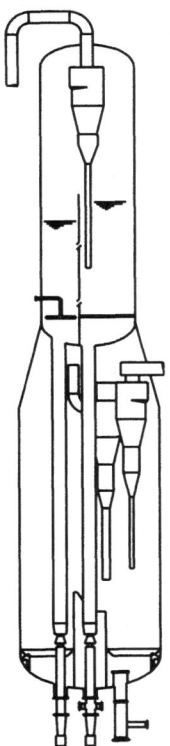

Figure 11. Kellogg Orthoflow A unit design (Courtesy of KBR)

By the mid-1950s and continuing into the 1960s, the Shell Oil Co. contributed multiple advances to FCC development. These developments include the first short residence time reactor in a two-stage fluid cracking mode. This included riser cracking together with a second stage conventional dense bed cracker. Other Shell developments include expander turbines for power recovery, catalyst fines recycle for improved particle size distribution control, FCC advanced process control, and stripper cyclones.

In 1953, Kellogg introduced its Orthoflow B design (Figure 13) in which the positions of the reactor and regenerator had been switched for improved utility usages. This design included feed injection directly into the dense reactor bubbling bed. In the early 1960s, Kellogg introduced its Orthoflow C configuration which was a return to the original reactor/regenerator configuration. This unit however was equipped with two risers, for fresh feed and recycle in order to make use of the more active zeolite based catalysts being developed. This unit retained the flexibility of utilizing bed cracking.

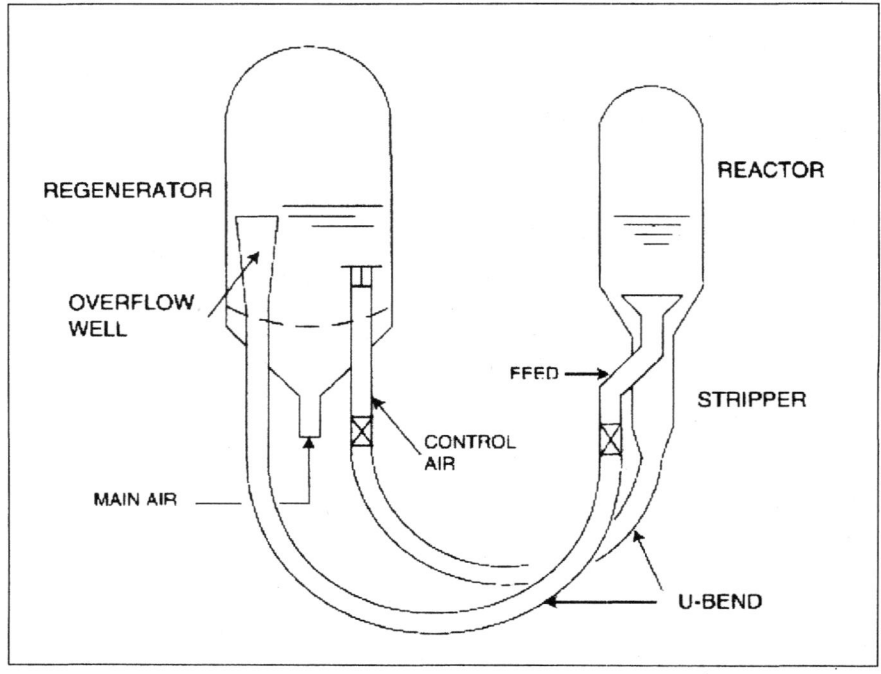

Figure 12. Esso Model IV unit design

Also in the early 1960s, Kellogg combined efforts with the Phillips Petroleum Co. to develop the first FCC unit to crack atmospheric tower bottoms. Kellogg modified its Orthoflow C with dense bed steam coils for heat removal at the Phillips Borger, Texas refinery.

An FCC Milestone: Zeolite-Based Catalysts

A major milestone in the history of FCC was initiated in 1962 when two innovators (Plank and Rosinski, Mobil Oil) began to experiment with the incorporation of zeolites in FCC catalysts. The substantial improvement in both activity and yield selectivities led to a quantum improvement in the FCC process. Zeolite containing FCC catalysts began appearing in FCC units in 1964. The first manufacturers of these catalysts were Filtrol and Grace Davison.

These new catalysts required fundamental changes to the design and operation of FCC units in order to adequately exploit the benefits of the improved activity and selectivities. It was quickly observed that the high levels

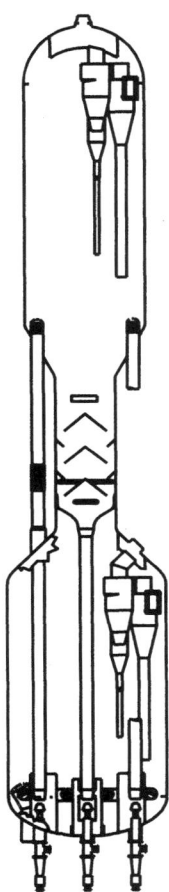

Figure 13. Kellogg Orthoflow C design (Courtesy of KBR)

of carbon on catalyst commonly practiced (0.5 - 0.7 wt%) at that time needed to be reduced in order to adequately regenerate these catalysts. This led to improved metallurgy in regenerator and cyclone construction enabling regeneration temperatures in excess of 1300°F. The second observation was that a much shorter hydrocarbon residence time was required in order to control coke formation. This rapidly led to the elimination of bed cracking in favor of riser cracking. The FCC designs subsequent to 1964 began to incorporate these advances (*26*).

UOP introduced its riser cracking configuration to the marketplace which featured the side-by-side vessel configuration. The reactor stripper vessel was elevated above the regenerator which enabled gravity flow of spent catalyst and permitted sufficient clearance for the new riser reactor. This unit presented the first annular stripper at the base of the reactor vessel.

In 1967, Texaco Inc. introduced its riser cracking design, which included two bent risers to enable simultaneous fresh feed and recycle cracking. This unit also retained bed cracking capabilities.

In 1971, Gulf Oil, introduced its riser based side-by-side configured FCC unit with a straight external vertical riser with feed injection at two levels. Also introduced in this design was the first countercurrent regenerator.

In the first half of the 1970s, rapid improvements in FCC catalyst quality were being made. One remarkable improvement was a level of hydrothermal stability enabling refiners to regenerate catalyst to levels of carbon on regenerated catalyst less than 0.1 wt%. Licensors then began to develop configurations with regenerator internals capable of operation consistent with these newly developed catalyst technologies.

In 1973, Kellogg introduced its Orthoflow F model (Figure 14) which maintained the stacked configuration but with a two-stage regenerator. Approximately 70% of the coke is burned in partial combustion within the first stage with the remainder of the coke being burned in full combustion mode in the second stage. This staged approach to regeneration minimizes the negative impact of vanadium on the equilibrium catalyst. The first stage included the first-ever spent catalyst distributor for improved regeneration. This unit was equipped with an external riser which entered laterally into the reactor vessel. This riser included two new features: a multiple nozzle feed injection system and inertial separation of hydrocarbon from the catalyst at its discharge into the reactor vessel.

In 1978, UOP introduced its side-by-side configured high efficiency regenerator (Figure 15) which utilized a high velocity combustor in place of the traditional bubbling bed regenerators. Catalyst velocities in the combustor were increased producing a fast fluid bed with both the catalyst and air flowing co-currently into a disengaging vessel situated above the combustor. Velocities in the combustor resulted in improved mixing of the spent catalyst and air thus enhancing the burning kinetics in the regenerator. This design permitted a smaller regenerator with a greatly reduced catalyst inventory.

In 1979, Exxon built its first Flexicracker which is a side-by-side configured FCC unit with an elevated reactor vessel. This unit contained a straight external vertical riser terminating in a proprietary separation system.

Also in 1979, Kellogg introduced its Ultra Orthoflow unit which combined the technologies of Kellogg's Orthoflow with Amoco's "Ultra cat regeneration". This unit consisted of a stacked arrangement with an external vertical riser terminating in a rough cut cyclone, an internal baffled stripper together with the standard plug valve control.

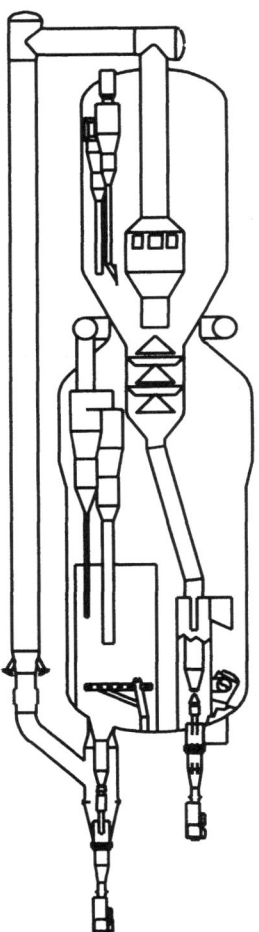

Figure 14. Kellogg Orthoflow F design (Courtesy of KBR)

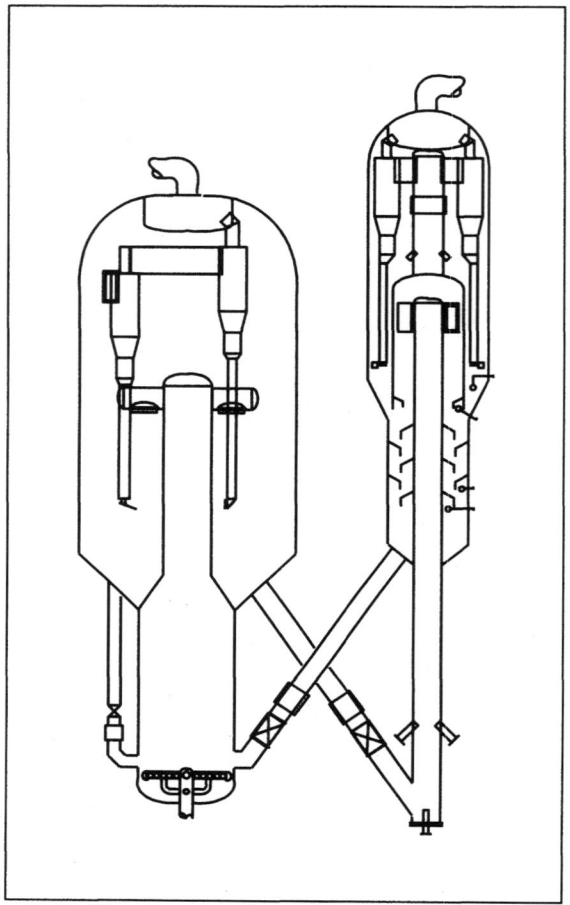

Figure 15. UOP High-Efficiency design^{TM} (Courtesy of UOP)

FCC Milestone #2: Residue Cracking

A second major milestone in the development of the FCC process was the inclusion of residue feedstocks into the FCC beginning in the early 1980s. Nearly 20% of the feedstocks currently processed today on the FCC are residues. The inclusion of these residues in the FCC feed slate initiated substantial modifications to FCC unit and required improved hydrothermal stability and vanadium resistantance in the FCC catalysts. Beginning in the 1980s, we see a flood of innovative new designs entering the marketplace to enable cost-effective residue cracking.

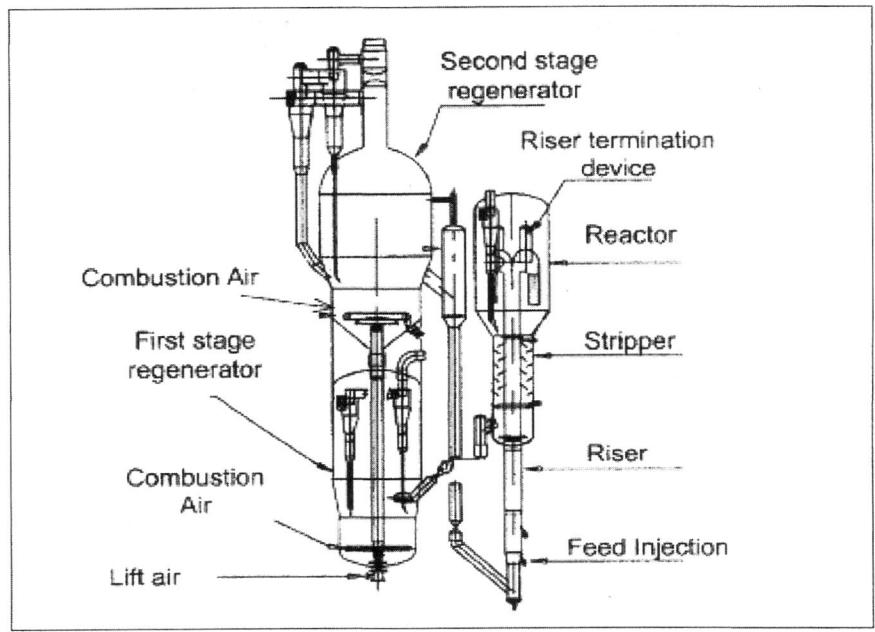

Figure 16. S&W/IFP R2R design (Courtesy of Shaw S&W)

. In 1981, Total Petroleum USA, developed its residue cracking unit (Figure 16) at Arkansas City, Kansas. This side-by-side configured unit with two-stage regeneration consisted of an external vertical riser with a proprietary feed injection system which enabled countercurrent or co-current feed injection. Flow of catalyst between regenerator stages is controlled by a plug valve. The second stage regenerator was equipped with external cyclones which were first seen in the Model I design. It is interesting to note that this technology represents the only entry in the FCC unit design field not associated with the original Catalytic Research Associates group. This technology is marketed by Shaw Stone & Webster/IFP.

Also in 1981, Kellogg introduced a modified stacked configuration for its heavy oil cracking unit (HOC). This unit was designed with a straight external vertical riser discharging laterally into a rough cut cyclone. The regenerator included heat removal via internal bed coils and external steam generation catalyst coolers.

UOP in collaboration with Ashland Petroleum developed the Residue Catalytic Cracking (RCC) process (27). The first RCC unit was brought on line in Catlettsburg, Kentucky, in 1983. The landmark design feature of this unit was the first external, dense phase FCC catalyst cooler enabling regenerator

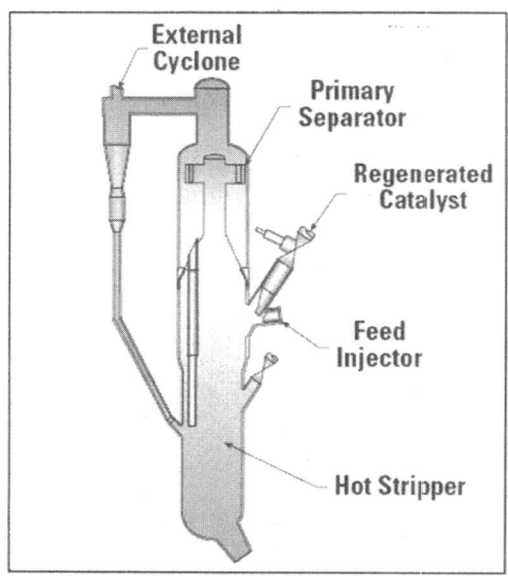

Figure 17. UOP MSCC^{TM} reactor design (Courtesy of UOP)

temperature control resulting from the high concarbon levels observed in residue feedstocks. This design utilized two-stage regeneration to better control the heat released by coke combustion. The first stage is placed above the second stage. The second stage regenerator was designed without cyclones, the flue gas from this stage passes into the bed of the first stage. This innovation resulted in the requirement for only one flue gas system.

The MSCC cracker (Figure 17) licensed by UOP went on-line in 1994 at Coastal's Eagle Point refinery. This unit achieves very short contact time cracking by injecting feed stock perpendicular to a shaped falling curtain of catalyst. Cracked products flow directly into external cyclones while spent catalyst gravity flows into a stripper and then is air lifted into a bubbling bed regenerator.

FCC Milestone #3: The FCC as Petrochemical Feedstock Platform

A third milestone in the advance of this technology has been the use of the FCC unit as a petrochemical feedstock platform Beginning in the mid-1990s rocketing propylene prices have led many refiners to specialize in operations producing maximum propylene. In addition, some refiners are beginning to selectively separate high valued chemical feedstocks from the FCC naphtha

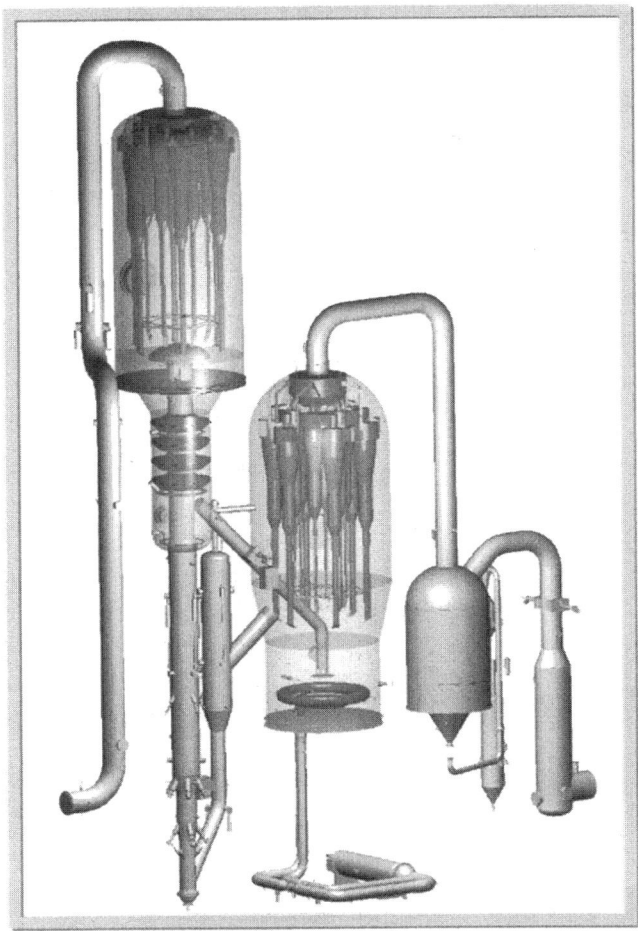

Figure 18. Shaw S&W DCC Unit (Courtesy of Shaw S&W)

stream. Several highly innovative designs have been brought to the marketplace. It is believed by the author that this step change is still in process.

The Research Institute of Petroleum Processing (Sinopec) together with Shaw Stone & Webster has launched their Deep Catalytic Cracking (DCC) process (Figure 18) which is based upon a traditional side-by-side configuration (*28*). With its emphasis on maximum propylene yield this design requires a higher heat of reaction. S&W designed a riser terminating in a bubbling dense bed to produce additional coke and for increased conversion to propylene. Slurry recycle into elevated nozzles is also provided to increase coke make. This unit operates at a lower operating pressure together with a lower hydrocarbon partial pressure achieved through additional steam injection nozzles into the riser.

In 2004, UOP launched a new unit design, RxCat (Figure 19), focused on low delta coke operations (light VGO and deeply hydrotreated feedstocks). This process recycles a portion of spent catalyst from the reactor stripper to a mixing vessel at the base of the riser where it is contacted and thoroughly mixed with regenerated catalyst. This technology is focused on adding additional conversion and selectivity to light VGO and deeply hydrotreated feedstocks. Furthermore, the RxCat technology is an integral part of UOP's PetroFCC unit design focused on producing maximum propylene yields.

In 2006, Kellogg Brown & Root started up its first Superflex unit design to crack naphtha into propylene and ethylene.

Further Specific Examples of Engineering Ingenuity

In addition to the remarkable ingenuity observed in successive unit designs, substantial creativity was brought to the market in many areas such as feed distribution, riser termination, spent catalyst regeneration, etc.

Feed Distribution

Very little attention was paid to feed distribution in the early days of cat cracking due to the very low conversion observed in the risers (29). In most cases, the feed injector was a "showerhead" consisting of a single four-inch to eight-inch pipe in the center of the riser just above the intersection of the regenerated catalyst standpipe. Some units had vertical introduction sections consisting of multiple feed injectors in the riser at an angle of about 30°. Steam was added to give a nozzle velocity at 50-60 fps.

An indication of how far feed nozzle design has advanced can be observed based on current "state-of-the-art" nozzles. Most units are now equipped with multiple elevated radial side entry nozzles which provide a spray pattern covering the cross-sectional area of the riser. Atomization devices used to generate droplet sizes less than 100 μm are common. Substantial effort is made by designers to ensure that catalyst has been accelerated using either steam or lift gas and that a fully developed plug flow regime is present prior to the catalyst entering the mix zone. High concentrations of dispersion steam are used to enhance feed vaporization, reduce hydrocarbon partial pressure and contact time. High pressure drops are used to enhance vaporization with residue feedstocks.

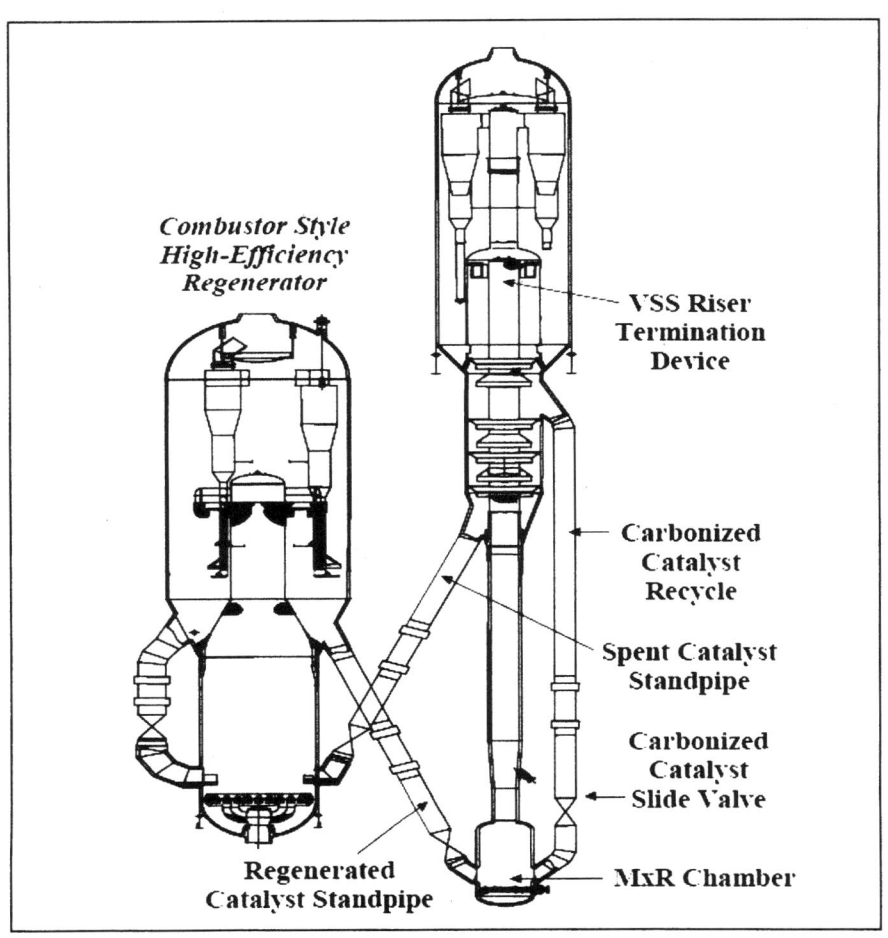

Figure 19. UOP RxCAT^{TM} design (Courtesy of UOP)

Riser termination

Due to the initial low catalyst activity, no attention was placed on a quick separation of catalyst and hydrocarbons in the early days of FCC. However, after the development of zeolite-based catalysis and riser cracking combined with improved catalyst stability, riser outlet temperatures of 970°F and greater were observed. It was observed that quick disengaging of hydrocarbons reduced dry gas and delta coke yields.

Some of the first improvements in riser termination devices were in the form of inertial separators (Figure 20). The inertial or ballistic separator was a device attached to the end of the riser which deflected the catalyst in cracked product 180° in a downward direction toward the base of the reactor vessel. Gravity was used to draw the catalyst into the stripper and density differences to draw the hydrocarbons into the cyclones.

Mobil Oil developed and commercialized a "closed cyclone system" in which a rough cut cyclone was directly to the riser termination which dramatically reduced the residence time of hydrocarbon vapors in the dilute phase of the reactor vessel thereby dramatically reducing "post-riser cracking".

There ensued a period of rapid innovation in which multiple designs were brought to the market place. The vast majority of cat crackers today operate with some form of quick disengaging of hydrocarbon and catalyst for improved yield selectivities.

Spent Catalyst Regeneration. Rapid development in spent catalyst regeneration technology coincided with the improvement in catalyst stability. The design of the early regenerators was limited by the capability of the synthetic catalysts available at that time. These catalysts lost substantial activity when exposed to temperatures greater than 1100°F. Regenerator temperature control was achieved primarily through control of combustion air. CO-to-CO_2 ratios averaged 1.0 and greater.

- *Afterburn.* The greatest concern faced by early FCC operators was the initiation of a runaway afterburn. Temperatures in excess of 1150°F exceeded the ignition point of carbon monoxide resulting in rapid combustion of carbon monoxide. Control of afterburn was achieved through operation in deep partial burn, use of catalyst coolers, and the use of elaborate spray systems in the dilute phase of the regenerator, the regenerator cyclones, and occasionally in the flue gas lines.

- *Full combustion.* Regenerator temperatures were increased to 1200°F with the improved regenerator metallurgy and eventually to 1350°F via cyclone construction utilizing stainless steel alloys. In the mid-70s, Amoco introduced complete combustion which led to

225

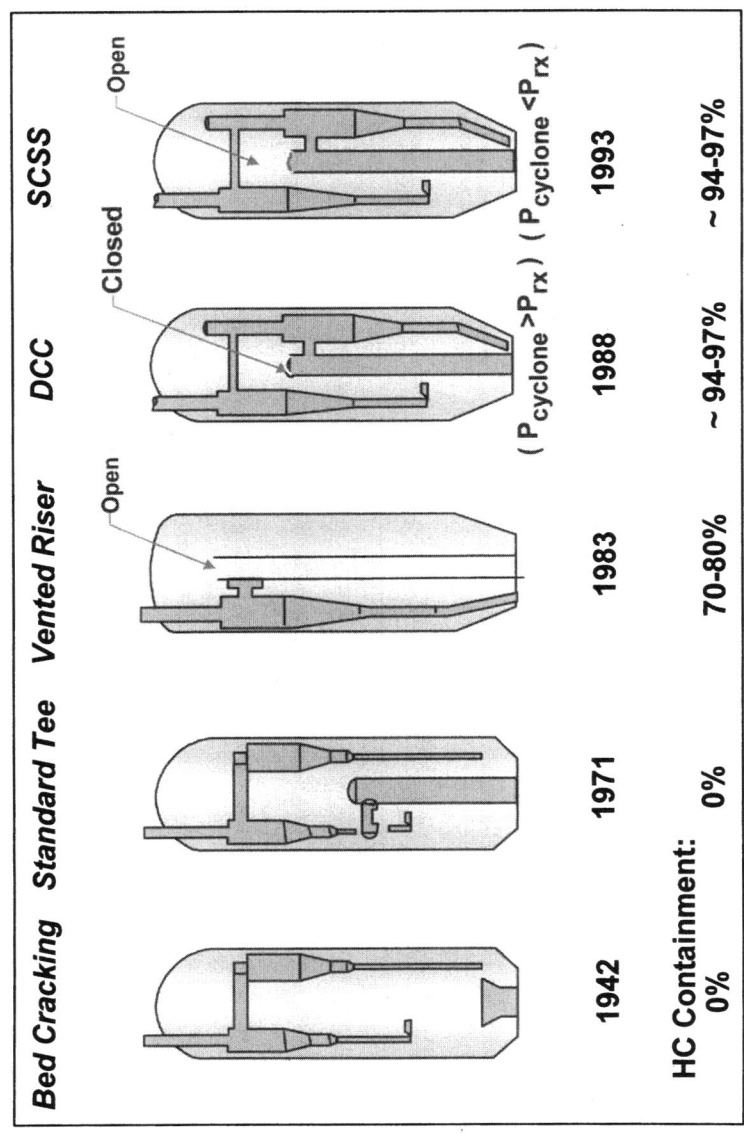

Figure 20. Advanced riser termination devices (Courtesy of UOP)

small amounts of excess oxygen present in the flue gas for the first time (*30*).

- *Air distribution.* During the first years of FCC, the predominant air distribution system was a perforated plate grid with multiple holes having pressure drops on the order of 1-3 psi to ensure adequate air and spent catalyst mixing. Eventually, air rings began to appear in regenerators. These air rings progressed from simple holes, to ferrules on the top of rings, to nozzles with an orifice at its inlet to ensure good air distribution. Nozzle lengths are carefully calculated to ensure that the venturi jet formed in the orifice is fully dissipated prior to entry into the catalyst bed.

- *Spent Catalyst Distribution.* A critical element in the design of regenerators is the observation that in well fluidized beds, catalyst mixes very well vertically but very slowly horizontally. The driving force for vertical mixing is the upward flow of combustion air through the bed. The driving force for horizontal mixing is the eruption of bubbles escaping from the bubbling bed (*31, 32*).

    The designers of today's regenerators have employed multiple innovative techniques to ensure good distribution of spent catalyst as it enters into the regenerator. Designs which include multiple troughs distributing the spent catalyst over the cross-sectional area of the regenerator and the side-by-side units with deep conical bottoms tend to result in evenly distributed spent catalyst.

- *Two-Stage Regeneration.* The advent of two-stage regenerators began with the introduction of residue in the FCC feed slate. Innovation was required to meet the challenge of processing feedstocks high in CCR and contaminant metals. In order to limit the deactivation rate of the catalyst during regeneration, the first stage regenerator operates in partial burn mode and burns approximately 70% of the coke. Most of the hydrogen in the coke is converted to steam during this initial burn. The lower temperature and partial burn conditions minimize the damage of vanadium to the zeolite crystals present within the catalyst.

    In some designs, the flue gas from the first stage is vented and the partially regenerated catalyst is transported to the second stage. The second stage regenerator operates in full combustion mode at higher temperatures to complete the regeneration. Due to the lower moisture content in the second stage atmosphere, higher temperatures can be tolerated by the catalyst with minimal hydrothermal deactivation.

- *Fast Flux Regeneration.* A further improvement in spent catalyst regeneration was developed by UOP in the mid-70s. This unique design stepped away from the prevalent bubbling bed regenerators to regeneration in a co-current, dilute phase, fast flux fluidized flow in a combustor vessel. The regenerated catalyst was then transferred into a disengaging vessel. This innovation resulted in improved regeneration, shorter catalyst regeneration residence times and lower regenerator catalyst inventories.

The Birth of FCC Catalysts

The development of FCC catalysts can be easily separated into two periods characterized by 1) synthetic catalysts and 2) zeolite based catalysts. The synthetic catalyst era began in 1942 and was made obsolete by Plank and Rosinski's (Mobil Oil) discovery of application of synthetic zeolites in 1964 (*33*). The inclusion of zeolites into FCC catalysts completely transformed the face of FCC within a relatively short period of time. This benchmark discovery resulted in an uncounted number of unit revamps in order to transition from bed cracking to riser cracking. The history of catalyst manufacturing plants is chronicled in Table IV.

Synthetic Catalyst Era

The origin of FCC catalysts may be traced to an observation made by Indians living in eastern Arizona used to solve a very practical problem. When the Indians sheared their sheep, they were left with a greasy fleece that had to be cleaned before it could be used. The Indians learned that they were able to absorb grease from the wool by applying clay they obtained from the ground. This clay would absorb grease if left in contact with the wool, and when brushed off, gave the Indians a clean fleece (*34*).

Many years later, a group of three entrepreneurs recognizing the chemical reactivity present in this bentonite clay, formed the Filtrol Company in 1922 to exploit this discovery (Figure 21). This clay formed the basis of an absorbent used in various applications including removal of color bodies and gums from petroleum oils.

During the late 1930s, a French chemist named Houdry discovered that heavy crude oil could be "cracked" into aviation gasoline using Filtrol's activated clay. Wartime demand for avgas resulted in many catalytic cracking units being built. Filtrol and Davison Division of W. R. Grace were a few of the very first suppliers of the catalyst.

The first delivery of Filtrol catalyst for use in the TCC or Thermofor Catalytic Cracking of petroleum was sold to Socony-Vacuum in the spring of

Table IV. History of Catalyst Manufacture

Company	Dates	Notes
Grace Davison	1832-present	• Company began in Baltimore, USA, "Grinders and acidulators of old bones and oyster shells" (1832) • South Gate catalyst plant started (1942-1985). Delivered a ground low alumina catalyst to PCLA#1 in 1942 becoming the first producer of synthetic catalyst. • Cincinatti, USA plant started (1947-1988) • First spray dried catalyst (1948) • Lake Charles plant started (1953) • Valleyfield, Canada plant started (1957) • Davison produces first catalyst to contain zeolite (XZ-15) in 1964 • Curtis Bay plant started (1981) • Worms, Germany plant started (1983)
Filtrol	1928 - 2001	• Early owners were Mobil Oil, US Filter, Ashland, Kaiser, Engelhard (1988), Akzo Nobel (1989) • Supplier of first FCC, PCLA #1, 1942
Nalco	1947-mid 70's	• Chicago, Il; ~25 Kmt/yr • Sydney, Australia, Nalco/ICI joint venture Catoleum, sold to Grace, closed in 1988
American Cyanamid	1948 - 1975	• Fort Worth, USA (~20 Kmt/yr) • Michigan City, USA (~15 Kmt/yr) started 1952 • Cyanamid continued their HPC & reforming catalysts later joining with Shell to form Criterion
UOP	1950–1960 1989-1992	• First period: Chicago, USA (~10 tpd) • Second period: new owner of Katalistiks (Delfzijl, Netherlands & Savannah, USA)
Crosfield	1953-1995	• Warrington, UK, licensee of Grace technology, ~20 Kmt/yr (1953-1993) • Haldia, India (1990-1995), ~5 Kmt/yr
Engelhard BASF	1952-present	• Attapulgus, USA (1952), moving bed catalysts for TCC process • Entered FCC technology (1972) filling the gap left by Cyanamid, UOP and Nalco. Invented "Insitu" technology.

Table IV. *Continued.*

Company	Dates	Notes
Engelhard BASF	1952-present	• Terneuzen, Netherlands (1989-1998) based Sisol technology. • Purchased Savannah facility from UOP (former Katalistiks) in 1992.
Ketjen, Akzo Nobel, Albemarle	1953-present	• Amsterdam, Netherlands (1953) • Pasadena, USA (1983) • Santa Cruz, Brazil (1985) • Acquisition of Filtrol (1989)
Russia	1960's-present	• Ufa (~15 Kmt/yr), Omsk (~5 Kmt/yr), Salavat (TCC, ~6 Kmt/yr) • Ishimbai (design capacity: 20 Kmt/yr); designed by CCIC and never started up • Omsk is the only surviving plant producing for the Omsk Refinery
Katalistiks	1980-1992	• Savannah, USA, Delfzijl, Netherlands (~25 Kmt/yr) • Sold to Union Carbide in 1984. Developed the high quality, high SAR, chemically treated zeolite, LZ-210. • Sold to UOP in 1989

1941, approximately 6 months before Pearl Harbor. This catalyst was a powder and was to be extruded into pellets by Socony for use in their semi-commercial catalytic cracking plant. In 1943, Filtrol began the manufacture and sale of small catalyst cylinders to the refining industry.

The manufacture of these clay-based catalysts consisted of crushing, drying and sizing the raw clay. The sized clay was then formed into slurry followed by washing. The sized clay was dewatered, dried, ground and sized as the final steps in the process (*35*). This clay catalyst could then be further formed into cylinders or beads depending on the final usage.

The market for natural clay catalysts began slowing down in the late 1950's and was replaced by more sophisticated catalysts, using alumina as a basic ingredient. These very early catalysts were used in a wide range of catalytic processes including: the fixed bed Houdry unit, TCC, and the FCC.

In the early days of catalytic cracking, the catalysts were highly temperature sensitive requiring regenerator temperatures to be limited to 1100°F. Temperatures in excess of 1150°F would ignite the carbon monoxide resulting in

Figure 21. Filtrol clay processing plant, Vernon, California, 1947

runaway temperatures leading to temperatures as high as 1800°F. Catalyst coolers and complex systems of spray water, cyclone quench, and flue gas line sprays were required to control after burn (*36*).

Another consequence of low regenerator temperatures was the low burning rate. Holding times in the regenerator were about 10-15 minutes compared to the 3-4 minutes in use today.

These synthetic catalysts had very low equilibrium activities of approximately 25-35 wt%. (Note: activities were reported in terms of "distillation plus loss", D+L.) There were few major quality differences between catalyst suppliers of the day and large refineries oftentimes used mixtures of catalyst from the various suppliers.

The preferred mode of operation in the early days of catalytic cracking for maximum gasoline was 1) minimum reactor bed height, 2) low bed temperatures, and 3) high recycle rates. Reactor temperatures of 900°F with recycle-to-feed ratios of 1.0 were common.

Most FCC units were designed with feed and recycle streams contacting hot regenerated catalyst (1050-1100° F, 0.5-0.7 wt% carbon) at the bottom of a sloped riser. The vaporized feed and recycle streams were transported with the catalyst upward through a short riser to plate grid fitted with punched holes designed to give a 1-3 psi pressure drop for good distribution. Nearly all early

FCC units utilized a plate grid for good distribution with the exception of the Orthoflow B unit which introduced feed directly to the reactor bed.

The first unit (Phillips Petroleum – Borger, Texas) operating with riser cracking was an Orthoflow C unit. This unit was also the first unit to be designed with bed coils for heat removal in order to process residue feedstocks.

The Zeolite Based Catalyst Revolution

Union Carbide discovered X & Y zeolites but it remained to Plank and Rosinski (1964, Mobil Oil) to apply these zeolites to clay or silica alumina-based FCC catalysts and thereby achieve a considerable boost to activity and selectivity. It is pleasing to note that Plank and Rosinski were introduced to the US patent office inventors Hall of Fame for this discovery! (Figures 22-24) Table VI documents of history of FCC catalyst manufacture.

The introduction of zeolites into synthetic catalysts brought about one of the most significant and widespread changes to the refining world. Nearly all catalytic crackers in the early 1960s (with the exception of the Orthoflow B) were configured with a very short riser which lifted the catalyst and oil mixture through a plate grid into a reactor bubbling bed. Once zeolites began appearing in catalyst commercially, these units required reconfiguration to prevent excessive coke make and over-cracking of the gasoline. This undoubtedly resulted in a significant source of income for the engineering companies of the day!

The unprecedented success of zeolite-based catalysts has been the result of the following characteristics (*37*):

1. High activity
2. Good activity stability
3. High selectivities to gasoline vs. coke and dry gas
4. Thermal and hydrothermal stability
5. Reasonable accessibility
6. Attrition resistance
7. Resistance to poisons (metals, nitrogen, etc.)
8. Acceptably low cost

Today's modern FCC catalyst is composed primarily of zeolite, active alumina, clay or filler, binder plus additional components for specific functions (Figures (25-27). One of the most significant differences found between various commercially available catalyst systems is related to the binding systems. These binding systems include: silica sol, alumina sol, alumina gel and insitu technologies. Insitu technologies are based upon the growth of zeolite crystal

232

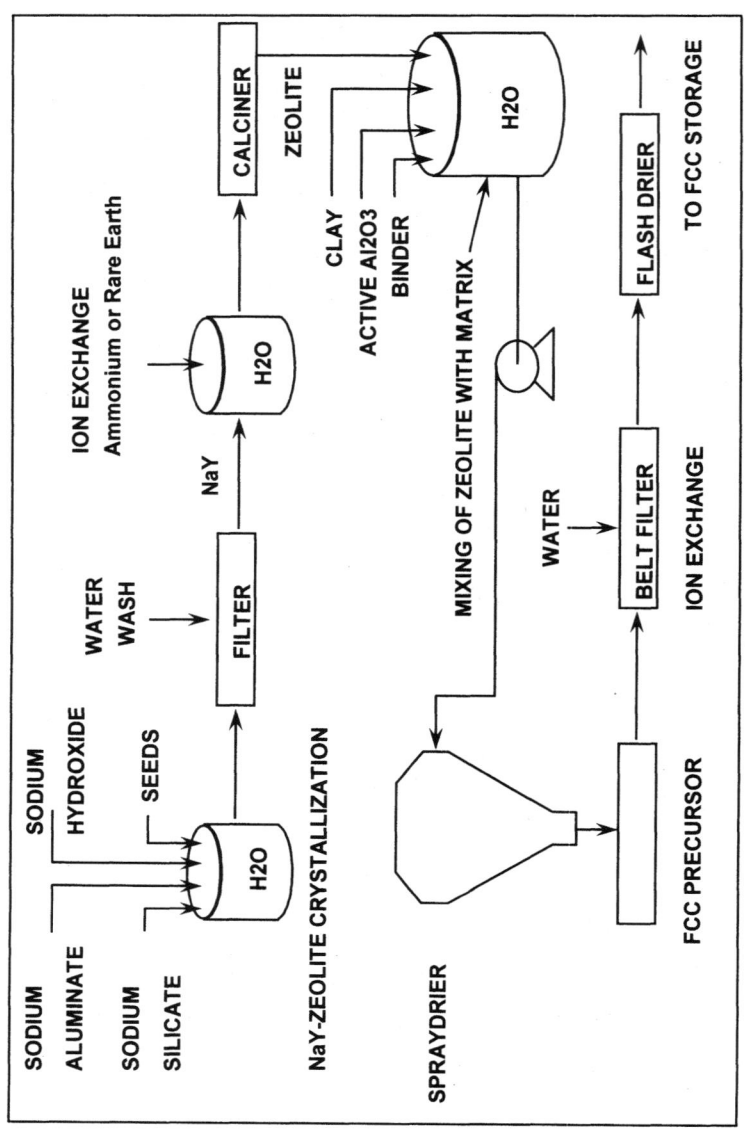

Figure 22. Typical FCC Catalyst Manufacturing Process

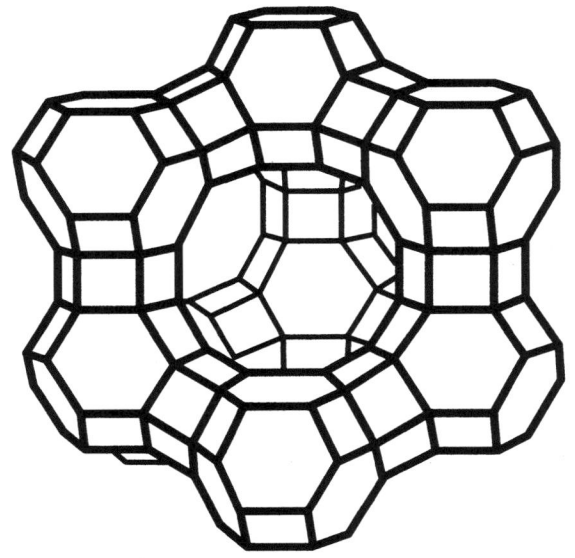

Figure 23. Schematic Y-zeolite crystal

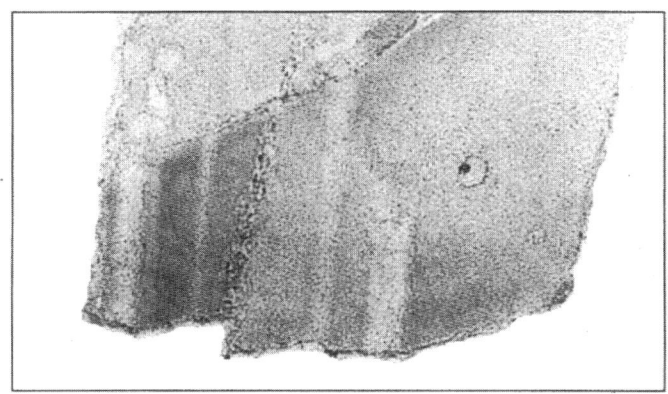

Figure 24. High-resolution TEM image of USY zeolite

234

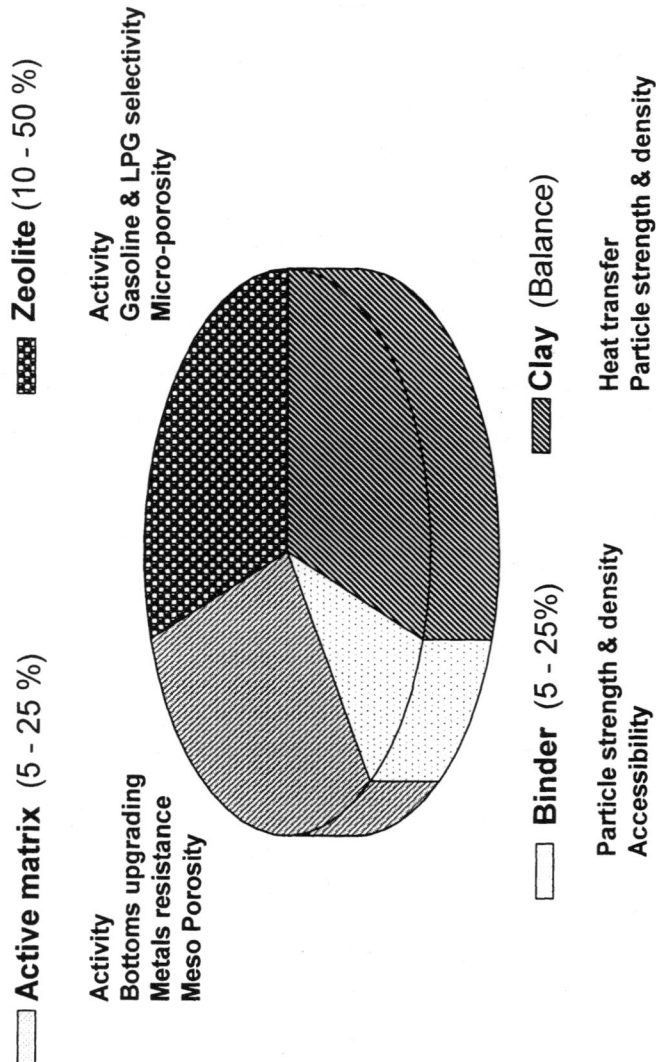

Zeolite (10 - 50 %)

Activity
Gasoline & LPG selectivity
Micro-porosity

Active matrix (5 - 25 %)

Activity
Bottoms upgrading
Metals resistance
Meso Porosity

Clay (Balance)

Heat transfer
Particle strength & density

Binder (5 - 25%)

Particle strength & density
Accessibility

Figure 25. Typical FCC catalyst compositions

235

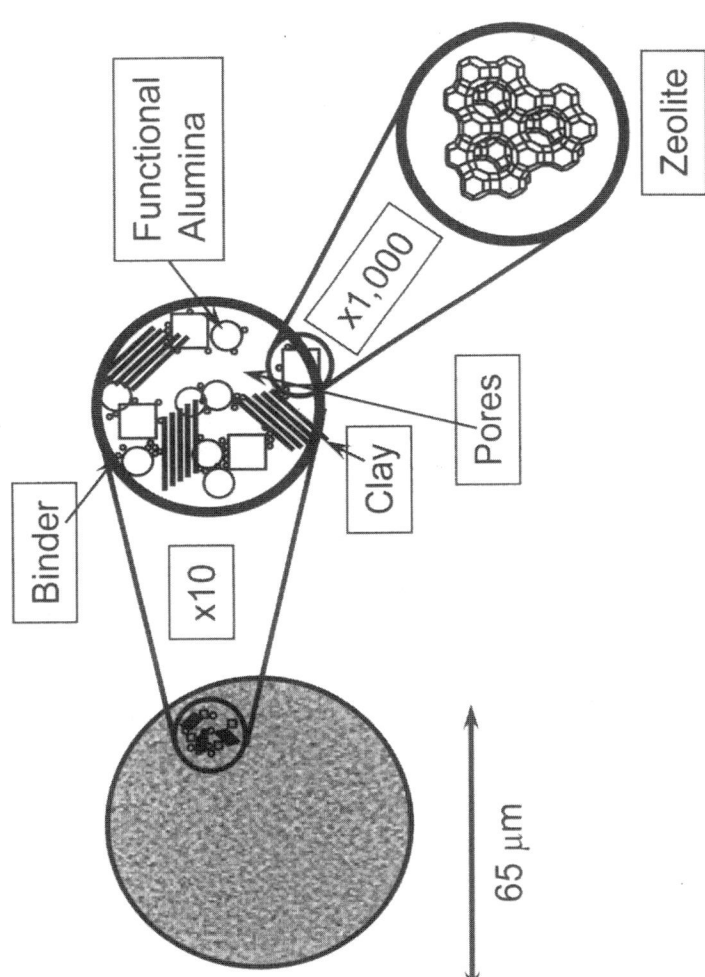

Figure 26. FCC catalyst is a heterogeneous composite of multiple ingredients

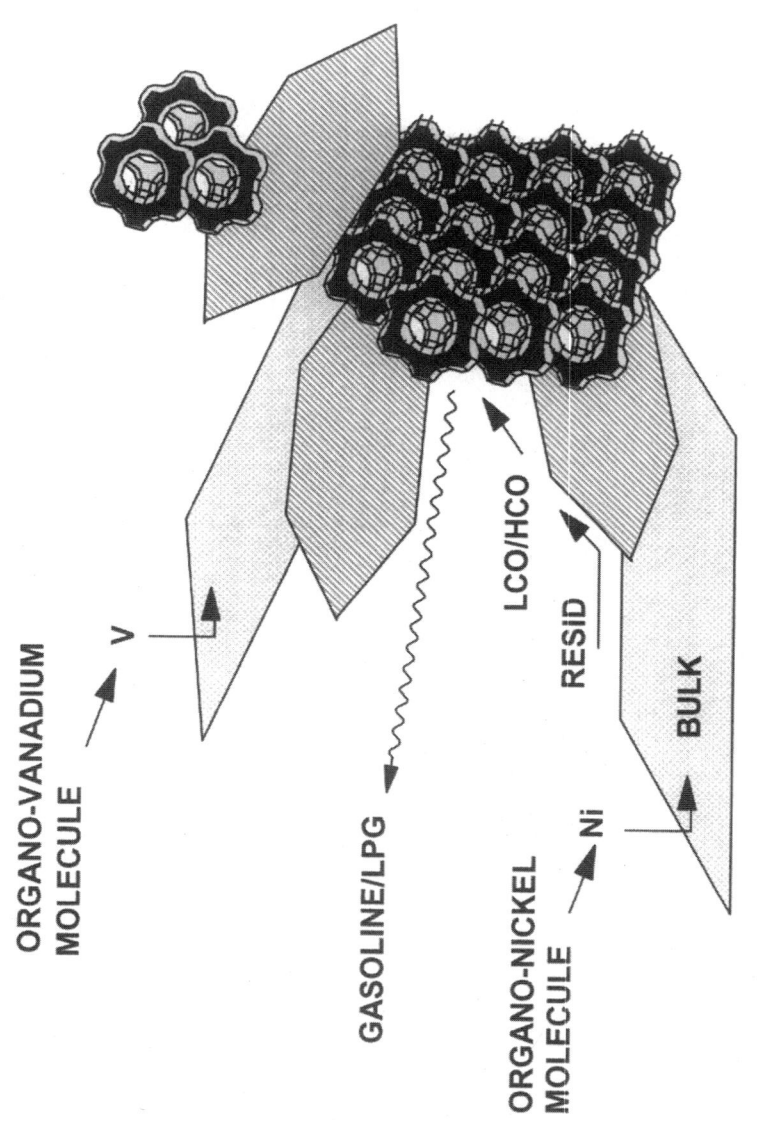

Figure 27. Catalyst Particle Diffusion

ORGANO-VANADIUM
MOLECULE

V

GASOLINE/LPG

LCO/HCO

RESID

ORGANO-NICKEL
MOLECULE

Ni

BULK

within spray dried calcined clay microspheres. The choice in the mode of binding may have significant impact on catalyst selectivities.

- *Zeolite.* Typical catalysts today contain 30-40% zeolite with some formulations reaching 50%. "Activity boosting" additives are currently available commercially with zeolite concentrations of 60% or more.
- *Zeolite type.* All Y zeolites begin as basically $Na_{56}[SiO_2]_{136}[AlO_2]_{56}*250H_2O$. They are produced via modification and optimization of synthesis conditions, treatment steps, and exchange agents and routes. The zeolite can be dealuminated thermally, chemically, or by employing both methods. Furthermore, the silica-to-alumina ratio can be optimized to provide maximum zeolitic activity (VGO operations), maximum zeolitic stability (residue cracking operations), or intermediate levels providing a balance of each.
- *Rare earth.* Rare earth (*38*) is added to the zeolite as an activity enhancer, selectivity modifier, and to improve hydrothermal stability. Rare earth on zeolite concentrations typically vary from zero to 16 wt%.
- *Active alumina.* Alumina is added to the catalyst to pre-crack high molecular weight molecules present in heavier portions of the feedstock. This pre-cracking function enables further cracking by the zeolite present in the catalyst increasing the yields of gasoline and LPG. The alumina technology present today includes aluminas capability of absorbing vanadium and encapsulating nickel which are catalyst poisons. Furthermore, catalyst manufacturers are now capable of modifying both the pore volumes and acidities of these aluminas enabling specific cracking of difficult feedstocks. Judicial use of alumina content is required as most aluminas increase coke and dry gas yields.
- *Zeolite-to-matrix ratio.* The relative composition of zeolite and alumina determines the yield selectivities observed by the refiner. A general guideline to catalyst formulation is shown in Table V below.

In 1984 Mobil Oil launched a revolutionary new additive, ZSM-5. ZSM-5 is a small pore zeolite which selectively cracks higher molecular weight olefinic molecules boiling in the gasoline range. ZSM-5 has found wide application in the refining industry and is used for multiple applications including LPG olefin maximization for the production of alkylate, improved gasoline octanes, and for petrochemical applications.

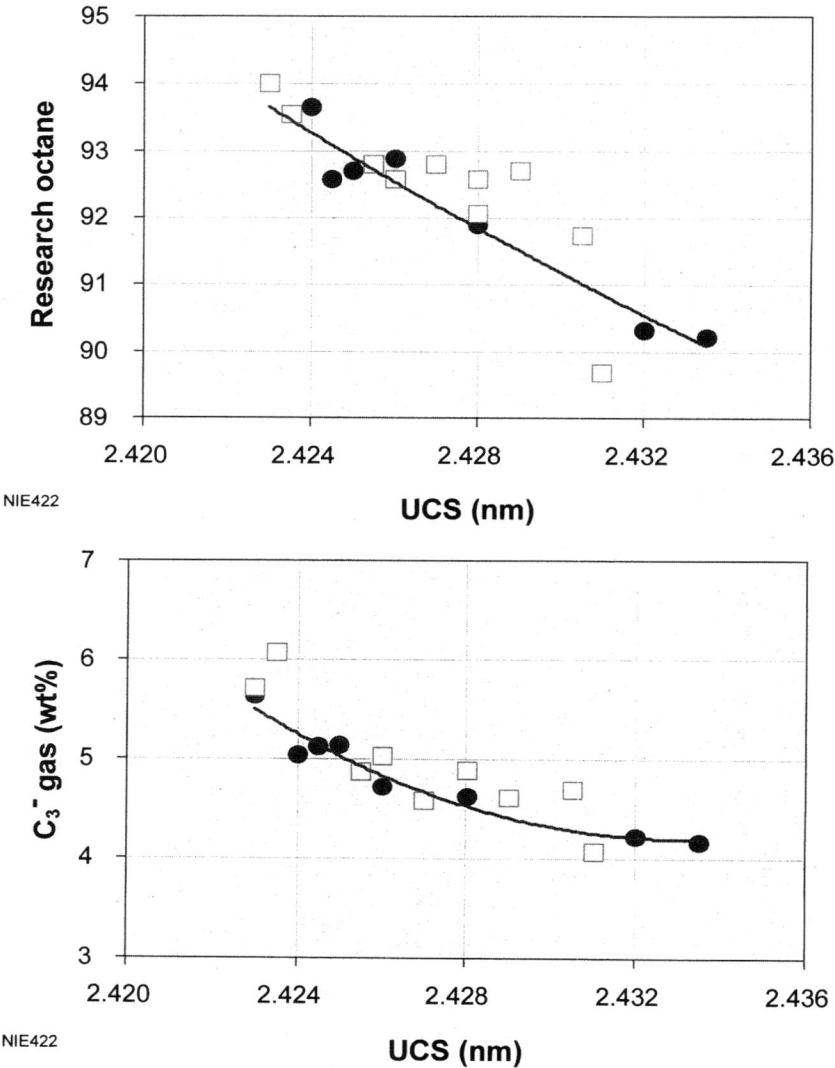

Figure 28. Unit cell size & rare earth impact on product selectivities

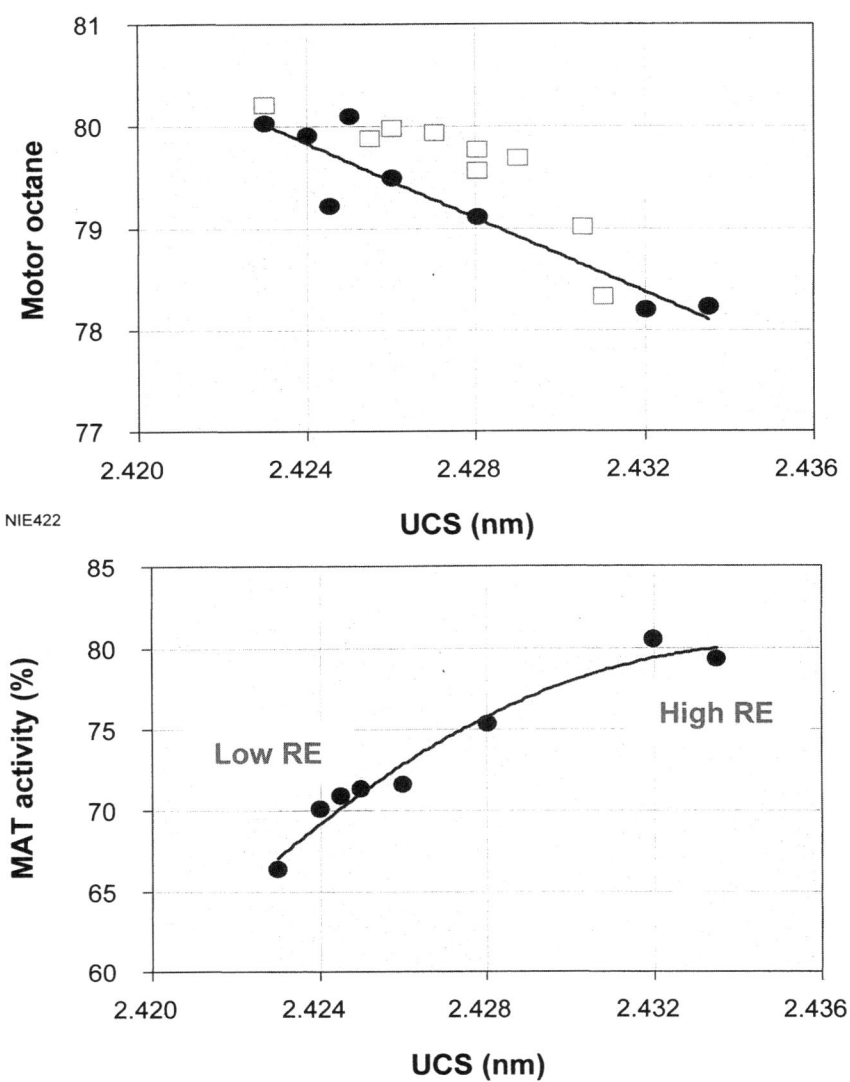

NIE422

Figure 28. Continued.

Table V. Catalyst formulation selectivity drivers

Desired selectivity change:	Typical catalyst formulation:
LCO-to-bottoms ratio	Active alumina content
Gasoline-to-LCO ratio	Zeolite-to-matrix ratio
LPG-to-gasoline ratio	Zeolite content and rare earth on zeolite
LPG olefinicity	Rare earth content

In addition to these basic components described above, the knowledge of FCC catalysis with respect to environmental additives has developed significantly over the last two decades. Refiners are now able to control SOx & NOx emissions, control CO emissions, reduce gasoline sulfur concentrations, improve bottoms cracking, improve fluidization characteristics, and enhance activity through controlled application of high-tech additives.

Residue Cracking Catalysts

The two greatest challenges facing the refiner cracking atmospheric residue include management of the increased levels of Conradson carbon levels and the contaminant metals in the feedstocks.

Higher catalyst makeup rates or more robust catalyst formulations can offset most of the effects of contaminant metals. Managing the higher CCR levels is best accomplished through the use of external catalyst coolers and possibly operation in a partial combustion mode.

There exist multiple approaches to the design of catalysts intended for residue cracking operations. In general, most approaches include careful control of both the rare earth on zeolite level and the zeolite-to-matrix (Z-to-M) level.

- *Rare earth.* The rare earth on zeolite level has a direct impact on catalyst stability and product selectivity. Directionally, increased rare earth levels on zeolite result in catalysts having enhanced hydrothermal stability. This results in a reduction in catalyst makeup rate. However, changes in product selectivities, especially in the LPG range, will result. Many refiners choose to compensate this loss in LPG olefinicity via the addition of ZSM-5 additive.

Table VI. High-tech additives augment FCC performance

Additive:	Function:	Typical additions (%):
Combustion promoter	Control of regenerator after burn	0.5 – 2.0 ppm
Non-Pt combustion promoter	Control of regenerator after burn with minimal NOx increase	1 to 4 ppm
SOx additive	Control of flue gas SOx emissions	1 to 5
NOx additive	Control of flue gas NOx emissions	1 to 5
ZSM5	LPG olefin and octane enhancement	2 to 10
Gasoline sulfur removal	Gasoline boiling range sulfur removal	5 to 15
Bottoms cracking	Selective cracking of slurry oil	5 to 10
Fluidization enhancer	Optimization of fluidization characteristics	5 to 10

- *Z-to-M level.* The zeolite-to-matrix ratio affords the refiner a second means to control bottoms destruction with residue feedstocks. Directionally, as the amount of active alumina present in the catalyst increases, the amount of slurry upgrading also increases. However, with this increased conversion comes an increase in dry gas and coke. Therefore, care must be given to ensure proper balancing between zeolite and active alumina levels.

An additional approach stresses the importance of tailoring matrix alumina type and pore size distribution (*39*):

- Large pores (>100 Å in diameter) with lower activity to control coke and gas make
- Meso pores (30-100 Å) with higher activity
- Small pores (<20 Å) with the highest activity

The meso pores are directly responsible for reducing bottoms yield with aromatic or naphthenic feedstocks, whereas the smaller pores and the zeolite are more important for paraffinic feedstocks. (*40*)

Akzo Nobel Catalysts pioneered catalyst particle accessibility in 1997. This was the result of the development of a quantified lab scale test designed to measure the accessibility of the catalyst particle. After a catalyst enters into an operating resid FCC unit, there is a gradual accumulation of contaminant metals on the surface of the particle. The presence of these contaminant metals accumulate in some residue units to the point that a resulting barrier to the diffusion of high molecular weight, sterically hindered oil molecules occurs. This discovery of catalyst particle accessibility coupled with proprietary catalyst manufacturing techniques has led to significant improvements in the accessibility of residue cracking catalysts and the profitable cracking of very heavy feedstocks in the FCC unit.

Metals passivation in the resid FCC unit may also be accomplished through the addition of metal traps. Tin, barium titanate, strontium titanate, magnesium oxide, manganese oxide, and specialized zeolite types and contents have all been used for vanadium trapping. In addition, zeolites coated with alumina and catalyst particles coated with rare earth have also been applied to resist vanadium poisoning. Antimony, bismuth, and specialized active high crystalline aluminas have all been used successfully to counteract the negative effect of nickel in the FCC unit.

Current Challenges in FCC

There have been few periods during the entire existence of Fluidized Catalytic Cracking in which there have been no challenges facing the FCC process. As the inheritors of this process we also are faced with particular challenges that can easily be transformed into opportunities. The following list, while not inclusive, lists several challenges to the innovative refiner through the next decade.

Propylene Maximization

The past decade has observed a strong demand for propylene as a feedstock for the chemicals industry. Many refiners have begun the addition of ZSM-5 to maximize propylene production within existing gas plant processing constraints (*41, 42*). Furthermore, specialty additives are being developed to minimize the dilution effects resulting from very high additive addition rates (in excess of 12%) and for units processing resid feedstocks desiring maximum propylene and minimum slurry production. The leading engineering firms have launched new unit designs for the refiner focused on delivering maximum propylene into this new marketplace. This trend is expected to continue.

Capacity Expansions. There has been a recent upsurge in announcements of new units to be built as a result of crude oil prices reaching rarely observed levels ($70 and higher per barrel) together with strong profit margins within the refining sector (Figure 29). At the time of this publication, there have been approximately 70 new cat crackers either in construction, in design, or having been recently announced. As a result, the historical overcapacity of the catalyst manufacturing sector is rapidly disappearing. It is anticipated that the supply and demand curves should meet within the next 4-5 years. Furthermore, given the historically low margins this industry has experienced there is a hesitance on the part of many investors to supply the capital necessary to build new world-class FCC catalyst manufacturing plants. This is anticipated to have a potential impact on both catalyst pricing and catalyst availability.

Globally Expanding Diesel Demand. The incremental growth of diesel has outpaced the incremental growth of gasoline (Figure 30) in many parts of the world including North America in recent years (*43*). This demand is being driven in large part by a growing awareness that the diesel engine is somewhat more "environmentally friendly" in terms of energy density and utilization than gasoline. There has been a widely recognized surplus of gasoline in Western Europe with a corresponding shortage of diesel. Many of the world's refined product markets are demanding more diesel than gasoline, such as in Europe, South America, and parts of Asia. This will be a growing challenge for refineries whose major conversion engine is the fluidized catalytic cracker since the FCC is a refining asset specifically focused on maximizing gasoline. This will be a challenge for many refiners who will eventually face this gasoline/diesel balance.

Environmental Regulations

Environmental regulations have been growing more stringent since the mid-70s. This trend is not expected to change. Continuing emphasis is expected to be paid to sulfur levels in gasoline and diesel. Benzene regulations are expected to become strict. Flue gas emission rates of particulates, SOx and NOx are expected to be steadily reduced. Continuing investments will be required for the "right to stay in business".

Future Challenges to the FCC Process

Humanity is entering an unprecedented era: population growth is exponential, access to clean drinking water for most of the world's population

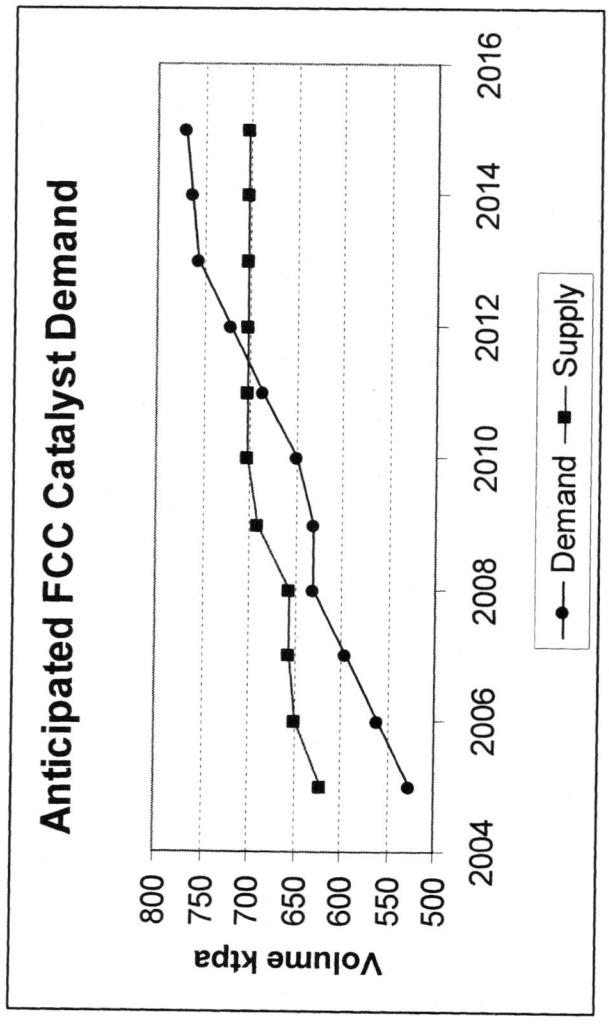

Figure 29. Anticipated Growth in Catalyst Demand

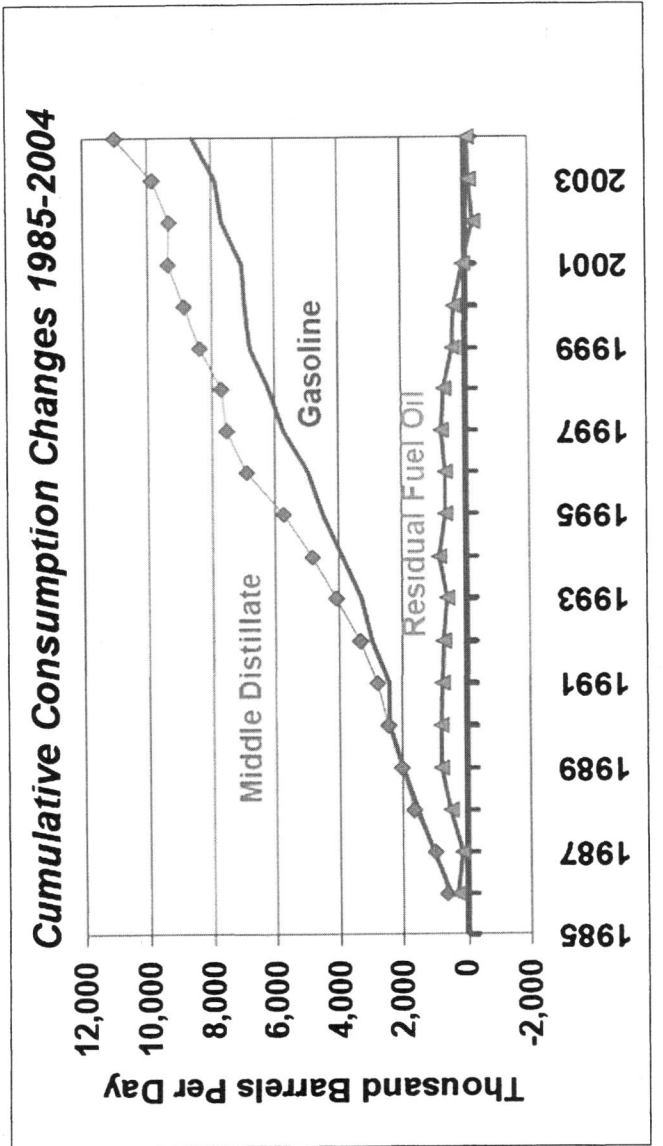

Figure 30. Incremental diesel demand outpaces gasoline (Source: BP world statistical review)

has never been lower, and a percentage of humanity remains undernourished. Furthermore, the global awareness of environmental degradation is at the highest point ever. The demand for clean air, renewable resources, and sustainable energy is increasing and is expected to grow ever faster.

The FCC process will be challenged in this era of wide-ranging change. The ability of the FCC to transform itself to meet these new demands and opportunities will be directly related to the continuing ingenuity of both today's and tomorrow's chemical engineers. Some of the challenges anticipated over the next 50 years include:

CO_2 *Emissions.* The impact of burning fossil fuels for transportation on the greenhouse effect is expected to grow as the global population increases. This pressure is expected to increase over the coming two to three decades as both China and India struggle to "get their fair share" of the lifestyle experienced by today's industrialized nations. Eventually however, it is expected that the awareness of the direct link between combustion of fossil fuels and average global temperature plus weather patterns will become widespread. Public awareness may inevitably lead to "political solutions" mandating reduction of CO_2 emissions by automobiles. It is anticipated that the percentage of fossil fuels devoted to transportation by individuals will gradually reduce exerting pressure on the FCC unit. The development of a CO2 "neutral" FCC unit will be a challenge for future engineers!

Declining Reserves of Easily Recoverable Crude Oil

It has already become evident that easily recoverable crude oil reserves are becoming more difficult to find. This is evidenced by the massive investments currently in progress to recover Alberta's tar sand reserves. This trend is expected to continue. This trend is expected to have a twofold impact on the refining industry as it exists today. The average price per barrel of crude oil is expected to increase as ever greater investments are necessary to bring to market newly discovered crude oil reserves. This should eventually lead to the economical exploitation of coal and possibly shale oil thereby limiting the impact of the FCC unit. The second anticipated driving force will be a reluctance to use this dwindling reserve on the personal family automobile. These two effects may combine to enforce changes to the operation of existing cat crackers as a preeminent engine to produce motor fuels.

Exploitation of Very Heavy Crude Oils

The exploitation of Alberta's bitumen oil from the vast fields of tar sands is currently at 1.5 million barrels per day. This production rate is expected to increase to 3.5 million barrels per day by 2015 (*44, 45, 46*). With a sharp increase in crude oil pricing and the apparent stabilization at these higher prices it is expected that other very heavy oil fields may become economic to exploit. The majority of the synthetic crude being produced in Alberta is coked followed by varying degrees of hydrotreating. These synthetic oils currently occupy a minor position in North America's FCC feed slates. The percentage of these crudes reaching the FCC is expected to increase over the next few decades. In spite of deep hydrotreating, these hydrocarbon streams remain highly cyclic and therefore pose a challenge to today's catalysts. Substantial improvements in the porosity and accessibility of today's "state-of-the-art" residue catalysts will likely be required.

Biofuels

Ever-growing volumes of biofuels as feedstocks to the FCC unit will likely be seen. Research focused on the specific requirements of these fuels will be required.

Integration with Petrochemical Plants

It is expected that by the middle of this century the world's remaining fossil fuel reserves will become much more precious given the growing global awareness of the impact of CO_2 on the environmental health of the planet and the declining availability of easily recoverable reserves. For this reason it is probable that the FCC will transform itself steadily into a feed preparation unit for existing and yet to be discovered chemical processes. The preeminence of today's FCC unit as an engine for the production of motor fuels will likely diminish as the decades pass.

The challenges that are being faced by the FCC process today and those anticipated over the coming decades can be viewed as problems or as opportunities. It is believed by the author that the industry may soon enter another period of extraordinary change requiring the FCC process to transform itself to the changing needs of society. This driving force will likely be gradual, yet not too dissimilar, to the driving force felt in World War II when society was seeking a plentiful supply of aviation fuel for the war effort. It is the author's hope that a new "Catalytic Research Associates" collaboration between tomorrow's FCC operators will converge to transform the very large existing FCC asset base into a process that continues to meet the needs of society well into the 22nd century.

References

1. Chemical Achievers—The Human Face of the Chemical Sciences, http://www.chemheritage.org/classroom/chemach/petroleum/houdry.html
2. Avidan, A., Edwards, M., Owen, H. OGJ Jan. 8, 1990, p 33-38
3. *Guide to Fluid Catalytic Cracking—Part One*, W.R. Grace and Co, 1993
4. *Handbook of Petroleum Processing*; Springer, p 245
5. Avidan, A. OGJ, May 18, 1992, p 59
6. Avidan, A., Edwards, M., Owen, H. OGJ Jan. 8, 1990, p 35
7. *Milestones in UOP FCC Technology*, internal UOP document
8. Chemical Achievers—The Human Face of the Chemical Sciences, http://www.chemheritage.org/classroom/chemach/petroleum/houdry.html
9. TheHoudryProcess, http://acswebcontent.acs.org/landmarks/landmarks/hdr/hdr_process.html
10. Avidan, A., Edwards, M., Owen, H. OGJ Jan. 8, 1990, p 33-38
11. *Guide to Fluid Catalytic Cracking—Part One*, W.R. Grace and Co, 1993
12. TheHoudryProcess, http://acswebcontent.acs.org/landmarks/landmarks/hdr/hdr_process.html
13. Avidan, A., Edwards, M., Owen, H. OGJ Jan. 8, 1990, p 43-45
14. Squires, A. M., *The Story of Fluid Catalytic Cracking – The First Circulating Fluid Bed Proceedings*; First International Conference on Circulating Fluid Beds, Technical University of Nova Scotia, Halifax, Nov. 18-20, 1985
15. Wrench, R., Wilson, J., Logwinuk, A., Kendrick, H. *Fifty Years of Catalytic Cracking;* M.W. Kellogg, 1986
16. Chen, N., Lucki, S., *Industrial Engineering Chemical Process Design Development*; Vol. 25, No. 3, pp. 814-820
17. Wrench, R., Wilson, J., Logwinuk, A., Kendrick, H. *Fifty Years of Catalytic Cracking;* M.W. Kellogg, 1986, pp. 4-6
18. Jahnig, C., Martin, H., Campbell, D., *The Development of Fluid Catalytic Cracking;* Chemtech, February 1984, pp. 106, 108-111
19. Jahnig, C., Martin, H., Campbell, D., *The Development of Fluid Catalytic Cracking;* Chemtech, February 1984, p.109
20. Reichle, D. OGJ, January 18, 1990, pp. 41-48
21. OGJ, Vol. 41, No. 5, June 11, 1942
22. Murcia, A. OGJ, May 18, 1992, pp. 68-71
23. Murphy, J. OGJ, May 18, 1992, pp. 49-58
24. Avidan, A., Edwards, M., Owen, H. OGJ Jan. 8, 1990, pp. 46-48
25. Wilson, J. In *Fluid Catalytic Cracking;* Penwell Books, Tulsa, OK, 1997, pp. 18-40
26. Warren Letszch, private conversations

27. Myers, D., Knapik, P., Lacijan, L., Brandner, K., *Use of Catalyst Coolers to Improve Resid Processing in FCC Units;* NPRA Annual Meeting, March 20-20 2, 1994

28. Myers, *Petroleum Refining Processes;* McGraw-Hill, New York, NY, pp. 3.101-3.112

29. Murphy, J. OGJ, May 18, 1992, p. 52

30. Horecky, C.J., US Patent 3,909,392, 1975

31. Milne, L.,Nienew, A., and Patel, K., *Lateral Mixing and Batch Beds of One or Two Components;* Fluidization Vol. VI, May 1989

32. Wilson, J. *FCC Regenerator Afterburn Causes and Cures;* NPRA, AM03-44

33. Wilson, J. In *Fluid Catalytic Cracking;* Penwell Books, Tulsa, OK, 1997, pp. 3-4

34. *Filtrol History*; Filtrol Facts, Filtrol Corporation, Vol. 2, No. 2, February 1950

35. "Salt Lake City Plant," Filtrol Facts, Filtrol Corporation, Vol. 3, No.2, February 1951

36. Murphy, J. OGJ, May 18, 1992, pp. 50-51

37. Venuto, P., Habib, E. *Fluid Catalytic Cracking with Zeolite Catalysts;* Marcel Dekker, Inc., New York, NY, 1979, p 30

38. Pine, L., Maker, P., Wachter, W. Journal of Catalysis, Vol. 80, No. 5, 1984, pp. 466-476

39. Avidan, A. OGJ, May 18, 1992

40. O'Connor, P., Gerritsen, L., Pierce, J., Desai, P. Humphries, A., Yanik, S., *Catalyst Development in Resid FCC;* Akzo Catalysts Symposium, May 1991

41. Couch, K., *FCC Propylene Production - Technology Integrations to Optimize Yields;* Albemarle Catalysts Symposium, Athens, 2007

42. Foskett, S., Edwards, M., *AFX: Catalyst Design and Application for Maximum Propylene in Resid FCC;* Albemarle Catalysts Symposium, Athens, 2007

43. Fletcher, R., *Producing Premium Quality Diesel from the FCC Unit;* Albemarle Catalysts Symposium, Athens, 2007

44. *Oil Sands Industry Update;* Alberta Economic Development, June 2006

45. *Oil Sands Technology Roadmap - Unlocking the Potential;* Alberta Chamber of Resources, January 30, 2004

46. Kelly, S., Wise, T. *Markets for Canadian Oil Sands Products;* NPRA Annual Meeting, March 19, 2006

Chapter 7

Purification of Automotive Exhaust through Catalysis

M. Rodkin[1], S. J. Tauster[1], X. Wei[1], and T. Neubauer[2]

[1]BASF Catalysts LLC, 25 Middlesex/Essex Turnpike, Iselin, NJ 08830
[2]BASF Catalysts Germany GmbH, Hanover, Germany

The worldwide effort to reduce the harmful effects of automotive exhaust has witnessed impressive progress during the past 30 years. Driven by the regulatory agencies, emissions have steadily declined and now represent a small fraction of their pre-control values.

Gasoline-powered passenger cars, which comprise a large majority of the US market, emit CO, unburnt HC and oxides of nitrogen (NOx). Starting in 1981, the automobile industry was mandated to sharply reduce the emissions of all three (previously only CO and HC had to be removed). This required "three-way-conversion" (TWC): the simultaneous oxidation of CO/HCs and reduction of NOx, a feat without precedent in the chemical industry. It can only be accomplished by keeping the exhaust gas composition extremely close to the stoichiometric point. Cerium oxide was soon recognized as indispensable to the success of the TWC catalyst, due to its ability to rapidly change oxidation states at the surface. This enables it to "store" and "release" oxygen in response to changes in the gas phase composition. The use of ceria was greatly expanded by the addition of zirconia; the ceria-zirconia was far more thermally stable than ceria itself and allowed the TWC catalyst to survive much higher temperatures.

Diesel exhaust differs from gasoline exhaust in important respects. It is always "lean" (i.e., net oxidizing) and three-way conversion is thus ruled out. The chief concern is particulate matter, which includes dry soot and a "soluble organic fraction" (SOF), comprised of mainly C_{20}-C_{28}. Prior to 2000, attention was mainly focused on SOF conversion, as dry soot emissions could be controlled within the standards by optimizing fuel delivery, air intake systems, and the combustion process. To meet the new emission standards proposed for 2007-2009 in the U.S, Europe and Japan, the diesel particulate filter (DPF) was created, in which the channel wall filters out the soot particles. The latter are burned off, at suitable intervals, by raising the temperature. DPFs were originally limited to trucks and buses, but their proven effectiveness has led to their planned use on passenger cars as well.

A final topic in diesel aftertreatment is NOx removal. A successful approach, already in place in Europe, is the use of urea, carried on-board as an aqueous solution. Urea hydrolyzes to release NH_3, which is a highly effective agent for converting NOx to N_2.

Introduction

The first catalytic converter was installed on a U.S. passenger car in the 1970s. This followed a lengthy period of academic studies, industry tests and Congressional testimony that focused on the feasibility of such an undertaking. On the one hand, the contribution of automobiles to the air pollution problem had been clearly established. A 1965 California study had concluded that cars were to blame for 80% of the unburnt hydrocarbons and 65% of the nitrogen oxides in the atmosphere. These gases were shown to be the principal components of "smog," to which the public had become sensitized. On the other hand, there was legitimate uncertainty whether the catalytic converter could be effective, and especially cost-effective, given the obstacles it faced. These included extreme temperature gyrations, poisons and severe mechanical stress (vibrations) in the exhaust system. There was also reluctance on the part of the car-buying public to pay for these new, improved devices.

With the benefit of hindsight, these concerns may seem exaggerated or unwarranted. Certainly, the catalytic converter has been overwhelmingly successful in meeting not only the original EPA targets, but a stream of ever-more-demanding limits as well. This does not mean, however, that the original reservations were unfounded. To make this point, consider the different situations confronting a typical industrial catalyst and an automotive catalytic converter. The former is carefully brought on stream and then operated with purified reagents under tightly controlled conditions. By contrast, an auto catalyst is "operated" (by a person who often does not know it is there) over a temperature range of many hundreds of degrees, in the presence of strong poisons. Both the temperature and the gas composition frequently undergo large changes over a time span of less than one minute. Whereas industrial catalysts operate in a stable mechanical environment, automotive converters face extreme vibration and pressure surges. Especially in the early days, it was feared that this vibration would pulverize the catalyst, which would then plug the exhaust system, creating a genuine safety hazard. These legitimate concerns were only outweighed by a strong and growing demand that something be done about air pollution.

Indeed, something has been done. Current production vehicles emit only a small fraction of the amount of pollutants released into the atmosphere in the pre-control days. As the allowable emissions have been constantly reduced, auto catalyst research has met the challenge. NOx limits, for example, stood at 3.1 g/mile in 1975/76; this was decreased to 1.0 g/mile in 1991, to 0.2 g/mile in 1996[1] and to 0.02 g/mile for the current Tier 2 Bin 4 federal emission standards. A major share of the credit belongs to the catalyst research and development community. Their achievements are the subject of this chapter. We will also acknowledge, although only briefly and superficially, the roles played by the ceramics industry and by the automotive engineering community. The former created, and has continually improved, the ceramic monolith, without which the auto catalyst would not be feasible. (Monoliths can also be fabricated from metal for a few occasions when higher heat conductivity is needed) Automotive engineers, using sophisticated electronic controls, made an essential contribution toward providing the catalyst with the correct gas composition, the critical importance of which will be described below.

This chapter will begin with the three-way-conversion (TWC) catalyst, which simultaneously converts CO, hydrocarbons and nitrogen oxides. TWC catalysts are currently used on all gasoline-powered passenger cars made in the US, as well as in many other world markets. The three milestones cited will be ceria, ceria-zirconia and close-coupled palladium. The first two of these bear on the central theme of gas-composition control, without which three-way conversion cannot be realized. The third milestone allowed the catalyst to become active at a much earlier point in the test cycle, leading to a major reduction in the amount of hydrocarbon emissions.

Then we will turn to diesel emissions control for passenger cars, trucks and buses. The problems here are very different from those that occur with the gasoline engine. Regulating the gas composition is not an issue. The emission of particulate matter, which is not a problem faced by the TWC catalyst, becomes a major challenge.

The Catalytic Converter

Catalytic converters were first installed on US passenger cars in 1975. For several years, their mandate was limited to the reduction of tailpipe concentrations of CO and unburnt hydrocarbon (HCs). Although oxides of nitrogen (NOx) were recognized as health and environmental hazards, the allowable limit was attainable by non-catalytic measures, such as recirculating a portion of the exhaust gas to cool the engine.

This situation changed in 1981 with the imposition of a 1.0g/mi limit on NOx. A catalyst now was clearly required, but it faced a seemingly impossible hurdle. Whereas the removals of CO and HCs are straightforward oxidations, NOx conversion can only be accomplished by reduction with a species such as CO or H_2. Oxidation and reduction are, of course, chemical reaction opposites and there was no precedent in industrial catalysis for combining the two in the same catalyst bed.

The initial approach was to divide the problem into its logical components: a front bed where the reduction would take place and a downstream bed for the oxidations. If the engine were tuned rich (i.e., net reducing), NOx conversion could proceed upstream and then, with the injection of air, oxidizing conditions would prevail in the rear bed, so that HCs and CO would be removed. From a chemical perspective, this was a rational strategy.

After being implemented for a few years, however, this plan was abandoned due to several factors. Running the engine rich lowered the fuel economy. The air pump was expensive and cumbersome. Neither Pt nor Pd proved effective front-bed catalysts, due to their tendency to reduce NOx to NH_3. This would only be re-oxidized in the rear bed to NOx. Ruthenium avoided this problem, but occasional periods of lean operation at high temperature led to its volatilization as RuO_2. The dual-bed strategy was discontinued.

The industry was forced to weigh the radical concept of three-way conversion, simultaneously oxidizing two species in the exhaust gas while reducing a third. There was no other way of complying with government standards. It was immediately clear that the key would be extremely tight control of the air/fuel (A/F) ratio. The gas composition at the catalyst surface would have to be kept very close to the stoichiometric point. Otherwise, either the oxidations or the reduction would suffer. Without sufficient O_2, some CO and HC would be emitted since their conversions require O_2. In the case of NOx, decomposition to N_2 and O_2 is thermodynamically possible. However, all known catalysts for this reaction are immediately poisoned by a reaction product, O_2 (N_2 is easily swept out), and this adsorbed oxygen must be removed by a reductant. When this occurs, the newly vacant surface site can adsorb NO or O_2, with these species in kinetic competition. Rh, more than Pt or Pd, shows a slight preference for NO, possibly related to its ability to form dinitrosyl adducts in organometallic chemistry. But this selectivity is limited, and, even with Rh, NOx conversion plummets as the A/F ratio moves into the lean (oxidizing) regime. Figure 1, probably the most familiar diagram in the TWC literature, shows the dependencies of CO, HC and NOx conversions on the A/F ratio (indicated as the lambda value, stoichiometic value is 1). Deviations from stoichiometry have severe impact. The rich-side decline for CO is steeper than that for HC, but both suffer. NOx conversion is even more acutely affected by small excursions away from stoichiometry.

Despite the difficult challenge inherent in Figure 1, the TWC catalyst has achieved extraordinary success during the past quarter century. As EPA standards have consistently become tighter, new formulations have allowed these targets to be met. This success was made possible by the industry's ability to control the A/F ratio at the catalyst surface, which is the essential precondition of TWC operation. Control has come from two sources. An elaborate technology has developed, incorporating sensors and electronically controlled fuel injectors, to constantly drive the engine-out A/F ratio back to the stoichiometric point. As an example, Figure 2 compares the engine-out A/F trace of a 1986 car with that of a 1990 vehicle.[2] During this brief span, central fuel injectors were replaced by multi-point injectors. Without these "hardware" advances, TWC improvements would have been difficult if not impossible.

But even more was needed. The extremely precise A/F control required meant that catalysis research had to shoulder part of the burden. A new concept, "oxygen storage," was put forward, first by H. Gandhi at the Ford Motor Co.[3] A large part of TWC research has been devoted to the need for improved oxygen storage systems and for better ways of measuring this critical function. We will describe this progress in detail. First, however, it is appropriate to take a closer look at the pre-TWC phase of auto catalyst research. During this relatively brief period, a repository of materials and techniques was laid down that TWC scientists were able to exploit.

The Catalyst Monolith

Catalysts used in the chemical and petrochemical industries are packed beds of particles, typically 1/16" to 1/4". The shapes vary, with spheres and cylinders common, but with a variety of other extruded forms found as well. In many cases, the entire pellet is catalytically active, while in other instances the active phase is coated onto an inert core.

In the early days of automotive catalysis, packed beds of catalyst beads were tried, but found completely unacceptable. The problem was the extreme level of vibration in the exhaust system, leading to the mutual attrition of neighboring particles. In a short time, this can reduce a significant fraction of the catalyst

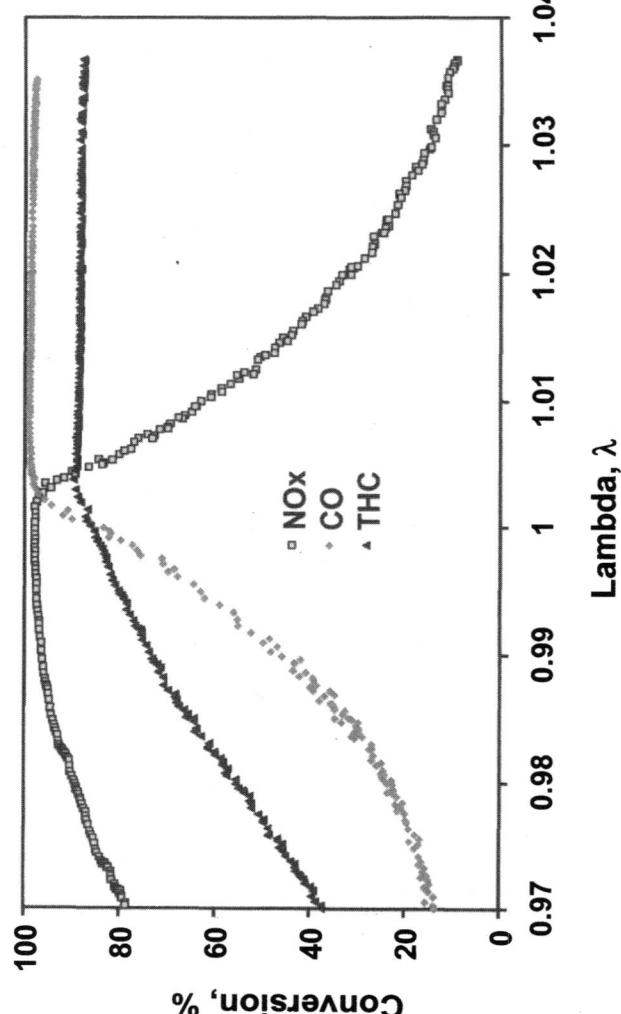

Figure 1. Impact of Air/Fuel ratio on CO/HC/NO conversion

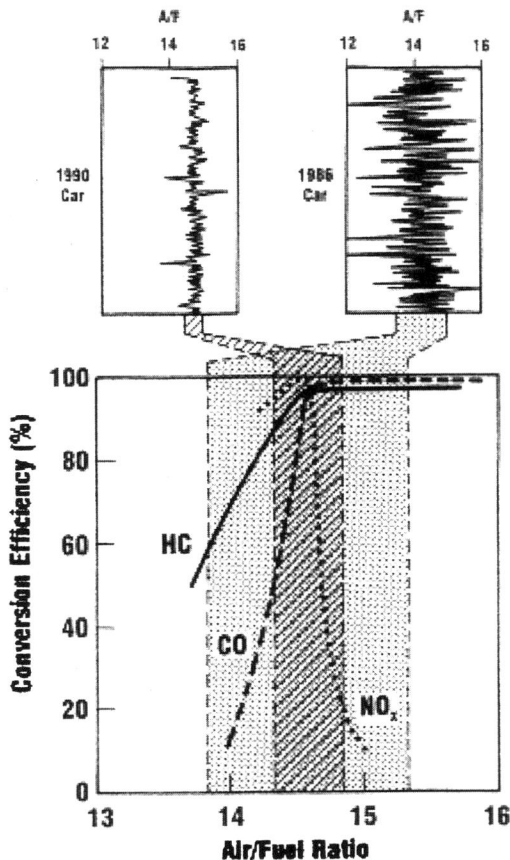

Figure 2. Comparison of engine Air/Fuel trace between a 1986 car and a 1990 vehicle[2]. (Reproduced with permission from reference 2. Copyright 2000 Elsevier.)

charge to powder, with disastrous consequences, including plugging of the exhaust system and dangerous pressure build-up.

It is thus no exaggeration to state that the auto catalyst owes its existence to the catalyst monolith, developed at Corning and NGK, the use of which was pioneered by the Engelhard Corporation. in the 1960s. This is a single-piece structure, usually ceramic, with longitudinal channels extending through its entire length. Viewed end-on, there are typically 400 or 600 channels per square inch, although both higher and lower values are sometimes used. Cordierite ($2MgO \bullet 5SiO_2 \bullet 2Al_2O_3$) emerged as the material of choice, due mainly to its low thermal expansion, an important attribute given the extreme temperature stresses to which it is subjected.

The catalyst is deposited on the walls of monolith channels as a slurry, which is then dried and calcined. Preparation of the slurry is an exacting procedure, with pH, particle size and solids content rigidly controlled. The optimal values of these parameters must be determined for each formulation. Without a high degree of adhesion to the channel wall, the catalyst can easily be turned to powder and blown out of the system, sometimes on vehicles in production. Such crises are not infrequent in the industry and must be attended to immediately. Sometimes a simple adjustment, such as changing the particle size, is effective. Several binding agents are commercially available and are often used.

TWC research inherited the ceramic monolith, together with well-developed procedures for applying the catalyst slurry and dealing with adhesion problems, should they occur. Along with this came an elaborate testing protocol. Thus, at the dawn of the TWC era, the "backbone" of the catalyst, plus methods of applying the active components to it, were in place.

Precious Metals

Before the TWC concept was even discussed, it had become clear that precious metals (the industry term for Pt, Pd, Rh) were indispensable ingredients in the auto catalyst. Despite their high cost and occasional problems of availability, the margin of superiority over base metals was too great.

Pt, Pd and Rh are usually referred to as "noble" metals, and their nobility is very relevant to their use as auto catalysts. As a simple illustration, we can compare the relative affinities of Pt and Ni for O_2. Both adsorb O_2, but the Pt-O interaction is much weaker. As a result, adsorbed oxygen can be removed from a Pt surface by reaction with H_2 at room temperature, whereas in the case of Ni, a temperature of 200°C is required. But the Federal Test Procedure (FTP), shown in Figure 3, starts with the vehicle under ambient conditions, and this "cold start" places a premium on activity at low temperature. A key metric used

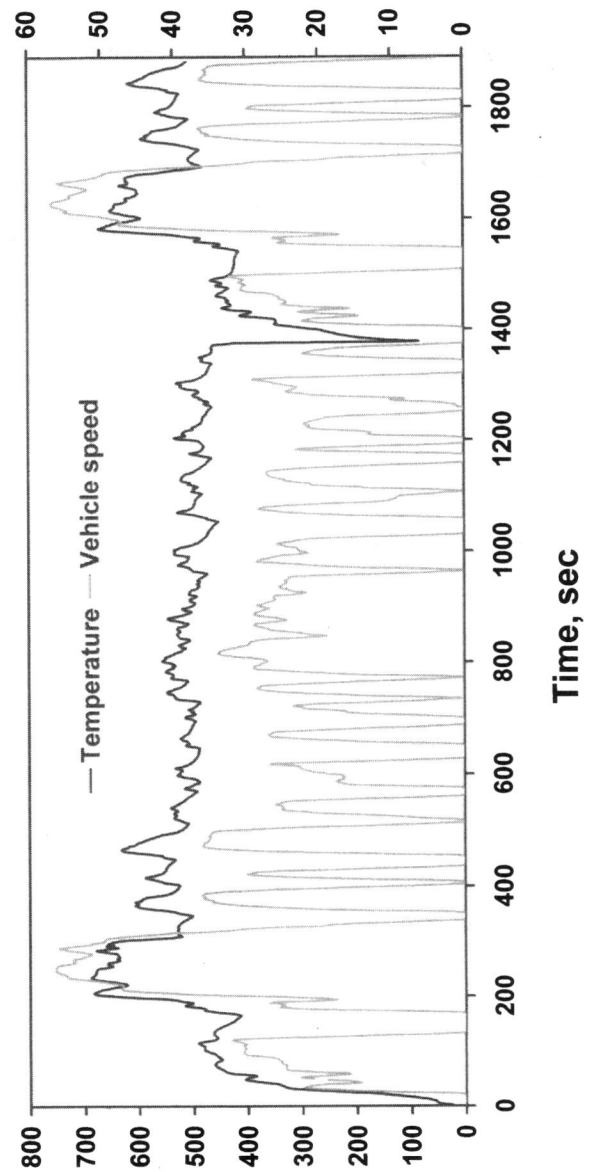

Figure 3. Vehicle speed and engine-out exhaust temperature trace in the Federal Test Protocol (FTP)

with auto catalysts is the "light-off" temperature. The temperature is gradually increased and the point at which 50% conversion occurs is recorded. The relative ease with which the Pt surface can be freed of adatoms and made available for catalysis is a crucial advantage. Similarly, nobility improves a metal's ability to tolerate sulfur, which is ubiquitous in auto exhaust. The formation of PtS is less favored, thermodynamically, than that of NiS. As a result, sulfur poisoning, which is a serious issue even with precious metals, is a significantly greater problem with base metals.

A final advantage of nobility involves interaction with the support, such as alumina. At high temperature, nickel will readily form nickel aluminate, whereas such formation does not occur with Pt or Pd. Rh is an intermediate case. Damaging interaction between Rh (as Rh_2O_3) and alumina occurs at high temperature under oxidizing conditions. This creates a severe impediment to the return to the metallic state (which is required for Rh's catalytic function) when stoichiometric/rich conditions are reestablished. Fortunately, methods have been devised (which will be mentioned later) to mitigate this harmful interaction. Thus, the nobility of the precious metals means that they will return to their active form far more readily than base metals. Since the FTP test places a premium on quick response and low temperature activity, nobility is indispensable for good performance.

Although precious metals have been shown to be necessary for acceptable TWC performance, their high cost creates a clear driving force to reduce or eliminate their use. This has been the target of much research, as yet unsuccessful. These economic incentives are magnified by the vast size of the catalytic converter industry. Rh is a striking example. Soon after its initial use in TWC formulations, this application accounted for the majority of Rh usage worldwide. Due to its relative scarcity, large price swings occurred: from $1,000/t-oz to $7,000 during the 1990s. Still, the industry was forced to accept this state of affairs.

TWC Catalysis

The Role of Oxygen Storage

We have discussed the fact that the TWC catalyst can only succeed if the gas composition stays close to the stoichiometric point. Control of the A/F ratio involves feedback. If the ratio is too high, as measured by an exhaust gas oxygen sensor, the fuel injectors are signaled to increase the fueling rate; if too low, to decrease it. The closer the set points are to stoichiometry, the better the control. But even with a time-averaged A/F ratio that is exactly stoichiometric, the A/F ratio will oscillate. If the perturbation frequency is 1 sec^{-1} (which is typical for actual vehicles) the engine-out gas will be slightly lean for ½ sec and slightly rich for the other ½ sec. From the standpoint of CO oxidation, e.g., there will always be a lean half-cycle during which surplus O_2 exits the system and a rich half-cycle when some CO goes unconverted for lack of this lost O_2.

The basic concept of oxygen storage is to choose an oxide with two (or more) stable oxidation states, and to use the interconversion between these valencies to buffer A/F changes in the gas. If, e.g., we have two oxidation states, AO and AO_2, during the lean half-cycle we can write

$$AO + \tfrac{1}{2}\,O_2 \rightarrow AO_2$$

If AO were not present, the superfluous O_2 would be lost, but now it is "stored" in the AO_2 molecule.

For oxygen storage to be effective, the stored oxygen must react with CO (or HC) during the rich half-cycle:

$$AO_2 + CO \rightarrow CO_2 + AO$$

The net result of these two equations is to capture superfluous O_2 and use it during the rich half-cycle to provide an additional increment of CO conversion. It is vital, however, that both steps proceed quickly enough to keep pace with the perturbation.

The Role of Ceria

Since there are many transition metal and rare earth oxides with two or more stable valencies, there would seem to be many candidates available for buffering the A/F oscillations in the gas phase. Yet only one, cerium oxide, has achieved widespread use. Were it not for ceria, the TWC strategy could never be implemented. The key, as noted above, is the rate of interconversion between oxidation states. Perturbation frequencies on cars are in the vicinity of 1 Hz. The OSC (oxygen storage capacity) component must keep up with this if it is to contribute to conversion. This means that both oxidation and reduction must be rapid.

The problem is not symmetrical. Oxidation of the reduced state is generally facile and nonactivated. Adsorption of O_2 is exothermic and, on most reduced oxides, occurs readily at low temperature. Reduction of the oxidized state, on the other hand, is temperature dependent and much more difficult. It is here that cerium oxide and, perhaps, the oxides of terbium and praseodymium, excel. Cerium oxide, however, has advantages in terms of cost and availability. It was first introduced as an oxygen storage component in 1981 and has been an essential part of the TWC catalyst ever since.

Ceria has a dramatic effect on the conversion of CO, and also of NOx. This is because NOx participates in the oxygen storage and release cycle, oxidizing Ce^{3+} and, in the process, being reduced to N_2. Figure 4 compares two Rh catalysts, one with ceria and one without, on an engine-laboratory test featuring repetitive changes in the gas composition. The catalyst with ceria is far better able to buffer these changes. Interconversion of Ce^{4+} and Ce^{3+} have a major effect on the catalyst's ability to remove CO and NOx.

The superior properties of cerium oxide can, in fact, be rationalized. CeO_2 has the fluorite structure, in which each Ce ion is surrounded by eight equivalent nearest-neighbor oxygen anions, each of which is tetrahedrally coordinated by four Ce cations. Importantly, this lattice can tolerate a high degree of nonstoichiometry. Thus, if the surface of CeO_2 is reduced, no major structural reorganization is needed. This is the key to the ability of ceria to respond quickly to A/F changes.

There is, in fact, a near-continuum of nonstoichiometric phases. As x in the formula CeO_x varies from 2 to 1.7, the lattice expands in a nearly linear manner due to the larger radius of Ce^{3+} compared with Ce^{4+}. Both metal and oxygen atoms remain face-centered, cubic-close-packed.

Further Studies of Oxygen Storage

Given the central role of oxygen storage in auto exhaust control, it is not surprising that considerable research has been directed at establishing its mechanism. Oxygen storage can be conveniently studied in the laboratory, provided that switching from oxidizing to reducing gas and back occurs at roughly the same frequency as used in vehicles. In a typical study, pulses of 2% CO and 1% O_2 are alternated at a frequency of 1 or 0.5 Hz. Gas composition is followed with mass spectrometry.

Using this arrangement, a variety of catalysts, some with ceria, some without, were investigated. There are two sources of CO_2 production. One is direct CO oxidation, i.e., resulting from the simultaneous presence of CO and O_2 in portions of the catalyst bed. This is unavoidable, despite the discrete pulses of O_2 and CO, because of gas phase mixing at these high switching frequencies. On changing from CO to O_2, this direct, non-oxygen-storage-related mechanism is responsible for whatever CO_2 formation is measured. At the same time, uptake of O_2 on the catalyst is observed.

On switching from O_2 to CO, the direct mechanism is again operative, but now there is an additional source of CO_2: reaction of CO from the gas phase with oxygen "stored" on the catalyst. This simple arrangement allows a wealth of information to be collected, including differences among catalysts and the effects of temperature and other experimental conditions.

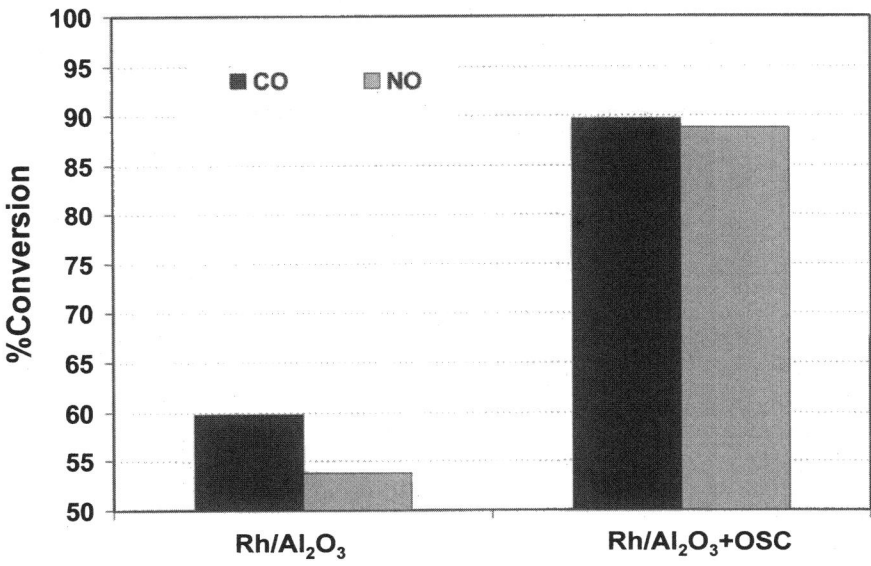

Figure 4. Effect of oxygen storage materials on CO & NO conversion

A significant result of these studies is that a synergy exists between the supported metal and ceria. This increases the rate of storage-related CO_2 production by a factor of roughly three to five.[4] With no metal, CO_2 results from a reaction on the ceria surface between CO and a stored O atom, perhaps via an Eley-Rideal mechanism, i.e., gas phase CO with adsorbed O. With metal present, the presumed mechanism involves CO adsorption on the metal surface and O_2 adsorption on ceria. An adsorbed O atom in the immediate vicinity of the metal island "spills over" onto the metal surface, where it reacts with an adsorbed CO molecule. This route apparently is favored over the ceria-only Eley-Rideal route.

Other studies confirm the oxygen-spillover mechanism. If CO is adsorbed on a metal/ceria sample at room temperature, subsequent heating leads to the formation of CO_2.[5] The added O atom is provided by the ceria. Well-ordered ceria phases, e.g., single crystals of (111) CeO_2 or (100) CeO_2, are much less active in this process than polycrystalline CeO_2. All in all, these several studies point to the metal-ceria junction as vital to effective oxygen storage. When an O atom spills over onto the metal surface, charge balance is maintained by reduction of two Ce^{4+} cations to Ce^{3+}, i.e.,

$$2 \, Ce^{4+} + O^{2-} \rightarrow O \, (\text{spillover}) + 2 \, Ce^{3+}$$

The spillover of O creates an oxygen vacancy which, as noted above, is easily accommodated in the fluorite structure. The Ce^{3+} ions that are formed will interact strongly with the metal. Several studies have demonstrated strong metal-support interaction behavior in metal/reduced ceria systems.[6] This interaction presumably aids the spillover process.

Ceria-Zirconia

The incorporation of ceria marked a breakthrough in auto catalysis. Its facile oxidation-reduction, augmented by the metal-ceria synergy just described, made three-way conversion a reality. It soon was incorporated in virtually all catalytic converters for gasoline-powered cars. For a time, research centered on optimizing the amount of ceria. Higher loadings improved oxygen storage, but were linked, under certain driving conditions, to the emission of H_2S. Although the latter was not regulated by the EPA, customer complaints made it an urgent matter. The use of transition metals, particularly Ni, which adsorbs H_2S and then releases it as SO_2 under lean conditions, proved to be a generally acceptable solution.

It soon became clear, however, that ceria had a fundamental limitation. Although it could be prepared with a reasonably high surface area, e.g., from the carbonate or nitrate, it could not withstand the severe conditions of auto exhaust. A surface area of 100-120 m^2/g collapsed to only a few percent of this value after exposure for prolonged periods to temperatures of 900°C. Furthermore, kinetic studies[7] on model systems showed that the ability of ceria to provide "spillover" oxygen to a metal particle (critical to OSC behavior) decayed even faster than the surface area. Since aging cycles (to which catalysts are routinely subjected before testing) often include portions at 900°C or higher, this was a serious drawback. Aging cycle temperatures were steadily being increased. This was due, in large part, to the EPA regulation that a converter maintain efficiency for a certain number of miles of vehicle operation. This was gradually increased from 50,000 to 120,000 miles for the federal Tier 2 Bin 4 standard and to 150,000 miles for the California PZEV standard (Partial Zero Emission Vehicle). It was obviously impractical to test a converter for 100,000 miles under normal driving conditions. Instead, a correlation was assumed between the required durability and the aging severity. As converters were required to last for longer and longer times, they were aged at higher and higher temperatures. But this drastically impacted the surface area, and the oxygen-storage efficiency, of ceria. Since the latter was central to the TWC concept, the situation was serious. This came at a time when non-catalyst components, such as sensors and fuel-injection systems had been significantly improved. But this in no way eliminated the need for effective oxygen storage by the catalyst.

The needed breakthrough came from mixing ceria with zirconia, forming a solid solution with far greater thermal stability than with ceria itself. Zirconia was no stranger to auto catalyst formulations. It had long been recognized for its ability to substantially reduce the harmful interaction between Rh and alumina that occurred under high temperature, lean conditions. Rh^{3+} cations are readily incorporated into the Al_2O_3 surface or sub-surface region. On returning to stoichiometric conditions, the Rh^{3+}-alumina complex has to be decomposed. This decomposition is not facile; it retards the reduction of rhodium to the metallic state which is required for catalytic activity. If ZrO_2 is present (using, e.g., co-impregnation of Rh and Zr salts followed by drying and calcination), the ZrO_2 spreads on the surface and forms a barrier that blocks the Rh_2O_3-Al_2O_3 interaction. Importantly, ZrO_2 is itself relatively unreactive toward Rh_2O_3, as it is toward ceria.

The passive nature of ZrO_2 makes it an ideal candidate for the structural stabilization, i.e., maintenance of surface area, of ceria. (Here again, ZrO_2 stands in contrast to alumina, which tends to react with ceria to form $CeAlO_3$.) If the ceria-zirconia system consists of ceria domains separated by zirconia domains, the growth of ceria particles will be hindered and its thermal stability will be increased. The structural stabilization of ceria by zirconia can be measured in several ways, including direct measurement of the oxygen storage capacity. OSC data has been collected for a series of co-precipitated ceria-zirconia samples, using a technique of alternating CO and O_2 pulses. The samples were calcined at 900°C before determining the OSC capacity at 400°C. Structural promotion was clearly in evidence. Compared with pure ceria, Ce_x $Zr_{1-x}O_2$ with x values between 0.6 and 0.7 exhibited four times the OSC capacity.[4] Interestingly, all the samples in this range had the fluorite structure. If one combines these results with surface area data, an important conclusion emerges. Whereas with ceria the OSC properties are limited to the surface, with ceria-zirconia at least one sub-surface layer contributes.

Early data with zirconia-stabilized ceria showed a major improvement over ceria by itself. After aging at 1050°C, FTP testing showed sharp decreases in tailpipe concentrations of CO and NOx, together with smaller ceria particle sizes. This argued for effective structural stabilization.

There is much evidence to suggest that zirconia-stabilized ceria is, in fact, a solid solution, rather than a mixture of ceria and zirconia particles. The solid solution leads to structural stabilization, i.e., smaller particle size, after high temperature aging. Surprisingly, it improves performance in another way as well: due to greatly increased solid state diffusivity, oxygen from the interior of the particle participates in the oxygen storage process. Incorporation of ZrO_2 into the CeO_2 distorts the lattice because of the different cationic radii (0.097 nm for Ce^{4+} vs. 0.084 nm for Zr^{4+}). This distortion increases the mobility of oxygen; it also lowers the coordination number of the Zr^{4+} to < 8 (EXAFS), whereas 8 is the expected value for the fluorite lattice. Comparing post-aging temperature programmed reduction (TPR) scans of unpromoted and zirconia-promoted ceria, the latter shows a low temperature peak that can be ascribed to oxygen emanating from the interior of the particle. This participation of deep subsurface oxygen means that even larger particles can contribute to oxygen storage. It should be noted, however, that sintering of ceria-zirconia is still harmful, since it serves as a support for precious metals.

Close-Coupled Palladium

The two major advances cited thus far have involved problems that were solved by the introduction of new materials. First, the fast oxidation-reduction properties of ceria (combined with sophisticated electronic A/F control) made three-way conversion a viable concept. Second, the replacement of ceria with ceria-zirconia vastly improved the ability of the oxygen storage function to survive exposure to high temperature.

In the case of close-coupled Pd, however, the source of progress was a new strategy that allowed existing materials to be used much more effectively. The objective was to reduce the amount of pollutants, especially HCs, emitted in the very early portions of the FTP test. As mentioned above, one of the most difficult challenges the auto catalyst faces is the need to become active almost immediately after the engine is started at room temperature. This is the so-called "cold start." Even with precious metals, significant conversion requires temperatures of about 300°C. But the FTP test makes no allowance for this. Tailpipe emissions are counted immediately after the engine is switched on. Although the FTP test lasts 1877 seconds, HC emissions during the first 60 sec can easily account for the majority of the total.

A strategy to cope with cold-start involves the location of the catalyst in the exhaust system. If the converter is "close-coupled", i.e., mounted at the engine manifold, the catalyst will heat up faster. Compared with the downstream "underfloor" position, a close-coupled converter can reach 300°C as much as 75 seconds earlier. This can easily cut HC emissions by 50%. But this accelerated warm-up comes at a steep price. The manifold position will experience significantly higher temperatures during the aging cycle to which the catalyst is subjected before being tested. Until recently, this trade-off was unfavorable and close-couple mounting was avoided. Starting in the late 1990s, however, new emission standards made it imperative to reduce the time required for light-off. The Ultra-Low Emission Vehicle (ULEV) target for HCs necessitated light-off within 50 sec. This could only be accomplished with manifold mounting and close-coupled catalysts became a focal point of research.

It should be mentioned that some measures promote light-off without increasing the heat burden on the catalyst during aging. These include retarding the spark during combustion in the engine and tuning the A/F ratio slightly lean during cold-start. The catalyst monolith has been modified as well: using higher cell density, thinner wall substrates, heat-up can be accelerated. Rather than the older 400 cells (channels) per square inch, cell densities of 600 and 900 per square inch are now common. This has been shown, in some situations, to reduce HC emissions by 25%. Thus, in confronting the challenge of close-coupled siting, catalysis research benefited from valuable outside assistance. But more was needed. It came not from a new material, but from a new strategy that reexamined the roles played by different parts of the catalyst system.

Although we have thus far spoken of the converter as a single entity, in most applications there are two modules ("bricks") in serial alignment. Most of the precious metal is usually loaded onto the front brick, where most of the conversions take place; the rear converter plays a back-up role, preventing breakthrough during high-space-velocity portions of the FTP test. To a large extent, the precise distribution of metals, or ceria, between the two bricks is not critical; whether, for example, 90% of the CO is combusted over the front brick and 10% over the rear brick, or, instead, the split is 80%-20%, is not important.

When the front brick is close-coupled, however, the situation changes. Now, not only must CO be converted; where this occurs becomes an issue. This is because CO oxidation is highly exothermic. As a rough approximation, converting an additional 1% of CO raises the catalyst temperature by 90°C. Since the close-coupled converter must contend with a large amount of heat from the engine, making it the site of CO conversion adds unnecessarily to this burden.

The simple expedient of removing ceria-zirconia from the close-coupled brick, and thereby shifting a large part of the CO conversion to the rear converter, proved to be a winning strategy. In one published study, two Pd catalysts, one with ceria and one without, were aged on an engine dynamometer for 100 hrs at 900°C. The Ce-free brick showed a highly significant 40°C light-off advantage during stoichiometric cold-start conditions; with a lean cold-start, the advantages are even larger: 70°C.[8] Several catalyst manufacturers introduced high Pd, non- (or low)-Ce front-brick systems, which gained commercial acceptance.[9] As a result, expensive options, such as electrically heated monoliths, which had been considered, were no longer required.

On-Board Diagnostics (OBD)

A final point to be noted in connection with the TWC catalysts involves on-board diagnostics. The EPA has stressed the need for a vehicle owner to be made aware if the auto catalyst has lost activity and needs replacement. Current efforts focus on the OSC function. A fresh catalyst will completely dampen the A/F perturbation in the gas coming from the engine, while a severely deactivated catalyst will lose this ability. By comparing pre-catalyst and post-catalyst A/F traces, a measure of deactivation can be derived which will trigger a warning light. Interestingly, the great progress that has been made in the development of thermally stable OSC component has complicated the OBD requirement. Oxygen storage materials that undergo a sharp drop in OSC activity in a narrow temperature range above 1100°C are highly desirable for enhanced on-board diagnostics.

Catalytic Purification of Diesel-Engine Exhaust: An Overview

Diesel constitutes a major segment of the world automotive market. In addition to large vehicles, such as trucks and buses, diesel passenger cars are gaining in popularity. At the present time, about half of all cars sold in Europe use diesel fuel. The driving force is economy: the fuel is often less expensive than gasoline and diesel motors far outlast gasoline motors.

The two types of engines are fundamentally different. With gasoline, the fuel-air mixture is a homogenous premixed gas which is ignited with a spark, while in the diesel engine, liquid fuel is injected into compressed air so that ignition is caused by compression. Their exhausts, too, differ significantly. Diesel exhaust is always lean, with A/F ratios typically greater than 20. This means that factors that are fundamental to TWC catalysis, such as perturbation, rich-lean transitions and oxygen storage, do not occur. Lean exhaust means that NOx conversion does not occur either. Fortunately, however, the much cooler operating temperatures and the lean combustion environment of diesel engines mean that engine-out concentrations of NOx (as well as of HC and CO) are much lower than found with gasoline engines.

This does not mean, however, that diesel exhaust requires no aftertreatment. A negative feature of diesel motors is that the flame front is cooled before all the fuel is oxidized. This leads to the generation of "particulate matter" (PM), a mixture of unburnt carbon, organic species, sulfates and oil additives. Small particles of PM (<2.5μm) have been shown to deposit in the lungs and to have significant adverse health consequences. PM reduction is, as a result, by far the most important goal of diesel after-treatment.

Particulate matter consists basically of three fractions, the relative amounts of which depend strongly on fuel quality and engine conditions. The first is dry soot, which is a low-hydrogen-content, polyaromatic species that, regardless of the catalyst, cannot be oxidized under normal diesel exhaust conditions. The second component is the soluble organic fraction (SOF), derived from unburnt fuel and lube oil, and consisting mostly of HCs in the C_{20}-C_{28} range. The SOF can be present as discrete droplets or as an adsorbed layer on a dry soot particle. In contrast to dry soot, the SOF can be oxidized and an effective diesel oxidation catalyst (DOC) must do this. But a DOC can be highly active for SOF conversion and still fail to qualify. This is because of the third PM fraction, which consists of droplets of sulfuric acid. These arise when SO_2 (derived from sulfur present in the diesel fuel) exiting the engine is oxidized to SO_3, which is then easily hydrated to H_2SO_4. While it is not clear whether H_2SO_4 is as harmful to human health as the organic constituents of diesel exhaust, the testing protocol guarantees that high H_2SO_4 will lead to poor results. This is because the test for PM is simply based on weight - everything collected from diluted diesel exhaust on an in-line filter at a temperature not in excess of 52°C. If H_2SO_4 droplets are present, they will adsorb on the dry soot or, perhaps, directly onto the filter, in either case increasing the recorded amount of PM. Thus, for much of the history of DOC development, the goal was a catalyst that combined a relatively high activity for the oxidation of SOF with a relatively low activity for the oxidation of SO_2 to SO_3.

Diesel catalysis is often divided into two branches: light-duty diesel (LDD), which pertains to passenger cars, and heavy-duty diesel (HDD), which refers to trucks, buses and other large vehicles. This distinction is useful, since for much of the recent past, they faced different challenges. These challenges, of course, come from the regulatory agencies, which, in the US, is the Environmental Protection Agency. Although car companies will on occasion pursue a "green" image, the car-buying public is, in general, more interested in styling, cost, performance and other features. The driving force for reducing emissions, whether from gasoline or diesel vehicles, has always been the increasingly difficult standards handed down by the agencies. Sometimes these changes are small, necessitating, perhaps, an increased amount of precious metal on the catalyst. On other occasions, regulatory revisions have led to major changes in catalyst technology. An example of the latter, which we have already discussed, was the establishment of the 1.0 g/mi limit on NOx emissions from gasoline vehicles. This led to a previously unknown concept in catalysis: three-way conversion. Diesel catalysis, too, has on occasion responded to changes in regulations with a break-through in technology. The best example of this is probably the need to remove dry soot; this will be described in a later section.

While the agencies, by toughening emissions standards, have often made the job more difficult, they have also significantly eased the burden on automotive catalysis R&D in other ways. First and foremost, the agencies have been the driving force behind the effort to reduce the fuel content of sulfur. This is true for both gasoline and diesel fuel, but for the latter, the changes have been larger and the impact more profound. In the sections that follow, the technologies must be understood in the context of the sulfur levels that existed at the time.

LDD Catalysis With High-Sulfur Fuel

Prior to 1994, diesel fuel contained more than 2000 ppm sulfur (compared with 300 ppm for gasoline). This greatly complicated the effort to develop a catalyst for the conversion of SOF. As a general rule, oxidation of SOF was accompanied by oxidation of SO_2 to SO_3, leading to H_2SO_4. Thus, good SOF conversion did not guarantee reduction of particulate matter.

It could be shown that the relative rates of oxidation of SOF and SO_2 could be altered, within limits, by simple measures. By sintering a Pt/alumina catalyst at 1050°C (increasing the metal crystallite size) the SO_2 oxidation rate decreased more than that of SOF, so that the effect on PM was favorable.[10] Similarly, replacing Pt with Pd increased selectivity.[10] These changes were not sufficient, but they spurred efforts to develop a selective SOF conversion catalyst. Two successful developments will be described in the following sections.

Vanadia-Promotion of Pt for Selective SOF Oxidation

A major effort was launched in the late 1980s to find a base metal additive capable of raising the selectivity of Pt for the conversion of SOF. Several oxides were able to lower SO_2 oxidation activity, as measured by the temperature required to achieve 50% conversion. Vanadia and chromia (added by impregnating precursor salts onto Pt/alumina) raised T_{50} by 218 and 165°C respectively. Many other oxides, including those of Zr, Mn and Ti, raised T_{50}, but by a much smaller amount (75°C or less). In addition to suppressing SO_2 oxidation, the promoted catalysts had to show retained activity for SOF conversion. Laboratory studies used CO, propene and decane as surrogates for SOF. Here, results were mixed, with vanadia performing well with CO and propene but poorly with decane.

Engine dynamometer tests, of course, are definitive, and here vanadia-promoted Pt stood out. Compared with unpromoted Pt, the effect of vanadia on SO_2 oxidation outweighed any effect on SOF. This was true for fresh catalysts as well as after 200 hours of aging. Other studies gave similar results[11,12] and vanadia-promoted Pt was widely used on European passenger cars during the 1990s.

A possible mechanism for the effect of vanadia is suggested by the observation that vanadium oxide severely reduces the propensity of a Pt/alumina catalyst to store sulfate. It seems reasonable to speculate that SO_2 oxidation on Pt/alumina involves adsorption of SO_2 onto the alumina surface, with migration ("spillover") of a sulfite species onto neighboring Pt crystallites and/or of oxygen atoms from the Pt onto the support. Coverage of alumina with high-valent vanadia can, because of its acidic nature, reduce affinity for SO_2 and thereby disrupt the mechanism.

Cerium Oxide for Selective SOF Oxidation

While most LDD research groups during the 1980s focused on improving the selectivity of Pt, another strategy was to abandon the use of Pt altogether. Since many transition metal and rare earth oxides are potent oxidation catalysts, it was not inconceivable that one would display better selectivity for SOF oxidation than any Pt-based system. After a lengthy screening process, a promising candidate, cerium oxide, emerged.[13,14,15] A preferred embodiment comprised bulk cerium oxide (stabilized with a small amount of alumina) mixed with a roughly equal amount of a second oxide, which could be alumina. Pt was either entirely absent, or present (to boost gaseous HC/CO conversion) in quantities much smaller than used in previous catalysts that relied on Pt for SOF conversion.

A simple laboratory test was devised in which a catalyst sample was mixed with lube oil, placed in a TGA/DTA instrument and heated in flowing air. The DTA signal was used to measure the total amount of lube oil oxidation as well as the temperature at which conversion began. These data clearly showed that ceria was superior to any other oxide tested. In other experiments, ceria was compared with Pt/alumina for SO_2 oxidation and found to be far less active. The combination of high activity for lube oil oxidation and suppressed activity for SO_2 oxidation made ceria an ideal candidate for selective SOF conversion. Engine dynamometer tests confirmed the laboratory results. Ceria went on to enjoy major commercial success during the 1990s.

To explain the high activity of cerium dioxide for SOF oxidation, recall that SOF consists mainly of high molecular weight paraffins. There is consensus that the important step in alkane oxidation is cleavage of a C-H bond. On oxide catalysts, most authors favor a heterolytic mechanism, with, for example, methane splitting to yield a proton and a methyl anion. An important observation in support of this comes from a comparison of methane and fluoromethane.[16] The fluorine atom would be expected to stabilize a transition state of the form $FH_2C^{-1}...H^{+1}$, and one finds that fluoromethane is an order of magnitude more active than methane in coupling reactions. This argues in favor of heterolytic C-H scission.

Oxides with high-valent cations, such as CeO_2, should favor heterolytic C-H cleavage. Basic oxides such as La_2O_3, will, however, be susceptible to poisoning, since CO_2, the product of paraffin oxidation, is weakly acidic. Of course, the more-strongly-acidic SO_2 will poison any basic oxide used for SOF oxidation. These observations suggest that cerium dioxide is an effective catalyst for SOF oxidation because its highly charged acidic surface promotes heterolytic scission while avoiding susceptibility to poisoning by acidic species.

Conversion of Gas Phase CO and HC: the Zeolitic HC Trap

In addition to their primary function of reducing particulate matter, diesel catalysts must convert gas phase CO and HC. Emission limits have come down sharply: for U.S. trucks, for example, 2009 standards allow only about 1/10 as much as HC as was permitted in 1994. For CO, 2003/2004 limits for Japanese heavy-duty trucks are about 1/3 of their values in 2000. The much leaner diesel exhaust, compared with gasoline exhaust, helps reduce CO and HC emissions, but diesel exhaust is generally cooler, which suppresses conversion. Low temperature activity is, accordingly, are research priority. Pt, or Pt/Pd, supported on alumina, is standard, but even small changes in thermal stability can confer a competitive advantage.

The introduction of zeolites into diesel catalyst formulations, starting in the late 1990s, play an important role in decreasing emissions of gas phase HCs. Zeolites are among the most important materials in industrial catalysis. They dominate the markets for catalytic cracking, hydrocracking and xylene isomerization as well as several smaller scale applications. Their uses stem from their strong acidity and their unique pore structure, which leads to size- and shape-selective effects.

Zeolites have been found to be effective additives to diesel catalysts, based not on catalytic activity, but on their properties as adsorbents. The extremely small pore diameters of zeolites (ca. 7Å even for "large-pore" zeolites) means that molecules adsorbed in their intracrystalline volume will be packed much more tightly than in non-zeolitic solids. This increases intermolecular binding, mainly through Van der Waals forces. As a result, hydrocarbons will condense in zeolites far more readily (i.e., at much higher temperatures) than in other solids. This property has led to the concept of a zeolitic "hydrocarbon trap." According to this idea, hydrocarbons will be adsorbed by the zeolite at low temperature, when the catalyst is not yet active. When the catalyst lights off, these "trapped" HCs will desorb and be converted.[17] This is potentially of great value, since high HC conversion no longer is predicated on low-temperature light-off. The underlying assumption, of course, is that desorption of HCs from the trap will not precede light-off. Otherwise, the released HCs will exit the catalyst without being converted and nothing will have been accomplished. This assumption is fulfilled with zeolites due to their unique pore structure.

The HC trap concept applies to gasoline cars as well as to diesel cars. But there is an important difference. Zeolites suffer from low thermal stability. At high temperature, their crystallinity breaks down and they become amorphous. Since their unique properties, either as catalysts or adsorbents, are directly linked to their pore structure (and, thus, to the crystallinity) zeolites are essentially denatured by exposure to high temperature. As already noted, diesel exhaust is significantly cooler than gasoline exhaust. Correspondingly, their aging temperatures are different. TWCs are often aged at 1000°C or more, but diesel catalysts must withstand much lower temperatures, typically 650-800°C. This allows the zeolite component in a diesel catalyst to be functional even after aging. As described in US 6,248,684,[18] the addition of zeolite to a diesel oxidation catalyst significantly increased the conversion of gas phase HCs without increasing SO_2 oxidation.

As sulfur levels in diesel fuel came down, to 500 ppm in 1994 and 300 ppm in 2000, H_2SO_4 formation receded in importance and promoted Pt was no longer necessary. Simple Pt, or Pt/Pd, was the catalyst of choice, based on its high activity for SOF oxidation. Zeolites continued to be used, since HC trapping had a role to play regardless of the sulfur content of the fuel. Currently, research is aimed at tailoring the HC trap to the specific vehicle cold start and to the light-off activity of the catalyst.

HDD Catalysis – Diesel Particulate Filter

Prior to 2000, heavy duty vehicles coped with the same problems as passenger cars: the need for high SOF conversion combined with as little H_2SO_4 production as possible. Ceria and vanadia-promoted Pt were used until the reduced sulfur content of the fuel rendered them no longer necessary. This, of course, mirrors the situation just described for LDD vehicles.

A drastic change occurred in 2000. An increasing amount of evidence indicated that dry soot was a greater health hazard than previously recognized. This was true not only for large particles (which give rise to the 'black smoke' discharges from trucks and bases) but also for particles smaller than 2.5 microns. Indeed, there was evidence that the smaller particles were actually more harmful, due to their greater ability to penetrate the lungs. There was an urgent need to reduce the emissions of dry soot from diesel vehicles. Since trucks and buses release far more than passenger cars, they were the targets of the first anti-dry-soot legislation. But it was not clear what could be done. It has already been pointed out that dry soot, unlike SOF, cannot be oxidized in a conventional monolith catalyst. The large size of the particles, combined with their chemically refractory nature, severely limit their conversion during the brief residence time afforded by a flow-through converter. Yet, from a public health standpoint, it made no sense to eliminate some components of diesel exhaust while doing nothing about what might be its most harmful constituent.

The solution came from an ingenious modification of the monolith converter that turned it into a filter capable of trapping soot particles. Starting with a conventional flow-through unit, the diesel particulate filter (DPF) blocks each channel either at its inlet or outlet end, in alternating fashion, as shown in Figure 5. This means that the exhaust must first pass through a ceramic wall before continuing down an adjacent channel and exiting the converter.

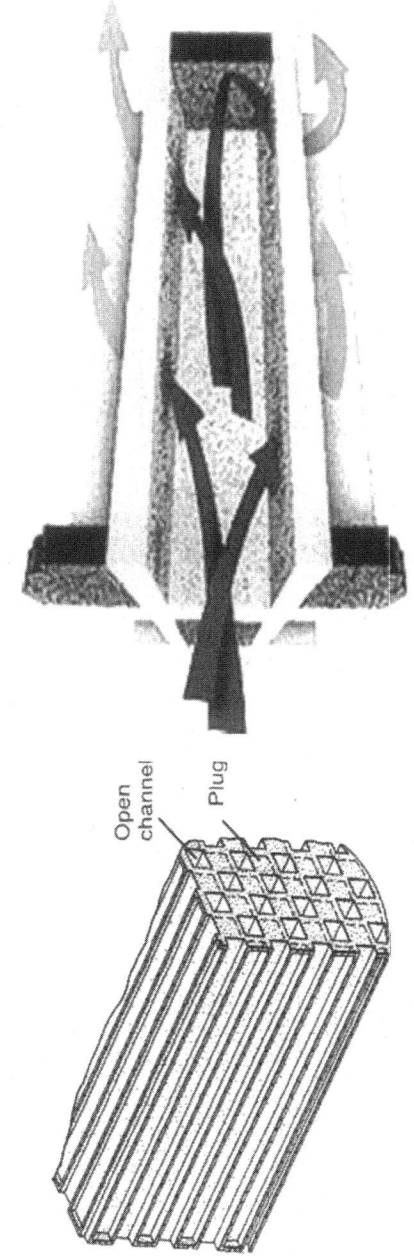

Open channel

Plug

Figure 5. *Structure and function of a diesel particulate filter (See page 1 of color insert.)*

An earlier version of the DPF has in fact been introduced in the 1980s and has been applied to off-road vehicles and low-speed urban buses. But extending its use to the entire heavy-duty fleet, including large trucks with huge exhaust volumes, was a different matter all together. A major re-design effort was launched, focusing on the key parameters of cell density, wall thickness, porosity and mean pore size. After much effort, a product emerged that was able to filter out soot particles while permitting the passage of the remainder of the exhaust, even though for large trucks. The fact that DPFs are now used on millions of vehicles worldwide should not obscure the boldness of this concept. Precise control of the channel wall porosity is key, and errors on either side will have severe consequences. If the pores are too large, the soot particles will not be captured and the device will fail. But if the porosity is too restricting, the result will be far worse. When a large truck accelerates, the engine-out flow can reach 10,000-20,000 liters per minute and this gas must make its way through a stone (i.e., ceramic) wall or a catastrophic pressure build-up will occur. The fact that the DPF has been developed to the point that it can avoid these pitfalls with essentially complete reliability is an impressive feat indeed.

The filtering action of the ceramic wall is complex. Depending on the pore size, soot particles will either collect primarily at the surface or deep within the wall. In the latter case, mechanisms such as diffusional deposition, inertial deposition and flow-line interception have been identified.[19] The filtered-out soot will have a much longer effective residence time in the converter, compared with the flow-through situation. This helps to overcome the intrinsically low reactivity of these particles. Currently cordierite, silicon carbide and aluminum titanate are being considered for DPFs. Modern DPFs are more than 95% effective in separating soot particles from gas phase diesel exhaust. Some soot will be continuously oxidized during normal engine operation. To facilitate this, the DPF can be coated with a catalytic layer, creating the catalyzed soot filter (CSF). Continuous conversion of filtered-out soot is termed 'passive regeneration' of the DPF. An important contribution to this is catalytic oxidation of NO to NO_2, where the latter is highly active for soot conversion:

$$2NO_2 + C = CO_2 + 2NO$$

The temperature of the exhaust gas, which varies widely among different vehicles, is a key determinant of the extent to which passive regeneration occurs. 'Active regeneration' is, therefore, often required. This involves periodically raising the temperature (e.g., by adding diesel fuel to the exhaust gas) to combust the accumulated soot. DPF regeneration is the focus of a great deal of research at the present time, with the major issues being dependability, pressure drop and the avoidance of excessive heat during the burn-off.

DPFs are now used on vast majorities of HDD vehicles. Today, the conventional system uses an upstream flow-through DOC and a downstream CSF. Conversion of gas phase HC/CO and SOF takes place over both monoliths, although primarily on the front catalyst. An additional task of the upstream converter is to assist regeneration of the CSF by combusting injected fuel to raise the temperature. The metal of choice for the CSF is Pt/Pd alloy, which offers greater thermal stability than Pt alone. This is important, since, during active regeneration, the measured temperature can reach 650°C with local "hot spots" created by the exothermic burn-off of soot.

Nor is the use of CSFs limited to trucks and buses. Starting in 2009, legislation calls for the control of soot emissions from passenger cars in Europe. As with HDD vehicles, cars will have upstream flow-through convertors followed by CSFs. Some car companies have, in fact, made this change in advance of the coming regulations.

Reduction of NOx in HDD Exhaust

As already pointed out, catalytic reduction of NOx in a highly lean environment, such as diesel exhaust, is extremely limited. The problem, of course, is that the reductant (CO or HC) is consumed by the relatively large amount of O_2, leaving virtually none for the reduction of NOx. The best selectivity has been reported for Cu/ZSM-5, but only above 350°C. Further limitations are susceptibility to poisons and poor hydrothermal stability.

The selectivity is dramatically improved, however, if the reductant is ammonia, rather than hydrocarbon. Selective Catalytic Reduction (SCR), using ammonia or a precursor thereof, is a well-established method of NOx abatement. It has been used for thermal power plants in Japan since the 1970s and in Europe since the 1980s. More recently, it has been applied to automotive diesel exhaust to reduce NOx emissions to levels that are presently contemplated.

The chemistry of SCR is straightforward. NOx reacts with NH_3, in the presence of O_2, to produce N_2 and H_2O. NH_3 can also react with O_2 to produce NO, which is an undesired side reaction. Thus, a good catalyst must be selective as well as active.

V_2O_5 supported on titania has emerged as one of the catalysts of choice while metal exchanged zeolites are also of high interest. Although Pt/Al_2O_3 is active and selective at 175-250°C, vanadia/titania is effective over a much broader range, roughly 200-450°C. The titania support has the further advantage of sulfur-resistance.

The application of SCR to diesel cars and trucks presents some obvious logistical issues. On-board storage of ammonia, either anhydrous or in aqueous solution, poses safety hazards. Aqueous solutions of urea are used instead. The solution is sprayed through an atomizing nozzle into the exhaust gas stream. The urea is easily hydrolyzed, releasing NH_3. Large-scale application of this technology requires a distribution infrastructure, which does not presently exist in the US, although it is currently in place in Europe. Establishment of this infrastructure in the US is planned for the next 2-3 years, in order to meet 2010 regulations. This investment is justified by the fact that SCR for heavy duty vehicles has achieved NOx reductions of 55-90%.

Another approach to the problem of NOx emissions from diesel vehicles is the use of NOx traps. NO_2 is acidic and is adsorbed by basic oxide, such as BaO, to form $Ba(NO_3)_2$. The NO_2 is released and reduced to N_2 during the periodic engine rich excursion. Basic oxides will, of course, bind sulfur oxides as well, and high temperature intervals with rich exhaust gas must be included in the engine control strategy to desulfate the catalyst. The lean-NOx-trapping concept is currently not in wide commercial use while development efforts are continuing.

Summary

Driven by the regulatory authorities, and in close collaboration with the ceramics and electronic control industries, the science of catalysis has fulfilled its promise to purify the vast quantities of exhaust released into the atmosphere by cars, trucks and buses. This effort has spanned several decades and has required large amounts of investment, but is easily justified by the contribution it has made to human health and welfare.

Progress has come from the efforts of many people and has often been slow and incremental, but a few landmark developments stand out. The ceramic monolith, a one-piece honeycomb structure with numerous channels that are coated with catalytic agents, launched the field of automotive catalysis, since it enabled the catalyst to survive the harsh vibrations and to have a low pressure drop in auto exhaust systems. With gasoline-driven cars, the early requirements of controlling CO and hydrocarbon emissions were met through the use of materials well known in other branches of catalysis, particularly the precious metals, such as Pt and Pd. But when it became necessary to promote the oxidation of CO and hydrocarbons, together with the reduction_of nitrogen oxides, new materials and concepts were needed. Three-way conversion (TWC) catalysis was born, based on the premise that if the gas composition stayed extremely close to the oxidation-reduction stoichiometric point, the necessary reactions would occur. This was accomplished through the combination of two factors: a sophisticated control system based on a device that responded to the oxygen content of the exhaust gas, kept the composition close to stoichiometric. But even tighter control was required, and it came from a special material, cerium oxide, that proved indispensable to TWC development. The key attribute of ceria was its ability to tolerate a large degree of non-stoichiometry (expressed by CeO_{2-x}) without significant structural rearrangement. This enabled it to undergo rapid changes in oxidation state and thereby compensate for deviations from stoichiometry in the gas phase. The term 'oxygen storage' pervaded the autocatalysis literature. Oxygen was "stored" by converting Ce^{3+} to Ce^{4+}. It was 'released' (to a molecule of hydrocarbon or CO) through the conversion of Ce^{4+} back to Ce^{3+}. Ceria was quickly established as an indispensable component of TWC catalysis.

But it suffered from a serious limitation. Exposure to high temperature drastically lowered its surface area and, with it, its ability to provide oxygen storage. Since the industry was steadily moving toward higher aging temperatures, the future of TWC catalysis was threatened.

The needed breakthrough came from the addition of zirconia to ceria, creating a system in which solid solution formation readily occurs. The surface area stability was significantly improved over that of pure ceria. Furthermore, the surprising mobility of oxygen ions in the ceria-zirconia lattice meant that the interior of the crystal could participate in the oxygen storage process. Together, these effects ensured that oxygen storage could survive the new, severe aging cycles. This was critical for the continued success of the TWC catalyst.

Diesel exhaust differs significantly from gasoline exhaust. Its constant leanness rules out conversion of NOx and oxygen storage. The chief concern has always been particulate matter, which includes soot, consisting of relatively large, unreactive carbon-rich particles and the soluble oil fraction (SOF) comprised mainly of C_{20}-C_{28} hydrocarbon molecules. SOF conversion is readily accomplished with standard oxidation catalysts, but if the fuel has a high concentration of sulfur, the catalyst will produce droplets of H_2SO_4, which by the testing protocol are counted as particulate matter. Thus, there was a need for a selective catalysts, able to convert SOF without promoting the oxidation of SO_2 to SO_3, the precursor of H_2SO_4. Promoting Pt with vanadium oxide was highly successful and this catalyst enjoyed commercial success during the 1990s. So, too, did cerium oxide. Both these systems were important while sulfur levels were high, but the large decrease in sulfur concentration that began in the mid-1990s ended their commercial reign. Pt, without promoters, was preferred because of its high activity for SOF conversion. Zeolites were often added to the catalyst because they provided a "hydrocarbon trapping" function; this was defined as the ability to adsorb hydrocarbon molecules at low temperature, when the catalyst was not yet active, and release them after light-off had occurred. Zeolites proved their ability to boost hydrocarbon conversion and gained widespread use.

Diesel catalysis faced its greatest challenge in 2000 when the regulatory agencies, armed with new data, re-opened the subject of soot conversion (Euro V and US 2007 standards). It was clear that, in contrast to SOF, the large, refractory soot particles could not be significantly reacted during the brief residence time within the convertor. Yet newly acquired public health information made it imperative that something be done.

The situation was resolved by a radical re-design of the ceramic monolith that turned it into a soot filter. The exhaust was no longer able to exit the convertor without first flowing through a channel wall. The porosity of the wall was adjusted to hold back the relatively large soot particles while allowing the remainder of the exhaust to pass through. Depending on the vehicle and the duty cycle, the trapped soot was combusted either under normally occurring conditions or during a programmed interval in which the exhaust temperature was artificially raised. Development work continues, but the success of the diesel particulate filter is no longer in doubt. Originally intended for heavy duty vehicles only, its performance has led the agencies to mandate its use, starting in 2010, for passenger cars as well.

A final topic covered in this chapter is the reduction of NOx to N_2 in diesel exhaust. Reductants such as CO are ineffective in the presence of excess O_2, but NH_3 converts NOx with high selectivity provided the proper catalyst is used. Urea, in aqueous solution, is used as a NH_3 precursor. It is stored on-board and an extensive infrastructure is required to ensure its availability. This exists in Europe today and efforts are being made to establish it in the U.S. Although costly and complex, the public has demonstrated that it rates clean air an important priority. The worldwide automotive catalysis community can take pride in the progress it has made toward this goal.

Acknowledgement

The authors wish to acknowledge the technical assistance received from Stan Roth, Ken Voss, Joseph Patchett and Sanath Kumar, in the preparation of this manuscript. A special appreciation is owed to Robert Farrauto, whose volume "Catalytic Air Polution Control – Commercial Technologies" served as an important source of information and whose personal consultations were valuble on many occasions.

References

1. Heck, R.M.; Farrauto, R.J. with Gulati, S.T. Catalytic *Air Pollution Control – Commercial Technology*, 2nd edition, Wiley-Interscience, **2002**; P70.
2. Shelef, M.; McCabe, R.W. *Catal. Today*, **2000**, *62*, 35.
3. Gandhi, H. S.; Piken, A. G.; Shelef, M.; Delosh, R.G. *SAE Paper 760201*, **1976**.

4. Duprez, D. and Descorme, C. *"Oxygen Storage/Redox Capacity and Related Phenomena on Ceria-Based Catalysts"*, in *Catalysis by Ceria and Related Materials*, eds A. Trovarelli, *Catalytic Science Series*, Vol 2, Imperial College Press, London, **2002**, page 243.

5. Overbury, S. H. and Mullins, D. R. *"Ceria Surfaces and Films for Model Catalytic Studies Using Surface Analysis Techniques"*, in *Catalysis by Ceria and Related Materials,* eds A. Trovarelli, *Catalytic Science Series*, Vol 2, Imperial College Press, London, **2002**, page 311.

6. Bernal, S.; Calvino, J.J. and Gatica, J.M. in *"Chemical and nanostructural aspects of the preparation and characterization of ceria and ceria-based mixed oxide-supported metal catalysts"*, in *Catalysis by Ceria and Related Materials*, eds A. Trovarelli, *Catalytic Science Series*, Vol 2, Imperial College Press, London, **2002**, page 85.

7. Shelef, M.; Graham, G. W.; McCabe, R. W. *"Ceria and Other OxygenStorage Components in Automotive Catalysts"*, in *Catalysis by Ceria and Related Materials*, eds A. Trovarelli, *Catalytic Science Series*, Vol 2, Imperial College Press, London, **2002**, page 343.

8. Williamson, W.B.; Dou, D.; Robota, H.J. *SAE paper 1999-01-0776*, **1999**.

9. Hu, Z.; Heck, R.M.; Rabinowitz, H.N. *US Patent 6254842*, **2001**.

10. Wyatt, M.; Roth, S.A.; Manning, W.A.; D'Aniello, M.J. *SAE paper 930130*, **1993**.

11. Osaka, K.S.; Osaka, K.U.; Himeji, Y.I.; Amagasaki, T.O. *US Patent 4,617,289*, **1986**.

12. Alzenau-Kaelberau, R.D.; Hanau, B.E.; Alzenau, E.K.; Zeiskam, V. H. *US Patent 5,157,007*, **1992**.

13. Farrauto, R.J.; Voss, K.E.; Heck, R.M. *World Patent WO 93/10886*, **1993**.

14. Farrauto, R.J.; Voss, K.E.; Heck, R.M. *US Patent 5,462,907*, **1995**.

15. Farrauto, R.J.; Voss, K.E. *Appl. Catal. B*, **1996**, *10*, 29.

16. 16. Burch, R.; *J. Mol. Catal. A*, **1995**, *100*, 13.

17. Hepburn, J.S.; Jen, H.-W.; Gandhi, H.S.; Otto, K *US Patent 5,772,972*, **1998**.

18. Yavuz, B.O.; Voss, K.E.; Deeba, M.; Adomaitis, J.R.; Farrauto, R.J. *US Patent 6,248,684,* **2001**.

19. Majewski, W.A. Diesel Particulate Filters, in *www.DieselNet.com*, **2001**.

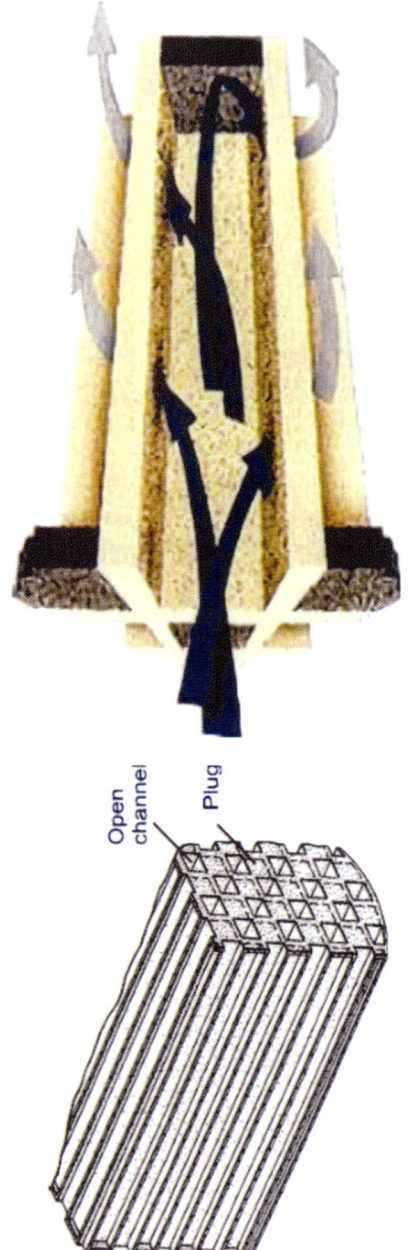

Figure 7.5. Structure and function of a diesel particulate filter

Table 11.I. I&EC ionic liquids symposia.

I&EC Symposium	Topics Covered
Green (or Greener) Industrial Applications of Ionic Liquids (*13*) April 1-5, 2001 San Diego, CA 221st ACS National Meeting http://bama.ua.edu/~rdrogers/sandiego (last accessed March 29, 2008)	• Ionic Liquids in Context • Separations and Engineering • Green Synthesis • Nuclear and Electrochemistry • Ionic Liquid Systems • Properties of Ionic Liquids • Catalysis I • Catalysis II • Structure and Photochemistry • High-Temperature and Other Systems
Ionic Liquids as Green Solvents: Progress and Prospects (*14*) August 18-22, 2002 Boston, MA 224th ACS National Meeting http://bama.ua.edu/~rdrogers/Boston (last accessed March 29, 2008)	• Ionic Liquid Tutorial • Manufacture and Synthesis of Ionic Liquids: Industrial and Academic • Characterization and Engineering • Novel Applications of Ionic Liquids • Separations • Biotechnology • Catalytic Chemistry • Non-Catalytic Chemistry • Electrochemistry • Photochemistry and Reaction Intermediates

Table 11.I. *Continued.*

I&EC Symposium	Topics Covered
Ionic Liquids III: Fundamentals, Progress, Challenges, and Opportunities  (*15,16*) September 7-11, 2003 New York, NY 226th ACS National Meeting http://bama.ua.edu/~rdrogers/New York (last accessed March 29, 2008)	• Ionic Liquid Tutorials • Fuels and Applications • Physical and Thermodynamic Properties • Catalysis and Synthesis • Spectroscopy • Separations • Novel Applications • Catalytic Polymers and Gels • Electrochemistry • Inorganic and Materials • General Contributions
Ionic Liquids: Not Just Solvents Anymore OR Ionic Liquids: Parallel Futures  (*17*) March 26-30, 2006 Atlanta, GA 231st ACS National Meeting http://bama.ua.edu/~rdrogers/Atlanta2006 (last accessed March 29, 2008)	• Why Are Ionic Liquids Liquid? • Structure-Activity Relationships/Modeling • Environmental Fate and Toxicity • New Industrial Applications of Ionic Liquids • Really New Ionic Liquids • Ionic Liquids and Education • Applications Based on Physical Properties • Functional Ionic Liquids/Ionic Materials • Analytical Applications of Ionic Liquids • Microengineering with Ionic Liquids

Chapter 8

Membrane Separation Technology: Past, Present, and Future

G. Glenn Lipscomb

Chemical and Environmental Engineering, The University of Toledo, Toledo, OH 43606–3390 (telephone: 419–530–8088, email: glenn.lipscomb@utoledo.edu)

The scientific underpinnings of the membrane separation industry date back to the eighteenth century. However, commercial products appeared less than 50 years ago. In the intervening half century, the industry has blossomed and is poised for continued strong growth. The evolution of the industry to its current state is reviewed here. The important role of the Office of Saline Water in the development of the industry is highlighted along with early patent literature that describes the manufacture, design, and operation of membrane systems – these patents still provide the technological base for the industry. The state-of-the-art in material selection, membrane formation, and module design and characterization is summarized. Finally, emerging future directions are identified in select areas including gas separation, biotechnology, water and wastewater treatment, nanofiltration, and alternative energy applications.

History

Origins of Membrane Science

The first report of osmotic phenomena by French Cleric Abbé Nollet [1] in 1748 might be considered the origin of membrane science. Nollet's work generated extensive interest in osmosis culminating ultimately in van't Hoff's [2] quantitative relationships in 1887.

During this time, Fick's [3] studies of dialysis with collodion (nitrocellulose) membranes, which led to his eponymous law, and Graham's exhaustive studies of liquid diffusion [4] and gas permeation [5] provided the basis for interpreting and analyzing membrane performance. Graham is credited with the first use of the term 'dialysis' to describe selective diffusion across semi-permeable membranes.

Bechold [6] reported one of the first careful studies of pressure to drive a membrane separation process. He developed a series of membranes from nitrocellulose with graded porosities and demonstrated how to characterize them through bubble tests [7]. The origin of the word 'ultrafiltration' is attributed to Bechold [8]. Subsequent improvements to Bechold's process led to the first commercially available microporous collodion membranes and the birth of the membrane industry [7].

Initially, the membrane industry served only niche markets. As with any new technology, adoption was hindered by concerns over reliability and cost. Cost improvements required dramatic increases in transport rates as well as increases in selectivity and how to achieve these changes was not obvious.

The Office of Saline Water

The rate of innovation in the industry increased dramatically with the creation of the Office of Saline Water (OSW). OSW was the successor to the Saline Water Conversion Program started in 1952 after passage of the Saline Water Conversion Act. The program was expanded in 1955 after amendment of the Saline Water Conversion Act and reorganized as the OSW [9]. One might argue that the funding and focus provided by the OSW is responsible for the emergence of the modern membrane industry.

The use of membranes specifically for desalination dates to at least the late 1940's. At The University of California at Los Angeles, Hassler proposed using synthetic membranes as biomimetic surrogates for cellular membranes [8]. His design utilized two membranes (cellophane sheets supported by a porous ceramic support) separated by an air gap that would permit the evaporation of water across the gap and subsequent condensation. The air gap itself was considered to be the semi-permeable membrane and the process was a forerunner of current membrane distillation processes.

With support from the OSW, Reid at the University of Florida pursued an alternative design in the mid 1950's based on filtration equipment available at the time. His design used pressurized air to drive water across polymeric films. Of the commercially available films, cellulose acetate was the most attractive due to its high salt rejections. Unfortunately, product water permeation rates were low. Since permeation rate was inversely proportional to film thickness, solution casting techniques were developed to form films as thin as possible [10]. Around the time of Hassler's and Reid's work, the term "reverse osmosis" was adopted to describe membrane desalination [8].

Reid and Breton [10] proposed one of the first transport models for the process based on hydrogen bonding between water and the carbonyl oxygen in cellulose acetate. Hydrogen bonded water filled membrane voids and prevented the passage of ions and non-polar molecules to impart high salt rejection characteristics. Water transport occurred by diffusion through the holes or voids in the membrane structure. This transport theory was challenged later by Lonsdale and co-workers with a solution-diffusion mechanism that did not rely on the presence of holes [11]. The commonality of solution-diffusion transport to a variety of membrane processes was demonstrated by Paul and co-workers [12] and is generally accepted as the transport mechanism for processes that utilize dense, non-porous membranes.

Membrane Manufacture

Low product water permeation rates prevented the development of commercial reverse osmosis membrane processes until the watershed discovery of the asymmetric membrane by Loeb and Sourirajan [13]. The serendipitous observation that desalination occurred only when one side of commercial cellulose acetate films contacted the saline feed led Loeb and Sourirajan to cast their own films. They cast thin films of cellulose acetate – acetone solutions on a glass plate using a doctor's blade. After casting the film, acetone was allowed to evaporate for a short period of time (10 – 100 seconds) before the film was immersed in cold water. The cast films possessed an asymmetric structure illustrated in Figure 1 in which a thin dense layer overlies an integrally attached underlying porous layer. The thin dense layer formed on the side of the membrane from which acetone evaporated. To enhance formation of the porous support, aqueous magnesium perchlorate was added to the casting solution.

The top dense layer of the asymmetric structure provided the selectivity of the membrane. Loeb and Sourirajan's process allowed the formation of virtually defect-free films that were ~10 times thinner than the free-standing films that

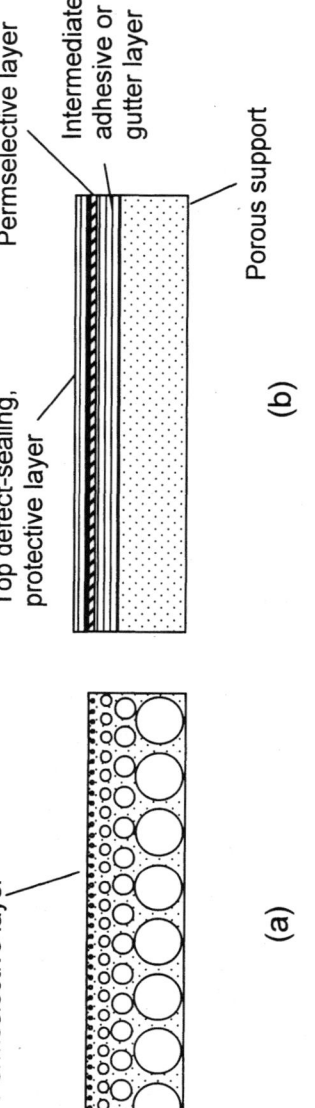

Figure 1. Typically asymmetric membrane (a) and composite membrane (b) cross-sectional structures.

could be produced using any other process. Thus, the membranes possessed permeation rates that were ~10 times higher than the best available materials. Moreover, the integrally attached porous layer provided the required mechanical support for the discriminating layer while posing little resistance to transport. Loeb and Sourirajan's invention was the breakthrough required for the development of economically viable membrane desalination. Additionally, modifications of the process led to the development of economical gas separation and ultrafiltration membranes and membrane separation processes.

A physical explanation of the Loeb-Sourirajan process was proposed by Strathmann et al. [14]. They depicted the process as compositional trajectories on a ternary phase diagram (water-acetone-cellulose acetate) as illustrated in Figure 2. The solid line in Figure 2 represents the binodal while the dashed line represents the spinodal for the mixture. The initial composition of the polymer solution is represented as point I. Two lines emanate from point I: one line represents the transient compositional changes that occur at the top surface from which acetone evaporation occurs while the second represents the composition at the opposite face. At the top surface, acetone evaporation occurs initially followed by acetone dissolution and water absorption in the water bath until the final concentration denoted by point II is reached. This concentration lies outside the two-phase region so the resulting structure is homogeneous or dense. At the opposite face, acetone loss does not occur until the nascent membrane is immersed in the water bath. Acetone loss and water uptake leads to a composition trajectory that passes into the two-phase region and therefore leads to the formation of a two-phase liquid structure consisting of polymer-lean and polymer-rich regions. Further loss of acetone to the water bath leads to solidification of the polymer rich regions into the porous support under the dense skin. The final composition is denoted by point III. The composition of points between the upper and lower surface of the membrane lie along the line connecting points II and III.

The Loeb-Sourirajan process often is referred to as diffusion induced phase separation (DIPS) to reflect the role of diffusion in forming the asymmetric structure. Liquid-liquid phase separation and the resulting asymmetric structure arise from diffusion of a solvent (acetone) out of the film and diffusion of a non-solvent (water) into the film. This physical interpretation provided the basis for the development of asymmetric membrane manufacturing processes for other polymer – solvent – non-solvent systems.

Commercialization of Loeb and Sourirajan's membrane required the development of membrane manufacturing processes and means for contacting the feed water with the membrane. Drawing upon the similarity between heat transfer devices (heat exchangers) and mass transfer devices, two forms for the membrane emerged: hollow fiber and flat sheet. Both forms were produced by extruding the precursor polymer into the desired shape, cooling, and removing any solvents or processing aids.

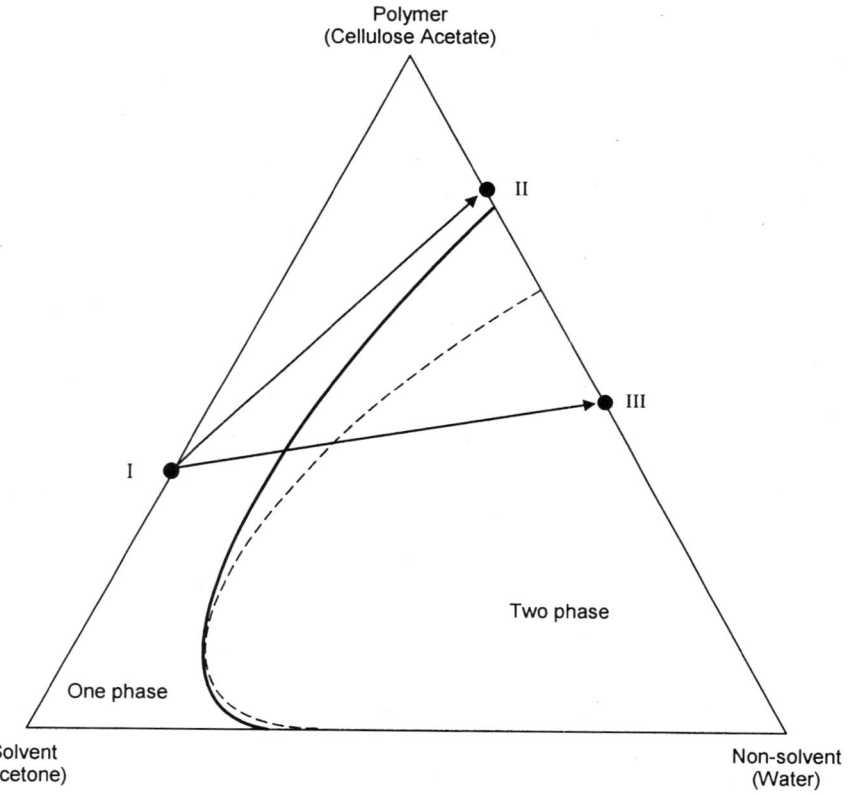

Figure 2. Ternary phase diagram illustrating compositional trajectories during asymmetric membrane formation. The solid line represents the binodal and the dashed line the spinodal. The initial casting solution composition is represented by I. The transient changes in concentration at the top surface are indicated by the top arrow emanating from I and ending in II. The changes at the bottom surface are indicated by the bottom arrow emanating from I and ending in III.

Fibers are manufactured using a melt (neat) or solution (polymer and solvent) spinning process [15]. A typical spinline is illustrated in Figure 3. The polymer in powder, flake or pellet form is fed to an extruder where it is melted or mixed with one or more solvents to form a solution. The resulting liquid is extruded through a die to form a hollow liquid cylinder. The interior of the cylinder is filled by a bore fluid (gas or liquid) that is fed to the die as well. This liquid cylinder may pass through a gas filled region (commonly air and referred to as an air-gap) to allow solvent evaporation before it enters one or more liquid baths (commonly water or aqueous solutions) to emulate the process of Loeb and Sourirajan. Although individual fibers may be spun, most commercial processes use dies that form up to ~100 filaments which pass through the process collectively; this group of fibers is referred to as a fiber tow. After formation, the fiber commonly is collected on a spool for storage until it is packaged into a device. Sheets are manufactured in a similar process that utilizes a slit die.

The patent and journal literature provides a description of the physics of these processes however conditions adopted for commercial manufacture are typically specified as ranges of composition, temperature, and contact times in gas or liquid baths. Optimal values are not disclosed.

Membrane Module Manufacture and Operation

Packaging of the membrane into a device or module followed the precedents set by past heat exchanger design. Initial desalination devices utilized membrane sheets packaged in a plate and frame device [8]. These devices quickly gave way to more compact hollow fiber and spiral wound module designs.

Both Dow [16] and DuPont [17] patented early hollow fiber designs reflecting the interest of the chemical industry in membrane technology for desalination and other separations. Dow's patent directly adapted the design of shell and tube heat exchangers for mass exchange applications. Individual fibers or tows of fibers are held between two fixed plates by potting the ends of the fibers in a holder. Potting involves simply pulling the fibers through the holder and filling it with an appropriate adhesive to hold the fibers together and affix them to the wall of the holder. Fiber lumens are opened by cutting through the potted end that extends out of the holder to form a tubesheet. The holders are placed in fixed plates on a regular array like tubes in a heat exchanger. The fibers are held under slight tension to align them along the axis of the module. The modules thus formed are claimed to offer an economical solution for desalination.

DuPont's patents are especially noteworthy in that they provide an unusually thorough description of module manufacture and insightful discussion of how design variables may affect performance. In particular, the patents

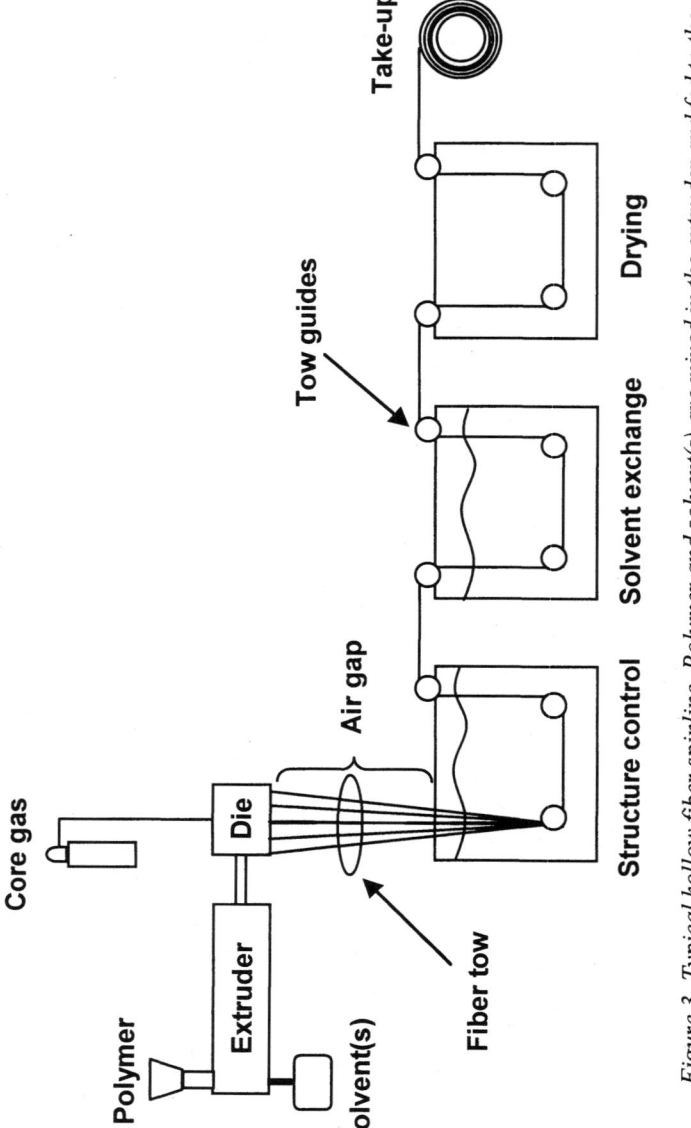

Figure 3. Typical hollow fiber spinline. Polymer and solvent(s) are mixed in the extruder and fed to the spinneret. The spinneret (die) creates hollow liquid filaments that pass through an air gap to allow solvent evaporation before the fiber tow enters a liquid bath to set the membrane structure. Subsequent baths may be used to replace solvents with water and dry the fiber before it is taken up on a spool.

succinctly describe sources of non-ideal performance that still confound manufacturers today and offer suggestions on how to mitigate them. Although the applications discussed are primarily gas separations, the design is claimed useful for all membrane separations. The patents are highly recommended reading for those new to membrane separations as well as those more experienced.

Figure 4 illustrates how a fiber bundle is formed in the DuPont patents. Tows of fiber are removed from a plurality of spools and wound to form a continuous shank of fiber. Individual shanks are covered by an elastic sock to hold them in a uniform spatial position. Figure 5 illustrates the formation of a fiber bundle from the individual shanks. A number of shanks are grouped together to form the fiber bundle which is encased in another elastic sock to hold the shanks in a uniform spatial position. The use of two elastic socks to maintain order at two length scales provided structure to the fiber bundle that is critical to creating uniform flow channels in the shell.

Each end of the fiber bundle is placed in a mold as illustrated in Figure 5. To create a tubesheet, the mold is filled with an adhesive, commonly an epoxy resin, similar to the Dow patent. However, a centrifugal potting process is used to fill the mold to minimize wicking of the epoxy into the fiber bundle. Figure 6 illustrates the process. The module is placed horizontally in a holder and the molds for forming tubesheets are attached at either end. The holder is rotated while the molds are filled with the potting material. Centrifugal forces reduce wicking of the potting material into the fiber bundle due to capillary forces. To open the fiber lumens in the tubesheet, a portion of the potted end is removed by cutting along the line indicated in Figure 6. Centrifugal potting produces a sharp interface at the edge of the tubesheet that lies inside the fiber bundle and allows precise control of tubesheet thickness. Consequently, it is the standard against which other potting techniques are compared.

The DuPont patents describe the use of hollow fiber modules for a number of gas separation applications. These applications include the use of staged systems with various inter-stage recycle schemes. One is illustrated in Figure 7. The two-stage scheme in Figure 7 is used to produce a high purity permeate, e.g., recover helium from a helium-nitrogen mixture. In operation, a high pressure feed is directed into the lumens of the first stage. As the gas flows through the module, helium preferentially permeates across the fiber walls into the shell side which is maintained at a lower pressure than the feed. This permeate product is enriched in helium. To further enrich the product, the first stage permeate is compressed and directed into the lumens of a second stage. The permeate produced by the second stage is the final enriched helium product. A similar staged system can be used to enrich oxygen from air.

To improve process efficiency, the non-permeate or retentate product from the second stage is recycled to the feed of the first stage. The retentate from the second stage possesses a higher helium concentration than the feed to first stage

290

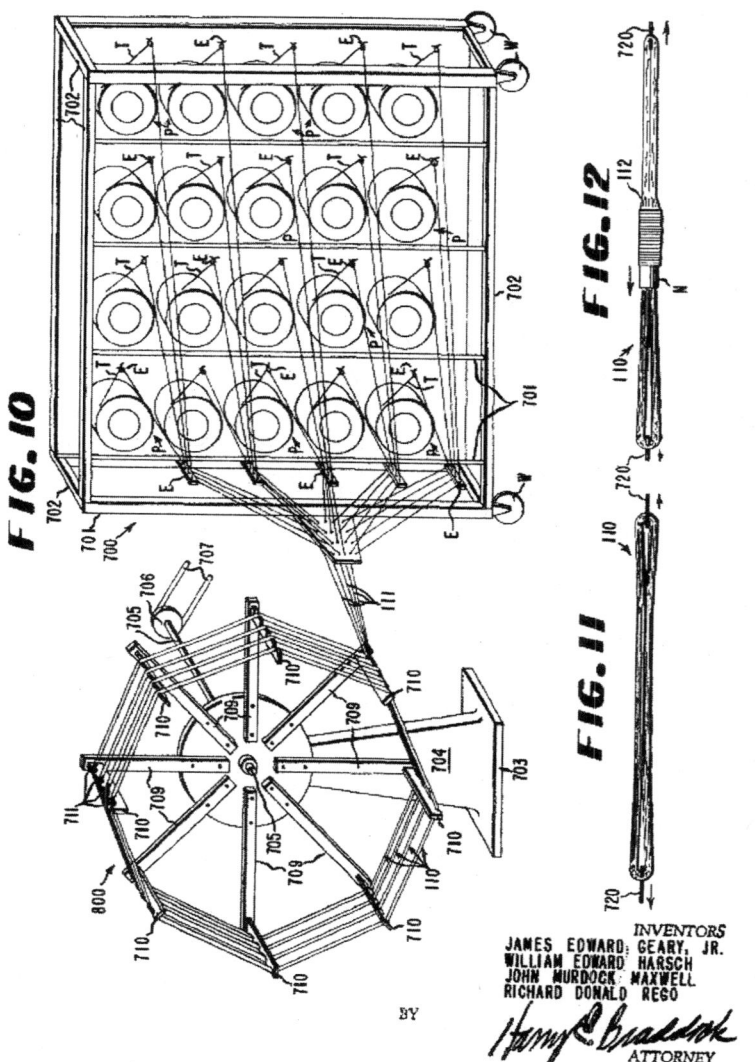

FIG. 10

FIG. 11

FIG. 12

INVENTORS
JAMES EDWARD GEARY, JR.
WILLIAM EDWARD HARSCH
JOHN MURDOCK MAXWELL
RICHARD DONALD REGO

BY

Harry E. Braddock
ATTORNEY

Figure 4. Formation of a hollow fiber bundle [17]. Continuous loops or shanks of fibers are formed by unwinding fiber from spools (Fig. 10). Individual shanks are covered by an elastic sock to hold fibers in uniform spatial position and facilitate subsequent handling (Fig. 12).

291

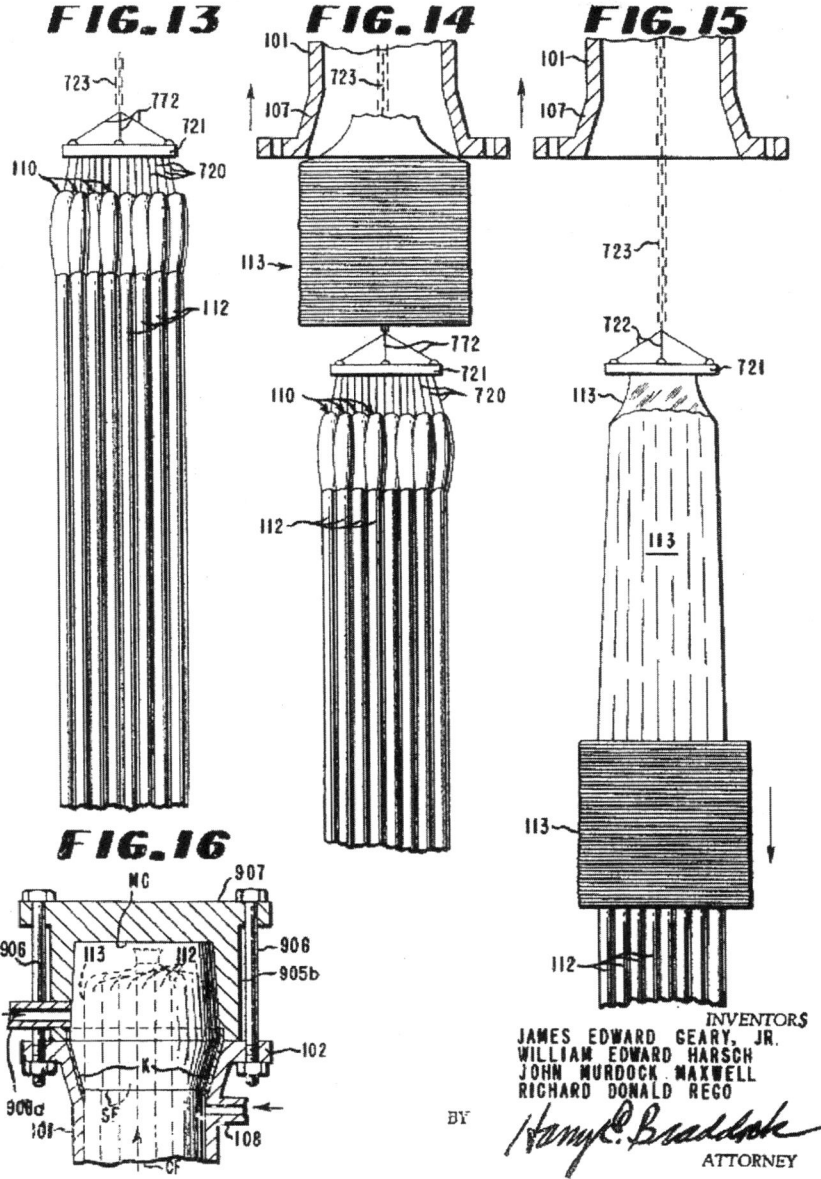

Figure 5. Formation of a hollow fiber bundle [17]. Sock covered fiber shanks are hung to form a large bundle (Fig. 13) which then is covered by an elastic sock to hold shanks in uniform spatial position (Figs. 14-15). Each end of the bundle is placed into a mold to form tubesheets (Fig. 16).

292

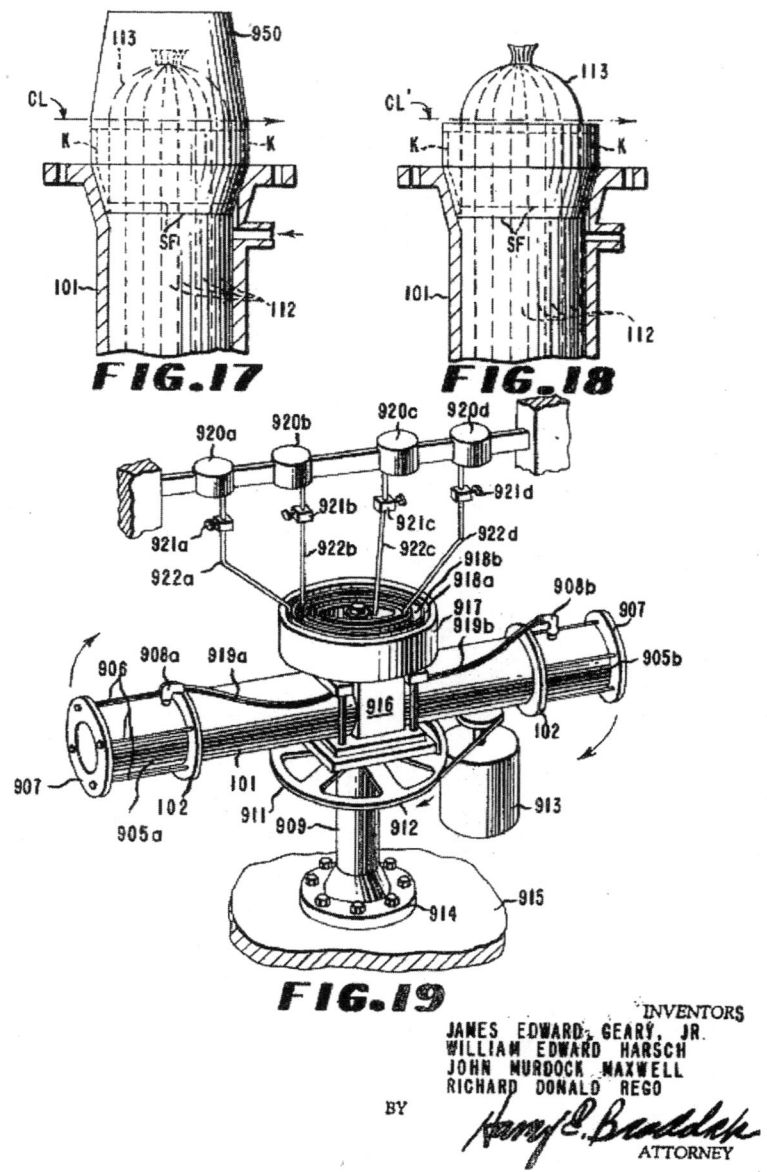

FIG.17 **FIG.18**

FIG.19

INVENTORS
JAMES EDWARD GEARY, JR.
WILLIAM EDWARD HARSCH
JOHN MURDOCK MAXWELL
RICHARD DONALD REGO

BY

ATTORNEY

Figure 6. Formation of a hollow fiber bundle [17]. Tubesheets are formed by centrifugal potting (Fig. 19). Once formed, fiber lumens are opened by cutting off the end of the tubesheet along line CL (Fig 17 and 18).

FIG. 6

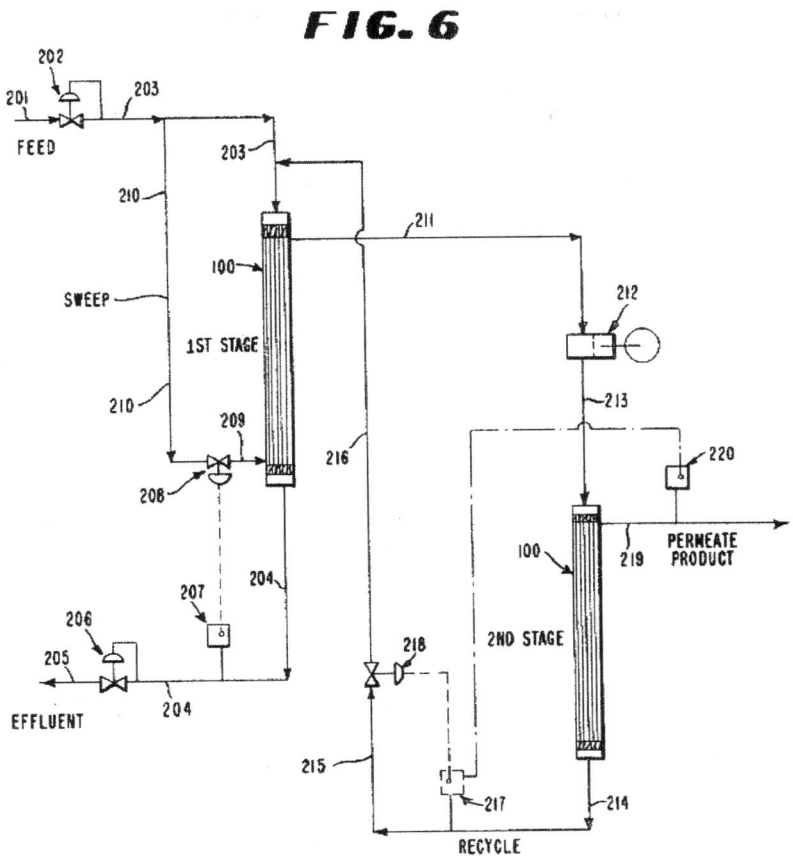

INVENTORS
JAMES EDWARD GEARY, JR.
WILLIAM EDWARD HARSCH
JOHN MURDOCK MAXWELL
RICHARD DONALD REGO

BY

Harry C. Braddock
ATTORNEY

Figure 7. Two-stage system for recovering helium from a dilute helium-nitrogen mixture [17]. The concentration of helium in the retentate product from the first stage, as measured by sensor 207, is controlled by adjusting the sweep to the first stage. Compressor 212 feeds the permeate from the first stage to the second. The concentration of helium in the retentate product from the second stage, as measured by sensor 217, is controlled by adjusting the retentate product flow rate. The permeate concentration from the second stage, as measured by sensor 220, is controlled by adjusting the set-point for the retentate product.

and upon mixing with it increases the helium concentration of the gas entering the first stage. This increases the driving force for helium permeation and thereby the amount of gas that the first stage can process. Additionally, a portion of the feed to the first is used to sweep the low pressure, permeate side of the first stage. The sweep can increase the permeation rate of helium and throughput for the system but limits the lowest concentration that can be reached in the retentate product stream.

Figure 7 also illustrates control schemes proposed to maintain uniform product purities. The concentration of helium in the retentate product from the first stage is controlled by adjusting the amount of the feed used as sweep – increasing the sweep reduces the effluent concentration. The concentration of the retentate product from the second stage is controlled by adjusting the flow rate of this stream – increasing the flow rate increases the concentration of helium. The concentration of the permeate product from the second stage is controlled by adjusting the concentration set point for the retentate product of the second stage – increasing the concentration set point increases the concentration of helium in the permeate product.

Introducing the high pressure feed into the fiber lumens (lumen feed) is recommended when the stage strips a more permeable component present at low concentration in the feed (less than ~25%). Shell-side feed is recommended when the more permeable component is present at higher concentrations (greater than ~80%).

The use of lumen feed is rationalized by noting that lumen feed results in more uniform gas contacting in the fiber bundle, i.e., flows in individual fibers are more nearly uniform and the permeate in the shell tends to flow more uniformly in a desirable counter-current manner. However, if fibers are plugged at either end of the module (due to poor cutting of the tubesheet to open the fiber lumens), performance deteriorates. Plugged fibers are filled with stagnant gas at the concentration of either the feed or the retentate product; if the plug is present at the end from which the retentate product is removed, the fiber is filled with the feed gas. All of the gas that enters a stagnant fiber will permeate into the shell. This will reduce the concentration of the permeate product since the concentration of the more permeable component in the feed and retentate product are significantly lower.

With shell-feed, the effect of plugged fibers is diminished greatly. If the plug is at the end from which the permeate product is removed, the fiber is dead-ended and cannot contribute to the permeate product. Therefore, such fibers simply reduce the effective surface area of the module. If the plug is at the end from which the retentate product is removed, the fiber performs identically to the other fibers in the module when a sweep gas is not used. If a sweep gas is used, it cannot enter the plugged fibers, but the purity of the permeate produced by the plugged fibers is not changed as dramatically as with lumen feed. Thus, shell-feed is preferred in stages where the feed contains a high concentration of the more permeable component and it is desired to produce a higher purity permeate.

The effect of variability in fiber size is similar to that of plugged fibers. The feed flow rate to a fiber decreases as fiber size decreases. For sufficiently small fibers, the feed flow rate is low enough that all of the feed will permeate into the shell like a fiber plugged at the retentate product end. With shell feed, the detrimental influence of smaller fibers is avoided as with plugged fibers. Fibers that are larger than average allow the feed to essentially by-pass the module with lumen feed. The gas contact time is reduced so the amount of gas that can permeate is reduced. However, shell side feed avoids this by-passing problem if flow channels inside the shell are uniform.

A widely used process for fixing broken fibers in the fiber bundle is described as well. A module is placed in a vertical position inside a case such that the top tubesheet may be covered by a liquid and the liquid will not drain off under the influence of gravity. A manifold is placed over the bottom tubesheet and connected to pressurized gas source to introduce gas into the fiber lumens. The top tubesheet also is covered by a manifold to control the pressure of the gas exiting the fiber lumens. To fix broken fibers, gas is forced to flow through the lumens, and the top tubesheet is covered by a solidifying liquid such as an epoxy. Gas exiting the lumens prevents the epoxy from entering fibers and solidifying therein. Covering the top tubesheet with a manifold and increasing the pressure in the manifold forces the epoxy into broken fibers since gas does not flow through such fibers. The gas flow is maintained until the epoxy sets and plugs the broken fibers. The tubesheet may be machined to remove excess epoxy that accumulates on the surface without removing the plug formed in broken fibers. To plug broken fibers in the bottom tubesheet, the module is flipped and the process repeated.

The detailed descriptions of module manufacture, process control, and the effects of non-idealities such as plugged fibers, fiber size variation, and broken fibers provided in the DuPont patents make them a great introduction to hollow fiber membrane technology. The manufacturers of the next generation of membrane products will benefit from this insight.

The evolution of modules that utilize membrane in sheet form is illustrated by the patents of Westmoreland [18] and Bray [19]. Westmoreland describes the simplest embodiment of the spiral wound construct. As illustrated in Figure 8, the module consists of four elements: 1) a membrane sheet, 2) a feed spacer, 3) a permeate spacer, and 4) a permeate collection tube. A membrane envelope is formed by layering one membrane sheet, the permeate spacer, and a second membrane sheet; the discriminating surfaces of the membrane sheets face away from the permeate spacer. Three edges of this envelope are bonded together using a suitable adhesive such as an epoxy. The fourth edge is bonded to the permeate collection tube such that the permeate spacer is in contact with the holes in the collection tube. Additionally, a fluid-tight seal is formed with the collection tube such that fluid can enter the channel created by the permeate

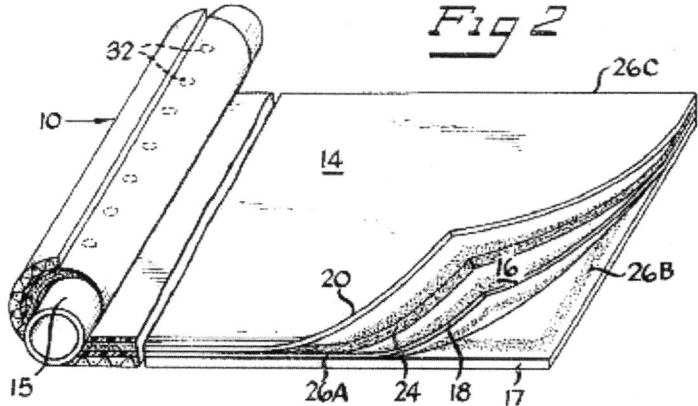

Figure 8. Construction of a single envelope spiral wound module [18].
A permeate spacer (24) is placed between two membranes (18 and 20) and the
envelope sealed by gluing along the periphery. This envelope is wrapped around
a permeate collection tube along with a feed spacer (17).

spacer only by permeation across the membrane. The feed spacer is placed either below or on top of the envelope.

In operation, feed water flows along the outer surfaces of the membrane envelope under pressure in a direction parallel to the permeate collection tube. Water that permeates across the membrane enters the channel created by the permeate spacer and flows perpendicularly to the feed (a crossflow contacting pattern) towards the permeate collection tube from which the product water is removed.

The module assembly is placed within a case that possesses manifolds to direct feed water along the membrane surface, withdraw product water from the collection tube, and remove the rejected water. A rectangular case might be used but a more compact unit is obtained by rolling the envelope and feed spacer around the permeate collection tube; the feed spacer defines a channel for feed water flow between successive layers of the envelope. A cylindrical pressure vessel is used to hold the spiral wound module thus produced.

A critical design concern is the permeate pressure drop within the envelope. The pressure at the end of the envelope will increase to the value required to drive the permeate into the collection tube. This pressure increase reduces the driving force for water permeation and limits the length of the envelope that can be used effectively. At a critical length, the pressure drop becomes large enough that water permeation will stop. Moreover, as the length of the envelope increases (or, equivalently, the module area increases), module efficiency decreases.

To overcome this limitation, Bray proposed the multi-envelope design shown in Figure 9. Fabrication of such a module is more complex and will not be described here. However, the design allows the construction of larger, more efficient modules by utilizing shorter envelopes. Notably, Bray acknowledges the support of the Office of Saline Water in the first paragraph of the patent.

State of the Art

From its origin in the work described in the previous section, the membrane industry has evolved steadily to its current state.

Although metallic and ceramic materials are used as membranes, polymeric materials account for the vast majority of commercial products. Polymer selection depends on a number of factors including intrinsic transport properties, mechanical properties, thermal stability, chemical stability (e.g., chemical resistance and biocompatibility), membrane manufacturability, cost, and patentability. The two most common types of polymers are glassy engineering thermoplastics and rubbery polysiloxanes.

To protect intellectual property investments, membrane companies often patent families of polymers. For example, in the gas and vapor separation area, a family of polyimides were patented by DuPont [20] while their competitors Dow and Air Products patented families of polycarbonates [21] and polyarylates [22], respectively.

Asymmetric membrane structures have been created from these materials using the diffusion induced phase separation process (DIPS) as well as a thermally induced phase separation process (TIPS) [23] that relies on temperature gradients to produce a gradient in phase separated domain size. Moreover, membranes formed by either process can be further modified by stretching or drawing to alter pore size and porosity.

The location of the discriminating layer can be manipulated through control of processing conditions. The discriminating layer can lie on either side of the membrane or somewhere in the middle [24-25]. Furthermore, the pore size distribution and morphology of the underlying support can be controlled. Excellent reviews of the phase separation process and the relationship between process variables and final membrane structure are available in the literature [26-27].

An alternative to the asymmetric structure is the composite structure illustrated in Figure 1 which compares the two. Like the asymmetric structure, the composite structure consists of a discriminating layer supported by an underlying support. In contrast to the asymmetric structure, the support generally is not made of the same material as the discriminating layer and hence is not integrally attached to it.

298

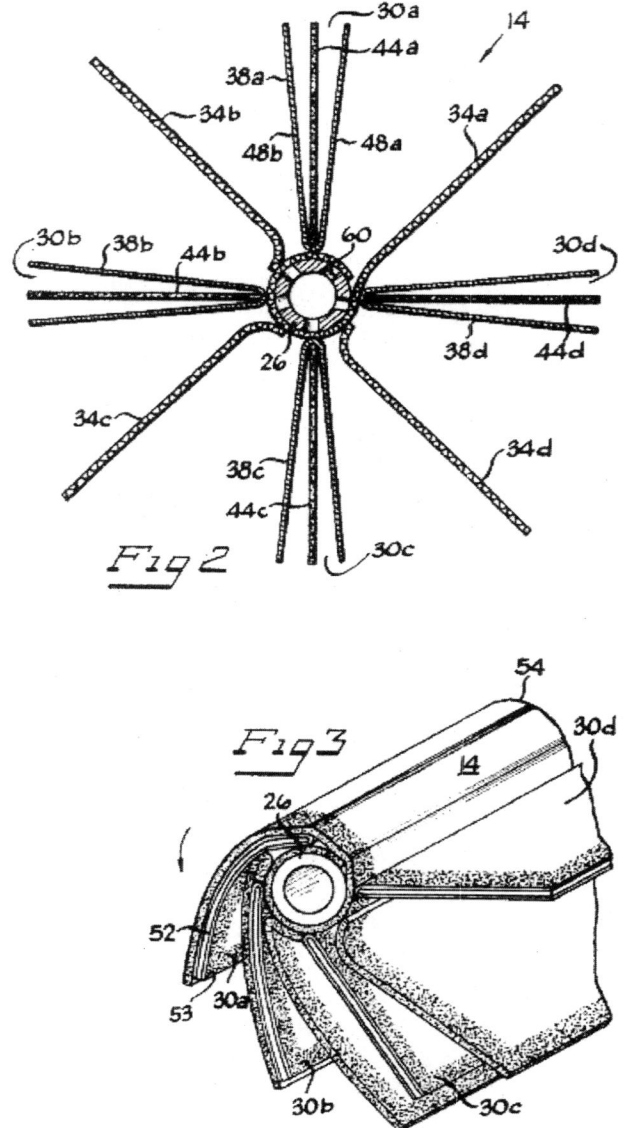

Figure 9. Construction of a multi-envelope spiral wound module [19]. Four envelopes are formed from membranes 38 (a-d) and permeate spacers 34 (a-d). Feed spacers 44 (a-d) separate the envelopes.

The porous support often is made separately by a phase inversion process followed by drawing to enhance porosity. The discriminating layer is added afterwards by dip or slip-casting [26] or interfacial polymerization [28]. Alternatively, the support and discriminating layers can be formed simultaneously by co-extrusion [29].

State of the art composite membranes for reverse osmosis consist of three layers: 1) the discriminating surface layer (commonly a polyamide produced by interfacial polymerization of m-phenylenediamine – trimesoyl chloride), 2) a supporting ultrafiltration layer (commonly polysulfone), and 3) a non-woven fabric that provides the majority of the mechanical strength [28, 30]. This tri-layer composite reduces the resistance to permeation in the supporting layers without compromising mechanical integrity.

The discriminating layer in state of the art asymmetric or composite membranes typically is less than 0.1 microns thick for reverse osmosis membranes [30] and gas/vapor separation [31]. This thickness is limited by the ability to produce defect free layers. Defects may appear due to particulate matter that extends through the discriminating layer or inherent variability in processing conditions that lead to variability in discriminating layer thickness. One may minimize the impact of such perfections by coating the membrane with a high permeability, low selectivity material to reduce permeation through imperfections [32-33].

The transport mechanism through the discriminating layer is dependent on whether the layer is porous (possesses large, fixed inter-molecular gaps) or dense (possesses small, transient intermolecular gaps). Porous materials are characterized by pore size using the definitions adopted by the International Union of Pure and Applied Chemistry (IUPAC) [34]. Unfortunately, the terminology adopted for pore size characterization by material scientists does not coincide with that used by membranologists [34-35]. Table 1 summarizes the terminology and primary transport mechanisms for each category.

Transport in dense discriminating layers is most commonly described using the well developed solution-diffusion theory [36]. The theory is based on the assumptions that: 1) the driving force for transport is a gradient in chemical potential, 2) at a fluid-membrane interface the chemical potential in the two phases are equal (i.e., equilibrium exists), and 3) the pressure within the membrane is constant and equal to the highest value at either interface. Baker [37] summarizes the application of the theory to a variety of membrane separation processes ranging from dialysis to gas separation.

For gas separations, solution-diffusion theory leads to the conclusion that gas permeation flux (J) is proportional to the difference in gas partial pressure across the membrane (Δp): $J=(P/l)\Delta p$. The proportionality constant is equal to the intrinsic permeability (P) for the membrane material divided by the effective membrane thickness (l). In turn, the permeability is equal to the product of a solubility (S) and diffusivity (D): $P=D*S$. The ability to separate two

Table 1. Pores size nomenclature and Transport Mechanisms [34-35].

Pore diameter (nm)	Material Classification	Membrane Classification	Transport Mechanism(s)
d>50	Macroporous	Microfiltration	Laminar flow, Bulk diffusion, Knudsen diffusion
2<d<50	Mesoporous	Ultrafiltration	Knudsen diffusion, Surface diffusion, Multi-layer diffusion, Capillary diffusion
d<2	Microporous	Nanofiltration	Activated diffusion
Molecular gaps	Dense	Gas/Vapor/Reverse Osmosis	Solution – diffusion

components of a gas mixture is reflected in the value of the intrinsic separation factor (α) defined as the ratio of the permeability for two components which may be expressed as the product of a solubility selectivity and diffusivity or mobility selectivity: $\alpha_{i,j} = P_i/P_j = (S_i/S_j)*(D_i/D_j)$.

Experimental measurements of permeability and selectivity for a large number of gas-polymer systems indicate that the mobility selectivity controls the intrinsic separation factor for nonpolar gases [27, 38]. To impart mobility selectivity, the polymer matrix must possess molecular size gaps sufficiently small and uniform in size to discriminate between molecules that may differ in size by less than 1 Å, for example oxygen and nitrogen differ in size by 0.2 Å.

The ideal membrane material possesses both high permeability and high selectivity. A significant body of work has sought to optimize transport properties through modifications of primary (composition) and secondary (chain conformation) molecular architecture. This work has led to the two rules of thumb for simultaneously increasing permeability and selectivity within a family of polymers [27]: 1) molecular changes that inhibit chain packing and reduce torsional mobility around the polymer backbone can increase permeability while maintaining selectivity and 2) molecular changes that reduce the mobility of segments of the polymer chain but do not reduce inter-segment spacing can increase selectivity without a loss of permeability.

Although these rules can lead to increased permeability and selectivity, an extensive compilation of the data in the literature by Robeson suggests an upper bound exists on transport properties – increases in permeability eventually lead to a decrease in selectivity and vice versa [38]. Figure 10 illustrates the upper bounds that exist for a number of gas pairs; such plots of selectivity versus permeability are referred as 'Robeson plots'. As one might expect, a similar upper bound exists for pervaporation membranes used to separate benzene-cyclohexane mixtures [39].

Freeman [40] demonstrates that solution-diffusion transport theory predicts the existence of an upper bound if diffusion is an activated process, the activation energy depends linearly on the square of the molecular diameter, and solubility depends exponentially on the Lennard-Jones temperature. Although one adjustable parameter is introduced to specify the dependence of activation energy on molecular size, a single value gives excellent predictions of the location of the upper bound for all gas pairs examined.

Mehta and Zydney [41] show a similar relationship exists for ultrafiltration membranes where transport through the membrane occurs by convective pore flow. A Robeson plot was created by taking the selectivity of an ultrafiltration membrane as the reciprocal of the protein sieving coefficient (the ratio of protein concentration in the permeate to that in the fluid adjacent to the membrane surface) and the permeability as the solvent hydraulic permeability. A plot of literature data for bovine serum albumin separation shows the existence of an upper bound. The location of the upper bound was predicted assuming the

302

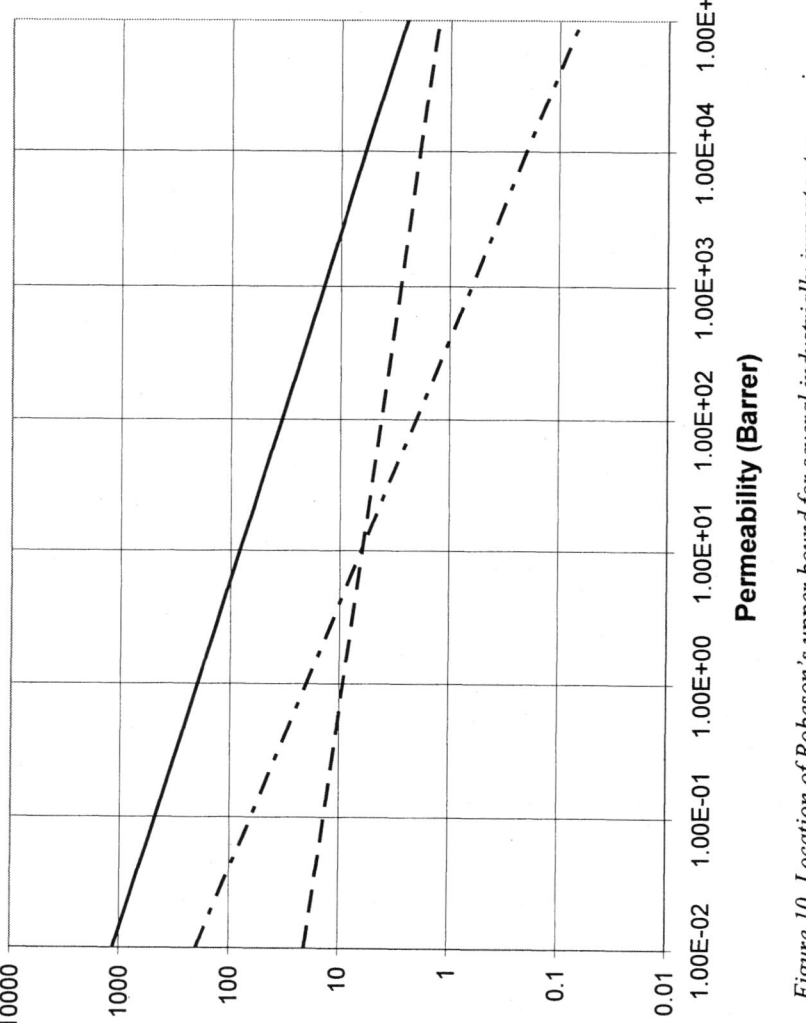

Figure 10. Location of Robeson's upper bound for several industrially important gas pairs: solid line CO_2/CH_4, dash-dot H_2/CO_2, dash O_2/N_2.

membrane pore structure consists of straight, parallel cylindrical pores log-normally distributed about an average pore size. To generate the upper bound, the average pore size was varied while holding constant the ratio of the coefficient of variation to average pore size as well as the ratio of membrane porosity to membrane thickness; values of both parameters were selected to provide the best estimate of the upper bound.

Our understanding of the hollow fiber spinning process has evolved significantly. McKelvey et al. [42] provide an excellent guide to the effect of hollow fiber spinning variables on the development of macroscopic membrane properties such as fiber size; size variance; circular, concentric cross-sections; and fiber breaks. As predicted by hollow fiber spinning theories [15], the primary variables that determine fiber shape are flow rates of the polymer and bore fluids, the draw ratio (i.e., the ratio of the final filament velocity to that at the exit of the spinneret), and spinneret dimensions. The dimensional changes in the draw zone also may be controlled by the formation of a large recirculating gas region in the fiber bore [43]. For a given spinneret geometry, recirculating regions appear below a critical bore-to-clad viscosity ratio and below a critical bore-to-clad flow rate ratio. Phase separation and polymer solidification rates also are primary factors. Theoretical models of momentum, heat, and mass transfer in the spinline provide useful tools for predicting asymmetric membrane structure formation and interpreting experimental results [44].

The literature also provides examples of current module design and manufacture. To improve the uniformity of fiber packing and facilitate the formation of the fiber bundle, individual fibers or tows of fibers may be woven into a hollow fiber membrane fabric [45]. As illustrated in Figure 11, this fabric can be wrapped around a supporting tube to form a module. Additionally, with proper control of the viscosity of the tubesheet forming adhesive, tubesheets may be formed by dipping the ends of the fiber bundle in a mold containing the potting compound. Such a process avoids the complexity and cost of centrifugal potting [17].

Innovations in spiral wound module design include the ability to incorporate a large number of membrane envelopes in one module [46], envelopes that allow counter-current contacting instead of cross-flow contacting [47], and envelopes that allow the use of a sweep stream [47]. An example of an envelope and module design that allows the use of sweep in counter-current contacting is illustrated in Figure 12 [47]. This design utilizes glue lines inside the envelope to direct the sweep and permeate counter-currently to the feed flow along the surface of the envelope. The benefits of using such a design must be balanced against the additional cost associated with creating the flow channels.

The importance of fluid mechanics in module performance and design has received a great attention. This includes experimental and theoretical studies of the development of velocity and concentration fields within modules as well as fouling and changes in membrane structure.

304

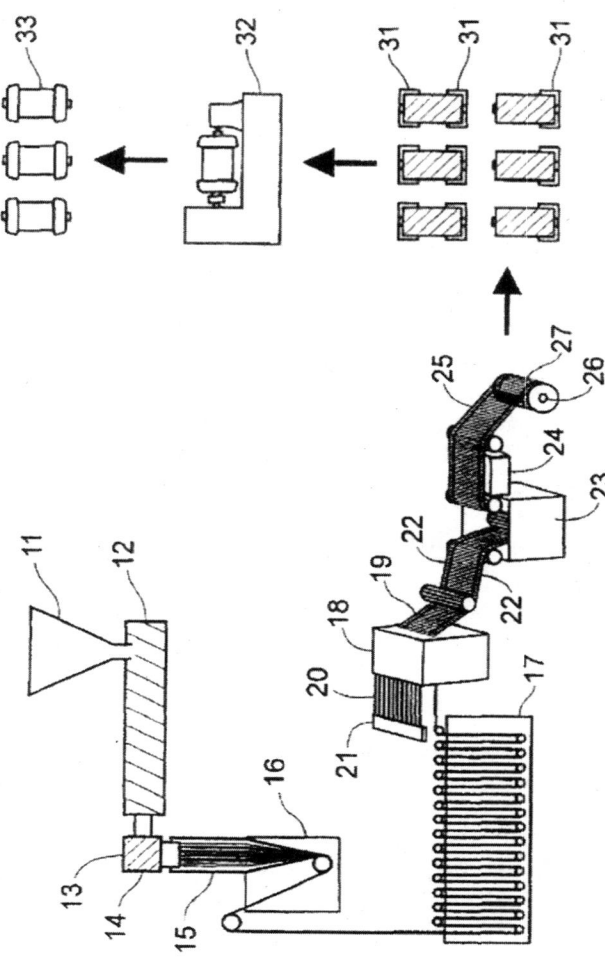

Figure 11. Hollow fiber module fabrication from a woven ho   v fiber fa `   `c [45]. Tows of hollow fiber are woven into a fabric which is subsequently wrapped around a supporting tube to form a bundle. Tubesheets are formed by dunk potting and fiber lumens opened be machining.

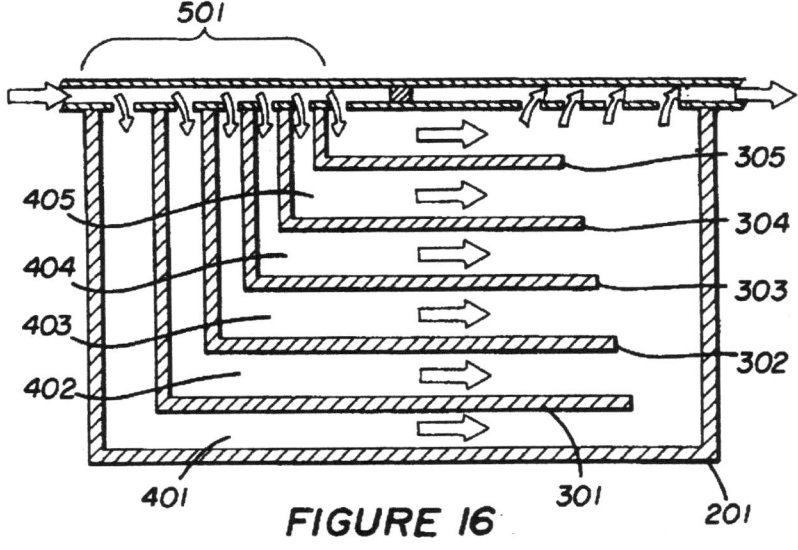

FIGURE 16

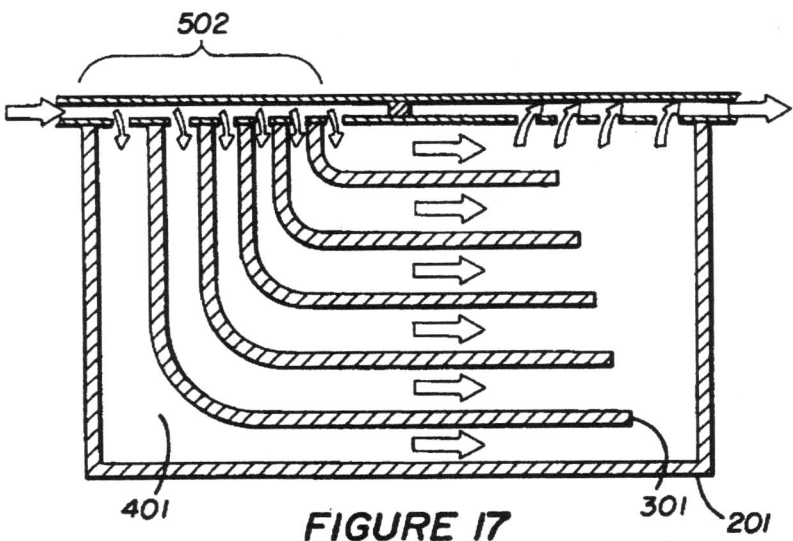

FIGURE 17

Figure 12. A spiral wound module envelop and permeate collection tube that permit counter-current flow between the permeate and the retentate [47]. Adhesive lines within the envelope direct permeate flow and the modified permeate collection tube allows introduction of a sweep stream.

Experimentally, non-invasive observations are possible with a wide range of optical techniques including: direct observation above the membrane, direct observation through the membrane, laser triangulometry, optical shadowgraph, refractometry, interferometry, photo-interrupt sensors, fluorescence, and particle image velocimetry [48-49]. Optical techniques are very effective for optically transparent systems, but such systems commonly require construction of special, idealized test systems. Non-optical methods are required for observations of more realistic systems and for use with commercial modules. Non-optical techniques include: impedance spectroscopy, ultrasonic reflectometry, magnetic resonance imaging, radio isotope labeling, x-ray computed tomography, small angle neutron scattering, electrochemical methods, and constant temperature anemometry [48-49]. These experimental techniques have been developed primarily for membrane filtration processes but have been applied to other separations including dialysis.

The use of computational fluid mechanics (CFD) to evaluate the performance of membrane separation processes has blossomed over the past decade. A recent review [50] summarizes this body of work which examines the complex interactions between momentum, heat and mass transfer. CFD allows prediction of concentration boundary layer and cake thickness along a membrane surface and throughout a module from which module performance can be determined. This tool has been used to evaluate a number of module design features and operational modes without having to construct and test a prototype including: module geometry, spacer design, pulsatile flow, gas sparing, and flow instabilities such as Dean and Taylor vortices. Although one would not rely solely on computational predictions, the results can be used to evaluate design modifications and identify the most promising for experimentation. Additionally, the results can be used to determine if a system is performing at its full potential.

A unique interaction between fluid mechanics and transport exists for filtration processes. Such processes perform better than expected based on the predicted impact of concentration boundary layers. The improvement in performance, a rare occurrence for membrane processes, arises from a combination of hydrodynamic diffusion and inertial lift [51]. Hydrodynamic interactions between particles or colloids that accumulate in the concentration boundary layer lead to shear-induced diffusion away from the membrane surface. Shear-induced diffusion can be significantly larger than molecular diffusion and thereby reduce surface concentrations. For sufficiently large particles at high shear rates, inertial lift becomes the dominant mechanism for particle movement away from the membrane.

The formation of a cake or gel layer on the membrane surface during filtration is highly dependent on filtration rate. A critical flux appears to exist below which fouling does not occur [52]. The critical value is dependent on particle-membrane interactions and device-specific hydrodynamic conditions.

Although the subject of continuing debate, it appears to provide a useful indication of when to expect fouling [53].

Recent work summarizes potential inefficiencies in module performance due to poor fluid distribution for hollow fiber modules [54-55]. The primary sources arise from either fiber bundle features that affect the uniformity of flows within the bundle or fluid distribution from inlet manifolds into the fiber bundle. These sources are illustrated in Figure 13.

The uniformity of flows within a bundle is controlled by the uniformity of the flow channels – some variation in fiber diameter and inter-fiber spacing is inevitable in module manufacture. Assuming the same pressure drop exists between inlet and outlet, flows will be higher in larger channels than in smaller channels since flow rate is proportional to the fourth power of the hydraulic diameter. If the variation in channel dimensions is large enough, the large channels allow fluid to by-pass the fiber bundle and have a significant detrimental impact on performance. The effects appear first at higher product purity and can be evident for variances as low as 10% of the average hydraulic diameter. The effects of variation in fiber transport properties (permeance and selectivity) are similar to those of flow channel dimension but are significant only for much higher levels of variance.

Distribution of fluid from inlet manifolds into the fiber bundle (lumen and shell spaces) also can impact performance. This effect arises from pressure drops within the manifold that lead to different pressure drops across the flow channels within the bundle.

Currently, the largest market for membranes is dialysis [56]. Driven by a global increase in the number of patients with end-stage renal disease, dialysis membranes are the leader in terms of surface area and sales revenue. Filtration membranes, ultrafiltration and microfiltration, are the next largest market with combined sales slightly less than that of dialysis membranes. Reverse osmosis and gas separation membranes round out the top five markets.

Future

During the course of the past decade, numerous reviews of membrane technology have been published as well as predictions for future growth. This section summarizes a number of these reviews and the conclusions contained therein.

Interest in membrane technology tends to increase with crude oil prices. Currently, many manufacturers of products are running at full capacity or expanding their capacity to meet increased demand. This correlation is due to the fact that membrane processes can offer significant energy reductions.

One of the most dramatic examples of the energy savings offered by membrane processes is desalination. Initial desalination plants relied exclusively

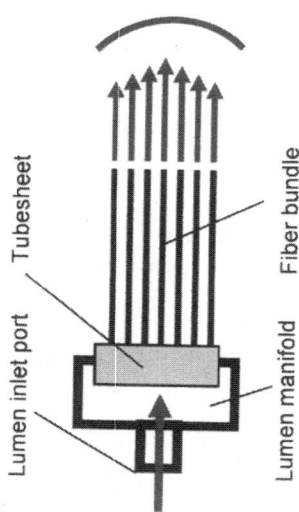

(a)

Lumen inlet port

Tubesheet

Fiber bundle

Lumen manifold

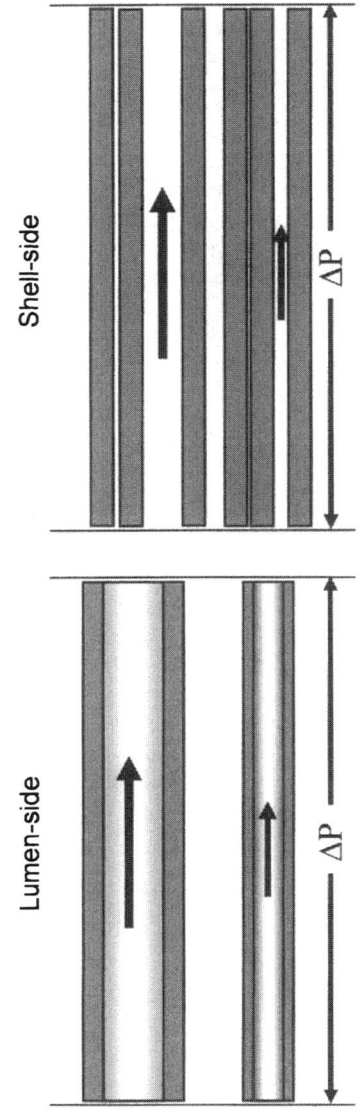

(b)

Figure 13. Sources of non-ideal flows in hollow fiber bundles: a) fluid distribution from the inlet manifold into the fiber bundle and b) fluid distribution within the fiber bundle. Fluid distribution from the lumen manifold into the fiber lumens (a, left hand side) can lead to higher flow rates in the fibers in the center of the bundle. Fluid distribution from the shell manifold (typically a fiber free collar that extends around the fiber bundle) into the shell can lead to higher flow rates near the inlet port than opposite it. Assuming a common pressure drop across all fibers, lumen flows are higher in larger fibers (b, left hand side). Similarly, assuming a common pressure drop along flow channels in the shell, shell flows are higher in larger channels.

on multi-stage flash and multiple effect distillation processes – processes that utilize energy as the separating agent. However, as membrane technology has evolved, initially driven by the OSW and later by the reverse osmosis industry, it has overtaken thermal technologies in installed capacity [57].

The economic driving force for this change is illustrated in Figure 14 which compares the energy consumption and cost of producing a cubic meter of water using thermal and membrane processes. Membrane processes offer almost an order of magnitude reduction in energy consumption *and* a 25% reduction in total cost. Moreover, new membranes and process designs offer the potential to reduce energy use to ~2 kWh/m^3 and cost below $0.5/m^3 [58-59].

The relative competitiveness of the two processes is determined primarily by the relative cost of energy and membranes. The dramatic drop in membrane costs evident in Figure 14 is due to economies of scale associated with mass production, improvements in membrane transport properties (water transport rates and salt rejection), and careful attention to feed pre-treatment and module design to reduce fouling and increase lifetime. These cost reductions in combination with the current cost of energy make membrane processes the preferred technology for future desalination plants.

Koros [60] refers to the replacement of thermal technologies with low-energy, non-thermal technologies such as membranes as the end of the thermal age of separations. Opportunities to save more than 240 TBtu/yr by replacing thermal separations have been identified in the four largest energy consuming industries: chemicals, petroleum refining, forest products, and mining [61]. Drioli and Romano [62] identify several specific areas for membranes ranging from water recycling in the textile industry to tomato juice concentration.

An area of current interest is the purification of biofuels produced from biomass. The fermentation of sugar from starch or cellulose typically produces an aqueous mixture of ~ 6-8% ethanol by weight. Other low molecular alcohols (e.g., butanol) may be produced at concentrations as low as 1% and the use of alcohol-resistant yeast can allow fermentation up to a concentration of 15%. Currently, fuel grade alcohol is produced by distillation of the aqueous mixture. However, membrane pervaporation [63], vapor permeation [64], and hybrid processes offer the potential for significant energy and cost savings.

The competition between distillation and membranes for biofuel purification is strikingly similar to that for desalination. The current state for biofuel purification corresponds to the state of desalination 40 years ago when membrane processes first appeared – membrane processes show potential and start-up companies are touting products but acceptance is low due to concerns over cost and reliability. However, experiences with desalination may dramatically reduce the time required for membrane processes to become as competitive in the biofuel industry as they are in desalination.

Future opportunities for the membrane industry in a few selected areas are described in the following sections. This list is not intended to be comprehensive and reflects in part the interests of the author.

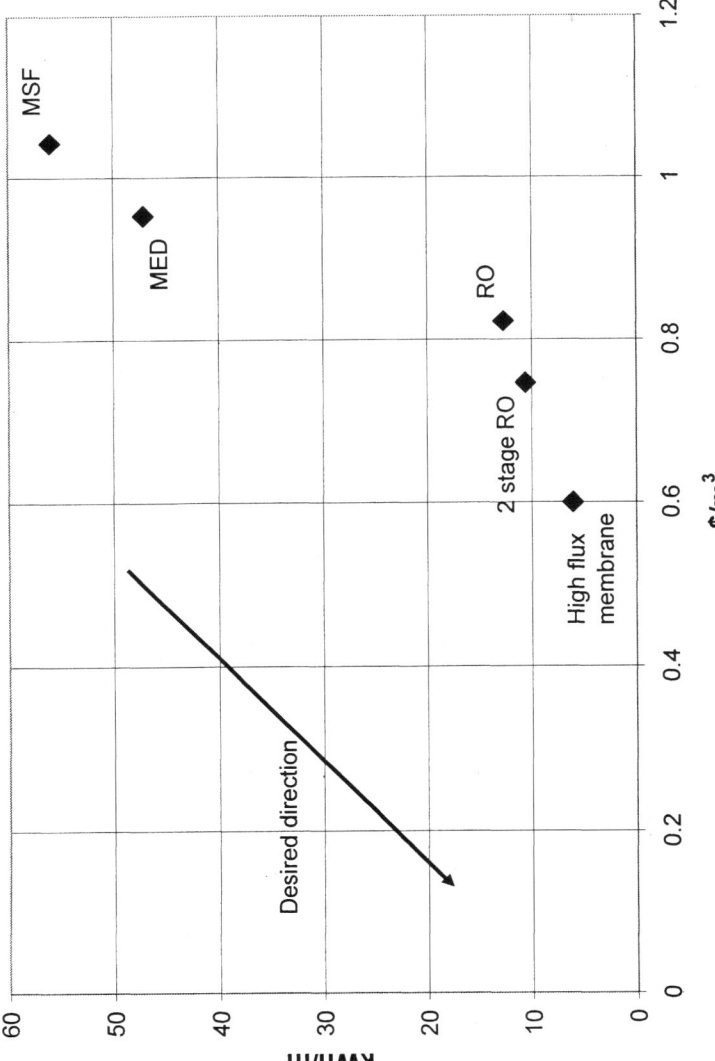

Figure 14. Energy requirement and cost of various desalination technologies. Data for multi-stage flash (MSF), multiple effect distillation (MED), reverse osmosis (RO), and two-stage RO (brine concentrator) taken from [57]. More recent data reported for a high flux membrane taken from [58].

Gas separation

Currently the largest market for membrane gas separation is nitrogen production from air. Baker [65] predicts this market will grow moderately but greater growth will occur in the refinery, petrochemical, and natural gas industries. Specific applications include hydrogen recovery, light hydrocarbon separations (e.g., propylene/propane), and natural gas treatment. Significant opportunities exist for a hydrogen-rejecting membrane and process that minimize fouling by light hydrocarbons. While polymeric materials are being used for some light hydrocarbon separations, facilitated transport or ceramic membranes with higher selectivities are needed for membranes to be competitive in most separations. Finally, natural gas treatment may become the largest membrane market in the next decade. Membrane processes are competitive options for carbon dioxide and nitrogen removal provided fouling from other contaminants in the as-produced gas can be avoided. Additionally, membranes may be used for natural gas liquids separation and dehydration.

Due to potential global warming effects, significant interest exists in technologies for post-combustion carbon dioxide recovery from flue gas. This application is particularly challenging due to the high temperature (~350 °C), low pressure (~ 1 atm), low concentration, and high flow rate of the feed. Moreover, DOE targets of 90% carbon dioxide recovery with no more than 10% cost increase are very aggressive.

Aaron and Tsouris [66] rank membrane separation processes as one of the most promising options. They anticipate advances in ceramic and metallic membrane technology will eventually lead to membrane processes that are more efficient than absorption – the best current option. Favre [67] argues that polymeric membrane technology is competitive today despite its dismissal in most studies. A techno-economic analysis suggests membranes are competitive when carbon dioxide concentration reaches ~20% (as found in the effluent from cement and steel factories) and vacuum is used to drive carbon dioxide transport. Although the subject of debate, the potential for membrane technology will continue to attract interest and research effort.

To surpass Robeson's upper bound, materials are emerging that rely on transport mechanisms other than solution-diffusion through glassy or rubbery polymeric materials. In particular, a number of materials have been developed that possess fixed microporosity (2 nm or less) in contrast to the activated, transient molecular gaps that give rise to diffusion in most polymers. These materials include amorphous and crystalline (zeolite) ceramics [68-69], molecular sieve carbons [70], polymers that possess intrinsic microporosity [71-72], and carbon nanotube membranes [73-76]. Transport in such materials is determined primarily by the average size and size distribution of the microporosity – the porosity can be tuned to allow discrimination between species that differ by less than one Angstrom in size. However, surface adsorption can dramatically influence transport rates for mixtures that contain a pore accessible, strongly adsorbing species.

Transport rates for single walled [75] and multi-walled [73-74] carbon nanotube membranes are particularly impressive. These membranes consist of aligned array of the nanotubes embedded in a solid polymeric or inorganic matrix such that transport is preferred through the nanotubes. Single component gas permeation rates through single walled nanotubes (1.3-2 nm in diameter) are 1-2 orders of magnitude larger than Knudsen flow as predicted by simulations [76]. Although gas permeation rates through larger multi-walled nanotubes (~7 nm in diameter) are consistent with Knudsen flow, liquid permeation rates are 4-5 orders of magnitude larger indicating slip occurs at the fluid-wall interface.

Pessimism exists about the ability to fabricate cost effective modules from microporous membranes. Specific concerns include the stability of the membranes (especially for silica membranes in humid, high temperature environments), the ability to produce large uniform membrane areas, the effect of thermal cycling (especially for applications that require operation at elevated temperature), the ability to form tubesheets, and the ultimate cost of manufacture.

These concerns can be addressed partially through the use of mixed-matrix membranes [77-79]. Dispersing the microporous material in the form of small particles within a polymeric matrix simplifies membrane formation dramatically. Mixed matrix materials possess transport properties intermediate between those of the polymer matrix and the microporous particle and operating temperatures are limited by the thermal stability of the polymer matrix. However, proper selection of the matrix, control of particle volume fraction, and development of a membrane formation process can yield materials with properties that approach those of the particles [77-78]. Special attention must be given to the particle-polymer interface. If the interface morphology is uncontrolled, the matrix may: 1) not wet the particle leaving a non-selective void around the particle, 2) enter the particle and block pores, or 3) rigidify around the particle and block access to it [79].

Dense ceramic or metallic membranes offer an additional alternative to solution-diffusion polymeric materials. Palladium alloy membranes [80] can be formed as thin films on metallic or ceramic supports by electroless plating, chemical vapor deposition, or physical sputtering. The polycrystalline metallic membranes thus formed selectively transport hydrogen due to the unique ability of hydrogen to dissociate and diffuse through the metal matrix. Commonly, palladium is alloyed with silver to avoid the effects of hydrogen embrittlement. Concerns also exist with changes in metal structure associated with metallic interdiffusion with the support or carbon deposited on the surface. High temperature perovskite type proton-conducting ceramics offer an alternative to palladium for hydrogen separation [68].

Oxygen ionic conducting perovskite type membranes permeate oxygen exclusively due to their unique combination of ionic and electronic

314

conductivities. Commercialization is being pursued by two large industrial teams: one led by Air Products [81] and the other by BP-Amoco-Praxair [68]. These teams are attempting to resolve difficulties associated with low oxygen permeances, instability in reducing environments, and stresses induced by oxygen vacancy gradients (arising from the dependence of density on oxygen concentration). Concerns also exist over the ability to form tubesheets and modules that will withstand the high operating temperatures required by the process and associated thermal cycling.

Biotechnology

Membranes are an attractive alternative for separations of labile species, such as those found in the biotechnology industry, due to their relatively low temperature, low pressure operation. Initially membranes developed for other industries were used but specialized materials and processes were developed as the industry grew.

Today, membranes perform critical separations in the production, purification and formulation of biotechnology products. Specific functions include sterilize filtration (both upstream and downstream of the reactor), virus removal (both upstream and stream), perfusion, medium exchange, harvest, and purification.

van Reis and Zydney [82] identify a number of opportunities in this expanding market. High performance tangential flow filtration (HPTFF) processes allow the separation of molecules nearly identical in size, for example the separation of protein variants that differ in a single amino acid residue [83] or removal of impurities comparable in size to a desired product [84]. This selectivity arises primarily from the use of charged membranes [85]. Such membranes rely on electrostatic interactions (instead of steric) to control transport and can surpass the upper bound on ultrafiltration membrane performance [41]. Additionally, buffer conditions that alter solute charge state can have a profound effect on selectivity [86]. HPTFF selectivity can be enhanced further through control of the trans-membrane pressure driving force for filtration and use of co-current filtrate flow to maintain a constant pressure difference along the membrane surface. HPTFF processes permit. Continued development of HPTFF processes may allow their use throughout the purification process and to combine several process steps (e.g., purification, concentration, and buffer exchange) into one unit operation.

Ultrafiltration unit operations would benefit from increased mass transfer coefficients. In addition to improvements in membrane properties, module designs that reduce concentration polarization in the contacting fluid, such as the use of a rotating membrane [87] or module vibration [88], are needed.

Adoption of membrane chromatography will require increases in binding capacity without an increase in cost. Moreover, if costs can be reduced sufficiently, single use columns could eliminate the need for some current chromatography steps [82]. Like chromatography, improvement in virus filters, especially for use in industry preferred normal flow filtration, will require increases in capacity/throughput and reduction in costs.

Exciting opportunities exist for increasing binding specificity through functionalized membranes. Separation of similarly sized molecules by binding with biological ligands (an affinity membrane) [89] remains a topic of great interest but limited commercial development. Moreover, enzyme immobilization [90] can impart bio-catalytic activity to the membrane.

Water and wastewater

The art of membrane manufacture has been perfected by a number of companies. Performance characteristics (water permeability as a function of salt rejection) of the standard 8 inch modules offered by several competitors are compared in Figures 15 and 16 for brackish water and seawater membranes, respectively.

Figure 15 clearly illustrates the similar characteristics of the brackish water modules. For 99.5% salt rejection, membrane productivity varies slightly from ~26 to 27 gallons/day/ft^2. At 99.7%, the spread increases to ~24-27. Somewhat larger differences are evident in the seawater offerings – rejections of 99.75-99.8% are accompanied by productivities ranging from ~17-22 gallons/day/ft^2. Manufacturers have had to match performance enhancements of competitors to the economic benefit of the end user.

While advances will continue in membrane manufacture, concentration polarization and fouling remain significant problems in membrane processes used to produce drinking water from saline or fresh water sources [91-92]. To control fouling, a variety of pretreatment, material design, and module design strategies have been studied.

Surface modification to control hydrophilicity, charge, roughness, and porosity can impact fouling dramatically [93-94]. A wide variety of surface modification techniques are available including: including: plasma treatment, chemical treatment, ion beam irradiation, physical adsorption, and graft polymerization [95]. Module array configuration (staging), geometry, and spacer design also impact fouling [96]. Although significant progress has been made in addressing fouling challenges, a universally applicable strategy to control fouling is not available.

An additional challenge in drinking water applications arises from the growing size of water plants. Plants can require hundreds or thousands of modules. Such large numbers complicate construction and maintenance.

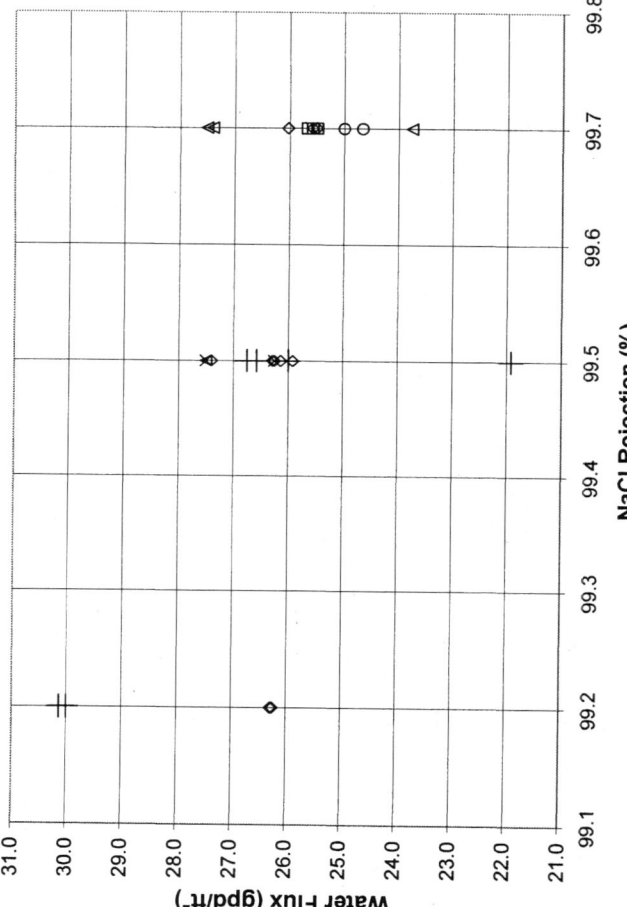

Figure 15. Permeate flow rate per unit membrane area (gallons/day/ft²) and NaCl rejection of brackish water membranes offered by: GE (◇), FilmTec/Dow (×), Koch (○), Toray (□), Trisep (+), and Nitto Denko/Hydranautics (△). All values taken from the manufacturers' web sites. Test conditions for all membranes were: 2000 ppm NaCl feed concentration, 225 psi feed pressure, 77 °F feed temperature, and 15% permeate recovery. Feed pH for test conditions varied slightly in the tests from 6.5-8. Note that a 1500 ppm NaCl feed was used to obtain the Nitto Denko/Hydranautics values; normalization to 2000 ppm NaCl reduces permeability by ~3%.

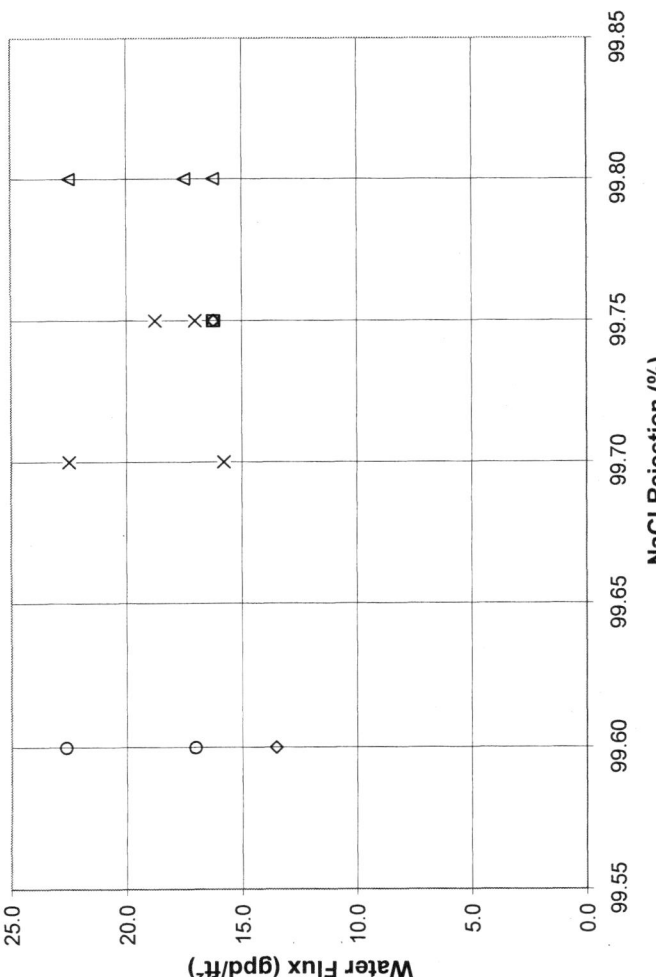

Figure 16. Permeate flow rate per unit membrane area (gallons/day/ft²) and NaCl rejection of seawater membranes offered by: GE (◇), FilmTec/Dow (×), Koch (○), Toray (□), and Nitto Denko/Hydranautics (△). All values taken from the manufacturers' web sites. Test conditions for all membranes were: 32,000 ppm NaCl feed concentration, 800 psi feed pressure, and 77 °F feed temperature. Permeate recovery varied slightly in the tests from 7-10% and feed pH varied from 6.5-8. Note that a 32,800 ppm NaCl feed was used to obtain the Koch values; normalization to 32,000 ppm NaCl increases permeability by ~2%.

Attempts to produce larger modules are underway but technical problems remain. Manufacturers also must adopt a standard size for larger units.

A number of alternatives to reverse osmosis are being considered. Two promising alternatives are membrane distillation [97] and forward osmosis [98]. Membrane distillation relies on vapor pressure differences across a membrane, arising from a temperature difference, to drive water transport. The process utilizes low temperature heat sources and operates at low pressure which can reduce operating costs relative to reverse osmosis. Forward osmosis relies on water permeation across a water selective membrane to a draw solution – the reverse of reverse osmosis. The water must then be separated from the draw solution but this may be less expense than reverse osmosis because the process operates at low pressure.

Tremendous opportunity exists for hybrid processes consisting solely of membrane processes or a combination of membrane and non-membrane processes. Of the large number of potential combinations, studies of several are reported in the literature including: nanofiltration with reverse osmosis [99]; nanofiltration with electrodialysis [100]; ultrafiltration with nanofiltration and reverse osmosis [101]; ultrafiltration with membrane distillation [102]; nanofiltration with reverse osmosis and a microfiltration membrane-based sorbent [103]; microfiltration with flotation [104]; microfiltration and ultrafiltration with ozone and activated carbon adsorption [105]; and membrane processes with photocatalysis [106-107]. Despite the activity in this area, a comprehensive approach to designing hybrid systems does not exist; future work would benefit from the development of such a design framework.

Submerged membrane bioreactors are revolutionizing waste water treatment [108]. These units dramatically increase the capacity of waste water treatment ponds while simultaneously producing a higher quality water by using an ultrafiltration membrane to remove treated water from the mixed liquor suspended solids (MLSS) produced by biological treatment. Such 'process intensification' is one of the hallmarks of applications where membrane processes have achieved commercial success (in addition to energy reduction and purification of labile compounds).

The growth of the industry is due to the development of robust, high permeability membranes in fiber or sheet form suitable for the MLSS environment. Additionally, modules and processes have been designed to reduce fouling of the membrane surface through a combination of aeration and backpulsing. Aeration or scouring of the membrane surface by bubbles minimizes the accumulation of suspended solids on the surface. Periodic backflushing, reversing the direction of flow through the membrane, helps remove what does accumulate and thereby increase membrane lifetime. Other currently used strategies for fouling reduction include periodically stopping flow (without flow reversal) and the addition of cleaning agents.

As with membranes for water production, fouling remains a significant problem [109]. Fouling increases the trans-membrane pressure drop required to maintain productivity or results in a productivity drop if the pressure difference is held constant. Development of enhanced anti-fouling coatings or membrane structures could have a dramatic impact on process economics. Likewise, new module designs or operational modes that reduce fouling are of interest.

The physics of MBR operation are poorly understood. Due to the nature of the measurement, the critical flux for MBRs is measured as an average along the length of the submerged fibers – the flux varies along the fiber length but local measurements of flux generally are not possible. This length average value is an apparent function of feed water characteristics (chemical oxygen demand, mixed liquor suspended solids, and viscosity), backflushing frequency and duration, gas sparging rate, compressibility of accumulated foulant, and observation time [110-113]. Optimization of such systems will require a better understanding of the physical origin of fouling and its control through system design and operation.

Nanofiltration

Nanofiltration emerged as a separate filtration membrane process with the development of low-pressure reverse osmosis membranes [114]. The rejection of monovalent salts by these low-pressure or loose reverse osmosis membranes was lower than that of high-pressure membranes. However, multi-valent salt rejection remained high as well as rejection of organic molecules of moderate molecular weight (greater than 300). Such separation characteristics along with the lower cost of low-pressure operation led to the use of nanofiltration in a wide range of applications in ground and surface water treatment, waste water treatment, and desalination pretreatment.

Ground and surface water applications include water softening [115], water disinfection by-product removal [116], natural organic matter removal [116-117], pesticide removal [118], and removal of a wide range of other pollutants [119]. Nanofiltration also can be used to remove microorganisms and viruses [120-121]. Such applications do not require the low molecular weight selectivity of reverse osmosis and are well suited for low-pressure nanofiltration.

Likewise, nanofiltration can be integrated into waste water treatment. Combined reverse osmosis/nanofiltration processes can offer higher water recovery than either process alone [122]. Moreover, nanofiltration can be combined with other membrane filtration processes [123], electrodialysis [124], or other waste water treatment processes such as ozonation [125].

Nanofiltration is used for water pretreatment in both thermal and membrane desalination processes. Desalination processes may rely solely on membrane processes (e.g., reverse osmosis in combination with nanofiltration, microfiltration or ultrafiltration [126]) or a combination of membrane and thermal (e.g., multistage flash [127]).

Fouling remains the biggest challenge with the use of nanofiltration membranes. Fouling occurs with the deposition of inorganic components, organic matter, colloids, bacteria or other suspended solids [114]. The foulants may deposit in membrane pores thereby reducing effective pore size/porosity or as a cake on the surface of the membrane. Both mechanisms lead to undesirable reductions in water permeation rates that require an increase in operating pressure to maintain productivity and a concomitant increase in membrane cleaning frequency.

Attempts to quantify the fouling propensity of a feed water such as the Silt Density Index (SDI) and Modified Fouling Index (MFI) have met with limited success [128] due to the complex interactions between membrane and foulant. Fouling by natural organic matter depends critically on solution pH, ionic strength, and the presence of divalent cations due to changes in macromolecular structure [129]. Techniques for monitoring biofilm and scale formation are summarized in the literature [130].

Transport through nanofiltration membranes is controlled primarily by electrostatic and steric interactions. The extended Nernst-Plank equation commonly is used with Donnan and steric partitioning to predict transport rates based on effective membrane charge density, pore radius, and thickness to porosity ratio [131-132]. Inclusion of solute-pore hydrodynamic interactions and a pore size distribution improves the predictive and correlative capabilities of the models [133].

Use of nanofiltration for non-aqueous separations is limited by membrane compatibility – a common material in composite nanofiltration membranes used for aqueous separations is polysulfone which possesses limited solvent resistance [134]. However, during the past two decades a number of materials have emerged with improved solvent resistance that have enabled a broad range of organic solvent nanofiltration (OSN) applications. These materials include polydimethylsiloxane, polyphenylene oxide, polyacrylic acid, polyimides, polyurethanes, and a limited number of ceramics. Commercial products are offered by Koch Membrane Systems, W.R. Grace, SolSep, and Hermsdorfer Institut fur Technische Keramik (HITK) [135].

Transport in OSN membranes occurs by mechanisms similar to those in membranes used for aqueous separations. Most theoretical analyses rely on either irreversible thermodynamics, the pore-flow model and the extended Nernst-Planck equation, or the solution-diffusion model [135]. To account for coupling between solute and solvent transport (i.e., convective mass transfer effects), the Stefan-Maxwell equations commonly are used. The solution-diffusion model appears to provide a better description of mixed-solvent transport and allow prediction of mixture transport rates from pure component measurements [136]. Experimental transport measurements may depend significantly on membrane preconditioning due to strong solvent-membrane interactions that lead to swelling or solvent phase separation in the membrane pore structure [137].

OSN has found applications in the food, pharmaceutical, and petrochemical industries [135, 137]. The process for edible oil production from oilseed commonly requires: 1) seed preparation, 2) solvent extraction, 3) solvent removal from the meal, 4) solvent recovery from the extract, and 5) further refining of the crude oil [138]. Energy requirements for cottonseed, corn, peanut, and soybean oil processing alone are nearly 65 trillion Btu/yr in the United States. Membrane processes offer the potential for reducing energy consumption by over 30%. The greatest potential exists in recovery of the extraction solvents and crude oil refining, especially degumming (removal of phospholipids and pigments) and deacidification (removal of free fatty acids). Other applications in the food industry include the recovery of amino acids and their derivatives (e.g., aspartame) [139] and the purification/concentration of bioactive compounds (e.g., nutraceuticals) [140].

Myriad applications for OSN exist in the pharmaceutical industry [135]. Specific applications include isolation and concentration of pharmaceutical intermediates and products [141], solvent exchange during pharmaceutical synthesis [142], diastereomeric resolution of chiral bases [143], and membrane based solvent back extraction [144].

Large scale OSN applications exist in the petrochemical industry [145]. One of the largest is lube oil dewaxing. The MAX-DEWAX^{TM} process has been used commercially since 1998 to recover solvents used in dewaxing. The use of OSN is driven by >20% reduction in energy consumption, increased product recovery, increased product quality, and increased throughput. Additional applications include solvent deoiling [145], homogeneous catalyst recovery [146], separation of phase-transfer agents [147], solvent exchange [148], and aromatics separation [149]. OSN membranes also can be used in a pervaporation process to remove sulfur compounds from hydrocarbon streams [150].

Nanofiltration will continue to be a rapidly growing segment of the membrane industry. To sustain this growth, a better understanding of transport is needed, in particular the differences between aqueous and organic systems. Additionally, tremendous opportunities exist in membrane and module development for separations that fall in the gap between reverse osmosis and ultrafiltration.

Alternative Energy

The mixing of fresh water from estuaries with seawater has the potential to produce more than 2.5 TW of power globally [138]. Recovering this energy has been discussed for the past half century using desalination processes operated in a reverse mode. Proposed alternatives include the use of reverse electro-dialysis (RED) [139], pressure retarded osmosis PRO [140-141], and vapor pressure differences [142].

RED and PRO are membrane processes that have received increased attention due to the high cost of energy. In RED, alternating cation and anion exchange membranes separate compartments filled alternately with concentrated (seawater) and dilute (fresh water) salt solutions. Diffusion of sodium from the concentrated solution to the dilute solution across the cation exchange membrane accompanied by simultaneous diffusion of chloride across the anion membrane leads to the generation of an electric current. In PRO, the concentrated and dilute solutions contact each other across a semipermeable membrane. The membrane, a reverse osmosis membrane, allows water transport and rejects salt. Without mechanically pressurizing the concentrated solution, water transport occurs from the dilute to the concentrated stream and can increase in the pressure of the concentrated stream. Expansion of this stream in a turbine generates electricity.

Experimental measurements of RED and PRO system performance are limited. However, a recent techno-economic analysis [143] suggests where the processes possess the greatest potential. RED is most attractive when using seawater as the concentrated salt solution while PRO is most attractive for more concentrated brines. An intriguing application is the use of PRO to recover energy from the concentrated brines produced by reverse osmosis desalination.

Additional work is required to realize the potential of these processes. Both would benefit from higher permeability membranes that increase power density (power per unit membrane area). Furthermore, the development of lower resistance spacers is critical for RED while a better understanding of water transport is needed for PRO.

The use of membranes to produce power is an exciting new area for the industry. Although intellectually appealing, commercialization will require significant manufacturing and system design advances.

Conclusions

Membrane separation processes have evolved into a multi-billion dollar industry. Industry growth was accelerated dramatically by funding provided through the Office of Saline Water. The processes developed to fabricate membranes and modules for water desalination have been adapted for virtually every membrane process developed afterwards.

The adoption of membrane processes is driven by economics. Membrane processes commonly are more capital intensive due to membrane costs. However, these costs are offset by either: 1) reduced energy consumption, 2) the ability to separate labile components, or 3) process intensification (increased throughput per unit volume). The energy savings provided by membrane processes will attract increasing attention as global competition for energy increases and costs rise.

The next generation of membrane technology will require surmounting the challenges of the current generation including how to:

1. overcome the upper bound that exists on transport properties for current materials – rigid, microporous materials offer promise for gas and vapor separations but manufacturing concerns exist;
2. eliminate fouling in membrane processes;
3. develop high-selectivity filtration membranes for both aqueous and organic solutions;
4. apply biomimetic concepts in synthetic membrane technology;
5. enhance transport properties through surface functionalization;
6. increase module size and optimize module performance; and
7. reduce manufacturing costs.

A new application area for membranes is energy production. Reverse electrodialysis and pressure retarded osmosis could provide significant quantities of energy from the mixing of fresh water with seawater or mixing the concentrated brine effluent of desalination plants with seawater. However, these applications will require significant reduction in membrane and module costs.

Opportunities for innovation in membrane technology are numerous, so membranologists should remain busy for the foreseeable future.

References

1. J.A. Nollet, Lecons de physique experimental, Hippolyte-Louis Guerin and Louis-Francios Delatour, Paris, 1748.
2. J.H. van't Hoff, Die Rolle osmotischen Drucks in der Analogie zwischen Losungen und Gasen, Z. Physik. Chem., 1 (1887) 481-508.
3. A. Fick, Uber diffusion, Ann. Phys. Chem. 94 (1855) 59-86.
4. T. Graham, Phil. Mag., Liquid Diffusion Applied to Analysis, 151 (1861) 183-224.
5. T. Graham, Phil. Mag., On the Absorption and Dialytic Separation of Gases by Colloid Septa, 156 (1866) 399-439.
6. H. Bechold, Kolloidstudien mit der Filtrationsmethode, Z. Physik. Chem., 60 (1907) 257-318.
7. R.W. Baker, Membrane Technology and Applications, 2nd Edition, Wiley and Sons, West Sussex, England.
8. J. Glater, The early history of reverse osmosis membrane development, Desalination, 117 (1998) 297-309.
9. Records of the Office of Saline Water, http://www.archives.gov/research/guide-fed-records/groups/380.html#380.2.
10. C.E. Reid and E.J. Breton, J. Appl. Polym. Sci., 3 (1959) 133.

11. H.K. Lonsdale, U. Merten, and R.L. Riley, Transport Properties of Cellulose Acetate Osmotic Membranes, J. Appl. Polym. Sci., 9 (1965) 1341.

12. D.R. Paul and D.J. Paciotti, Driving Force for Hydraulic and Pervaporation Transport in Homogeneous Membranes, J. Polym. Sci., Polym. Phys. Ed., 13 (1975) 1201.

13. S. Loeb and S. Sourirajan, Sea Water Demineralization by Means of an Osmotic Membrane, in Saline Water Conversion – II, Advances in Chemistry Series, No. 38 (28?), American Chemical Society, Washington, DC, 1963, pp. 117-132; S. Loeb and S. Sourirajan, High flow porous membranes for separating water from saline solutions, US Patent 3,133,132, 1964.

14. H. Strathmann, P. Scheible, and R.W. Baker, A Rationale for the Preparation of Loeb-Sourirajan-type Cellulose Acetate Membranes, J. Appl. Polym. Sci., 15 (1971) 811.

15. G.G. Lipscomb, The melt hollow fiber spinning process: steady-state behavior, sensitivity and stability, Polym. Adv. Tech., 5 (1994) 745-758.

16. H.I. Mahon, Permeability separatory apparatus, permeability separatory membrane element, method of making the same and process utilizing the same, US Patent 3,228,876, 1966.

17. J.M. Maxwell, W.E. Moore, R.D. Rego, Fluid separation process and apparatus, US Patent 3,339,341, 1967; J.E. Geary, W.E. Harsch, J.M. Maxwell, and R.D. Rego, Method of manufacture of fluid separation apparatus, US Patent 3,442,002, 1969; J.E. Geary, W.E. Harsch, J.M. Maxwell, and R.D. Rego, Method of repairing leaks in fluid separation apparatus, US Patent 3,499,062, 1970.

18. J.C. Westmoreland, Spirally wrapped reverse osmosis membrane cell, US Patent 3,367,504, 1968.

19. D.T. Bray, Reverse osmosis purification apparatus, US Patent 3,417,870, 1968.

20. R.A. Hayes, Polyimide Gas separation membranes, US Patents 4,705,540, 1987; 4,717,393, 1988; 4,717,394, 1988; 4,838,900, 1989; and 4,880,442, 1989.

21. J.N. Anand, D.C. Feay, S.E. Bales, and T.O. Jeanes, Semipermeable membranes consisting predominantly of polycarbonates derived from tetrahalobisphenols, US Patent 4,818,254, 1989

22. A.D. Surnamer and C.-F. Tien, Membranes formed from rigid polyarylates, US Patent 5,013,332, 1991.

23. A.J. Castro, Methods for making microporous products, US Patent 4,247,498, 1981.

24. E.S. Sanders, D.O. Clark, and J.A. Jensvold, Semipermeable membranes with an internal discriminating region, US Patent 4,838,904, 1989.

25. E.S. Sanders, J.A. Jensvold, D.O. Clark, F.L. Coan, H.N. Beck, W.E. Mickols, P.K. Kim, and W. Admassu, Semipermeable membranes with a non-external discriminating region, US Patent 4,955,993, 1990.

26. M. Mulder, Basic Principles of Membrane Technology, Kluwer Academic Publishers, Dordrecht, 1991.

27. W.J. Koros and G.K. Fleming, Membrane-based gas separations, J. Membr. Sci., 83 (1993) 1-80.

28. J.E. Cadotte, Interfacially synthesized reverse osmosis membrane, US Patent 4,277,344, 1981.

29. O.A. Ekiner, R.A. Hayes, and P. Manos, Novel multicomponent fluid separation membranes, US Patent 5,085,676, 1992.

30. C. Fritzmann, J. Lowenberg, T. Wintgens, and T. Melin, State-of-the-art of reverse osmosis desalination, Desalination, 216 (2007) 1-76.

31. L.M. Robeson, Curr. Op. Solid State Mat. Sci., 4 (1999) 549-552.

32. W.R. Browali, Method for sealing breaches in multilayer ultrathin membrane composites, US Patent 3,980,456, 1976.

33. J.M.S. Henis and M.K. Tripodi, Multicomponent membranes for gas separations, US Patent 4,230,463, 1980.

34. K. Keizer, R.J.R. Uhlhorn, V.T. Zaspalis, and A.J. Burggraaf, Transport and related (gas and vapor) separation in ceramic membranes, Key Eng. Mat. 61 & 62, (1991) 143-154.

35. W.J. Koros, Gas separation, in Membrane Separation Systems, R.W. Baker, E.L. Cussler, W. Eykamp, W.J. Koros, R.L. Riley, and H. Strathmann, Eds., Noyes Data Corporation, Park Ridge, NJ, 1991.

36. J.G. Wijmans and R.W. Baker, The solution-diffusion model: A review, J. Membr. Sci., 107 (1995) 1.

37. R.W. Baker, Membrane Technology and Applications, 2nd Edition, Wiley, West Sussex, England, 2004.

38. L.M. Robeson, Correlation of separation factor versus permeability for polymeric membranes, J. Membr. Sci., 62 (1991) 165-185.

39. S.J. Lue and S.H. Peng, Polyurethane (PU) membrane preparation with and without hydroxylpropyl-β-cyclodextrin and their pervaporation characteristics, J. Membr. Sci., 222 (2003) 203-217.

40. B.D. Freeman, Basis of Permeability/Selectivity Tradeoff Relations in Polymeric Gas Separation Membranes, Macromolecules, 32 (1999) 375 - 380.

41. A. Mehta and A.L. Zydney, Permeability and selectivity analysis for ultrafiltration membranes, J. Membr. Sci., 249 (2005) 245-249.

42. S.A. McKelvey, D.T. Clausi, W.J. Koros, A guide to establishing hollow fiber macroscopic properties for membrane applications, J. Membr. Sci., 124 (1997) 223-232.

43. Y. Su, G.G. Lipscomb, H. Balasubramanian, and D.R. Lloyd, Theoretical Observations of Recirculation Regions in the Bore Fluid during Hollow Fiber Spinning, AIChEJ, 52 (2006) 2072-2078.

44. H. Balasubramanian, Heat-Mass-Momentum Transfer in Hollow Fiber Spinning, Dissertation, University of Texas, 2004.

45. P.E. Alei, J.C. Schletz, J.A. Jensvold, W.E. Tegrotenhuis, W. Allen, F.L. Coan, K.L. Skala, D.O. Clark, and H.V. Wait, Loom processing of hollow fiber membranes, US Patent 5,598,874.

46. J.L. Bartlett, Reverse osmosis membrane module, US Patent 4,235,723

47. D.R Reddy, Spiral-wound membrane separation device with feed and permeate/sweep fluid flow control, US Patent 5,096,584.

48. V. Chen, H. Li, and A.G. Fane, Non-invasive observation of synthetic membrane processes – a review of methods, J. Membr. Sci., 241 (2004) 23-44.

49. J.C. Chen, Q. Li, and M. Elimelech, In situ monitoring techniques for concentration polarization and fouling phenomena in membrane filtration, Adv. Coll. Int. Sci., 107 (2004) 83-108.

50. R. Ghidossi, D. Veyret, and P. Moulin, Computational fluid dynamics applied to membranes: State of the art and opportunities, Chem. Eng. Proc., 45 (2006) 437-454.

51. G. Belfort, R.H. Davis, and A.L. Zydney. Review: The behavior of suspensions and macromolecular solutions in crossflow microfiltration, J. Membr. Sci., 96 (1994) 1-58.

52. R.W. Field, D. Wu, J.A. Howell, and B.B. Gupta, Critical flux concept for microfiltration fouling, J. Membr. Sci., 100 (1995) 259-272.

53. P.R. Neal, An examination of the nature of critical flux and membrane fouling by direct observation, PhD Thesis, The University of New South Wales, 2006.

54. L. Bao and G.G. Lipscomb, Mass transfer in axial flows through randomly packed fiber bundles, in New Insights into Membrane Science and Technology, D. Bhattacharyya and D.A. Butterfield, Eds., Elsevier, 2003.

55. G.G. Lipscomb and S. Sonalkar, Sources of non-ideal flow distribution and their effect on the performance of hollow fiber gas separation modules, Sep. Pur. Meth., 33 (2004) 41-76.

56. H. Strathmann, Membrane processes for sustainable industrial growth, Membr. Tech., 113 (1999) 9-11.

57. Review of the Desalination and Water Purification Technology Roadmap, The National Academies Press, Washington, D.C., 2004.

58. N.M. Wade, Distillation plant development and cost update, Desalination, 136 (2001) 3-12.

59. M. Busch and W.E. Mickols, Reducing energy consumption in seawater desalination, Desalination, 165 (2004) 299-312.

60. W.J. Koros, Evolving beyond the thermal age of separation processes: membranes can lead the way, AICHEJ, 50 (2004) 2326-2334.

61. Materials research for separation technologies: Energy and emission reduction opportunities, Final Report prepared for DOE Contract DE-AC05-00OR22725.

62. E. Drioli and M. Romano, Progress and new perspectives on integrated membrane operations for sustainable industrial growth, Ind. Eng. Chem. Res., 40 (2001) 1277-1300.

63. L.M. Vane, A review of pervaporation for product recovery from biomass fermentation processes, J. Chem. Tech. Biotech., 80 (2005) 603-629.

64. Y. Hayashi, S. Yuzaki, T. Kawanishi, N. Shimizu, and T. Furukawa, An efficient ethanol concentration process by vapor permeation through asymmetric polyimide membrane, J. Membr. Sci., 177 (2000) 233-239.

65. R.W. Baker, Future Directions of Membrane Gas Separation Technology, Ind. Eng. Chem. Res., 41 (2002) 1393-1411.

66. D. Aaron and C. Tsouris, Separation of CO_2 from Flue Gas: A Review, Sep. Sci. Tech., 40 (2005) 321-348.

67. E. Favre, Carbon dioxide recovery from post-combustion processes: Can gas permeation membranes compete with absorption?, J. Membr. Sci., 294 (2007) 50-59.

68. Y.S. Lin, Microporous and dense inorganic membranes: current status and prospective, Sep. Pur. Technol., 25 (2001) 39–55.

69. E.E. McLeary, J.C. Jansen, and F. Kapteijn, Zeolite based films, membranes and membrane reactors: Progress and prospects, Microporous Mesoporous Mat, 90 (2006) 198-220.

70. S.M. Saufi and A.F. Ismail, Fabrication of carbon membranes for gas separation—a review, Carbon, 42 (2004) 241-259.

71. P.M. Budd, P.M. Budd, K.J. Msayib, C.E. Tattershall, B.S. Ghanem, K.J. Reynolds, N.B. McKeown, and D. Fritsch, Gas separation membranes from polymers of intrinsic microporosity, J. Membr. Sci., 251 (2005) 263-269.

72. N.B. McKeown and P.M. Budd, Polymers of intrinsic microporosity (PIMs): organic materials for membrane separations, heterogeneous catalysis and hydrogen storage, Chem. Soc. Rev., 35 (2006) 675-683.

73. B.J. Hinds, N. Chopra, T. Rantell, R. Andrews, V. Gavalas, and L.G. Bachas, Aligned Multiwalled Carbon Nanotube Membranes, Science, 303 (2004) 62-65.

74. M. Majumder, N. Chopra, R. Andrews, B.J. Hinds, Nanoscale hydrodynamics: enhanced flow in carbon nanotubes, Nature, 438 (2005) 44.

75. J.K. Holt, H.G. Park, Y. Wang, M. Stadermann, A.B. Artyukhin, C.P. Grigoropoulos, A. Noy, and O. Bakajin, Fast Mass Transport Through Sub-2-Nanometer Carbon Nanotubes, Science, 312 (2006) 1034-1037.

76. D.S. Sholl and J.K. Johnson, Making High-Flux Membranes with Carbon Nanotubes, Science, 312 (2006) 1003-1004.

77. R. Mahajan and W.J. Koros, Mixed matrix membrane materials with glassy polymers. Part 1, Polym. Eng. Sci., 42 (2002) 1420-1431.

78. R. Mahajan and W.J. Koros, Mixed matrix membrane materials with glassy polymers. Part 2, Polym. Eng. Sci., 42 (2002) 1432-1441.

79. T.S. Chung, L.Y. Jiang, Y. Li, S. Kulprathipanja, Mixed matrix membranes (MMMs) comprising organic polymers with dispersed inorganic fillers for gas separation, Prog. Polym. Sci., 32 (2007) 483-507.

80. S.N. Paglieri and J.D. Way, Innovations in palladium membrane research, Sep. Pur. Meth., 31 (2002) 1-169.

81. P.N. Dyer, R.E. Richards, S.L. Russek, and D.M. Taylor, Ion transport membrane technology for oxygen separation and syngas production, Solid State Ionics, 134 (2000) 21-33.

82. R. van Reis and A. Zydney, Bioprocess membrane technology, J. Membr. Sci., 297 (2007) 16-50.

83. M.F. Ebersold and A.L. Zydney, Separation of protein charge variants by ultrafiltration, Biotech. Prog. 20 (2004) 543-549.

84. R. van Reis, J.M. Brake, J. Charkoudian, D.B. Burns, A.L. Zydney, High performance tangential flow filtration using charged membranes, J. Membr. Sci. 159 (1999) 133-142.

85. R. van Reis, Charged filtration membranes and uses therefor, US Patents 7,001,550, 2006; 7,153,426, 2006.

86. N.S. Pujar and A.L. Zydney, Electrostatic effects on protein partitioning in size exclusion chromatography and membrane ultrafiltration, J. Chromatogr. A 796 (1998) 229-238.

87. B. Hallstrom and M. Lopez-Leiva, Description of a rotating ultrafiltration module, Desalination 24 (1978) 273.

88. W.B. Krantz, R.R. Bilodeau, M.E. Voorhees and R.J. Elgas, Use of axial membrane vibrations to enhance mass transfer in a hollow tube oxygenator, J. Membr. Sci., 124 (1997) 283-299.

89. E. Klein, Affinity membranes: a 10-year review, J. Membr. Sci. 179 (2000) 1-27.

90. D.A. Butterfield, D. Bhattacharyya, S. Daunert, L. Bachas, Catalytic biofunctional membranes containing site-specifically immobilized enzyme arrays: a review, J. Membr. Sci., 181 (2001) 29-37.

91. S.S. Sablani, M.F.A. Goosen, R. Al-Belushi, M. Wilf, Concentration polarization in ultrafiltration and reverse osmosis: a critical review, Desalination, 141 (2001) 269-289.

92. M.F.A. Goosen, S.S. Sablani, H. Al-Hinai, S. Al-Obeidani, R. Al-Belushi, D. Jackson, Fouling of reverse osmosis and ultrafiltration membranes: a critical review, Sep. Sci. Tech., 39 (2005) 2261-2297.

93. J. Gilron, S. Belfer, P. Vaisanen, M. Nystrom, Effects of surface modification on antifouling and performance properties of reverse osmosis membranes, Desalination, 140 (2001) 167-179.

94. J.E. Kilduff, S. Mattaraj, M. Zhou, G. Belfort, Kinetics of membrane flux decline: the role of natural colloids and mitigation via membrane surface modification, J. Nanoparticle Res., 7 (2005) 525-544.

95. N. Singh, Z. Chen, N. Tomer, S.R. Wickramasinghe, N. Soice, and S.M. Husson, Modification of regenerated cellulose ultrafiltration membranes by surface-initiated atom transfer radical polymerization, J. Membr. Sci., 311 (2008) 225-234.

96. J. Schwinge, P.R. Neal, D.E. Wiley, D.F. Fletcher, A.G. Fane, Spiral wound modules and spacers – Review and analysis, J. Membr. Sci., 242 (2004) 129-153.

97. K.W. Lawson and D.R. Lloyd, Membrane distillation, J. Membr. Sci., 124 (1997) 1-25.

98. T.Y. Cath, A.E. Childress, and M. Elimelech, Forward osmosis: Principles, applications, and recent developments, J. Membr. Sci., 281 (2006) 70-87.

99. R. Rautenbach and A. Groeschl, Separation Potential of Nanofiltration Membranes, Desalination 77 (1990) 73-84.

100. V. Geraldes and M. Norberta de Pinho, Process water recovery from pulp bleaching effluents by an NF/ED hybrid process, J. Membr. Sci., 102 (1995) 209-221.

101. R. Rautenbach and R. Mellis, Hybrid processes involving membranes for the treatment of highly organic/inorganic contaminated waste water, Desalination, 101 (1995) 105-113.

102. M. Grytam, K. Karakulski, and A. W. Morawski, Purification of oily wastewater by hybrid UF/MD, Water Res., 35 (2001) 3665-3669.

103. S.M.C. Ritchie and D. Bhattacharyya, Membrane-based hybrid processes for high water recovery and selective inorganic pollutant separation, J. Haz. Mat., 92 (2002) 21-32.

104. C. Blöcher, J. Dorda, V. Mavrov, H. Chmiel, N. K. Lazaridis, and K. A. Matis, Hybrid flotation – membrane filtration process for the removal of heavy metal ions from wastewater, Water Res., 37 (2003) 4018-4026.

105. B. Schlichter, V. Mavrov, and H. Chmiel, Study of a hybrid process combining ozonation and microfiltration/ultrafiltration for drinking water production from surface water, Desalination, 168 (2004) 307-317.

106. R. Molinari, L. Palmisano, E. Drioli, and M. Schiavello, Studies on various reactor configurations for coupling photocatalysis and membrane processes in water purification, J. Membr. Sci., 206 (2002) 399-415.

107. P. Le-Clech, E.-K. Lee, and V. Chen, Hybrid photocatalysis/membrane treatment for surface waters containing low concentrations of natural organic matters, Water Res., 40 (2006) 323-330.

108. S. Judd, The MBR book: Principles and applications of membrane bioreactors in water and wastewater treatment, Elsevier, Oxford, 2006.

109. P. Le-Clech, V. Chen, and A.G. Fane, Fouling in membrane bioreactors used for wastewater treatment – A review., J. Membr. Sci., 284 (2006) 17-53.

110. J. Kim and F.A. DiGiano, Defining critical flux in submerged membranes: Influence of length-distributed flux, J. Membr. Sci., 280 (2006) 752-761.

111. N. Hilal, O.O. Ogunbiyi, N.J. Miles, and R. Nigmatullin, Methods Employed for Control of Fouling in MF and Membranes: A Comprehensive Review, Sep. Sci. Tech., 40 (2005) 1957-2005.

112. P.R. Bérubé and E. Lei, The effect of hydrodynamic conditions and system configurations on the permeate flux in a submerged hollow fiber membrane system, J. Membr. Sci., 271 (2006) 29-37.

113. G. Guglielmi, D. Chiarani, S.J. Judd, and G. Andreottola, Flux criticality and sustainability in a hollow fibre submerged membrane bioreactor for municipal wastewater treatment, J. Membr. Sci., 289 (2007) 241-248.

114. N. Hilala, H. Al-Zoubia, N.A. Darwishb, A.W. Mohammadc, M. Abu Arabid, A comprehensive review of nanofiltration membranes: Treatment, pretreatment, modelling, and atomic force microscopy, Desalination, 170 (2004) 281-308.

115. J. Schaep, B. Van der Bruggen, S. Uytterhoeven, R. Croux, C. Vandecasteele, D. Wilms, E. Van Houtte and F. Vanlerberghe, Removal of hardness from groundwater by nanofiltration, Desalination, 119 (1998) 295-302.

116. P. Fu, H. Ruiz, K. Thompson and C. Spangenberg, Selecting membranes for removing NOM and DBP precursors, J. AWWA, 86 (1994) 55-72.

117. I.C. Escobar, S. Hong and A. Randall, Removal of assimilable and biodegradable dissolved organic carbon by reverse osmosis and nanofiltration membranes, J. Membr. Sci., 175 (2000) 1-17.

118. B. Van der Bruggen, K. Everaert, D. Wilms and C. Vandecasteele, The use of nanofiltration for the removal of pesticides from ground water: an evaluation, Water Sci. Technol.: Water Supply, 1 (2001) 99-106.

119. M. Thanuttamavong, K. Yamamotob, J. Oh, K. Chood and S. Choi, Rejection characteristics of organic and inorganic pollutants by ultra low pressure nanofiltration of surface water for drinking water treatment, Desalination, 145 (2002) 257-264.

120. P. Laurent, P. Servais, D. Gatel, G. Randon, P. Bonne and J. Cavard, Microbiological quality before and after nanofiltration, J. AWWA, 91(10) (1999) 62-72.

121. M. Otaki, K. Yano and S. Ohgaki, Virus removal in a membrane separation process, Water Sci. Tech.., 37(10) (1998) 107-116.

122. R. Rautenbach, T. Linn and L. Eilers, Treatment of severely contaminated waste water by a combination of RO, high-pressure RO and NF —potential and limits of the process, J. Membr. Sci., 174 (2000) 231-241.

123. I. Koyuncu, An advanced treatment of high-strength opium alkaloid processing industry wastewaters with membrane technology: pretreatment, fouling and retention characteristics of membranes, Desalination, 155 (2003) 265-275.

124. V. Geraldes and M. de Pinho, Process water recovery from pulp bleaching effluents by an NF/ED hybrid process, J. Membr. Sci., 102 (1995) 209-221.

125. B. Schlichter, V. Mavrov and H. Chmiel, Study of a hybrid process combination ozonation and membrane filtration — filtration of model solution, Desalination, 156 (2003) 257-265.

126. E. Drioli, A. Criscuoli and E. Curcioa, Integrated membrane operations for seawater desalination, Desalination, 147 (2002) 77-81.

127. O.A. Hamed, Overview of hybrid desalination systems — current status and future prospects, Desalination, 186 (2005) 207-214.

128. S. Boerlage, M. Kennedy, A. Bonne Paul, G. Galjaard and J. Schippers, Prediction of flux decline in membrane systems due to particulate fouling, Desalination, 113 (1997) 231-233.

129. J. Vrouwenvelder, J. Kappelhof, S. Heijman, J. Schippers and D. Kooij, Tools for fouling diagnosis of NF and RO membranes and assessment of the fouling potential of feed water, Desalination, 157 (2003) 361-365.

130. S. Hong and M. Elimelech, Chemical and physical aspects of natural organic matter (NOM) fouling of nanofiltration membranes, J. Membr. Sci., 132 (1997) 159-181.

131. W.R. Bowen, A. Mohammad and N. Hilal, Characterisation of nanofiltration membranes for predictive purposes use of salts, uncharged solutes and atomic force microscopy, J. Membr. Sci., 126 (1997) 91-105.

132. J. Schaep, C. Vandecasteele, A.W. Mohammad and W.R. Bowen, Evaluation of the salt retention of nanofiltration membranes using the Donnan and steric partitioning model, Sep. Sci. Technol., 34 (1999) 3009-3030.

133. W. Bowen and J. Welfoot, Modelling of membrane nanofiltration – pore size distribution effects, Chem. Eng. Sci., 57 (2002) 1393-1407.

134. L.S. White, Development of large-scale applications in organic solvent nanofiltration and pervaporation for chemical and refining processes, J. Membr. Sci., 286 (2006) 26-35.

135. P. Vandezande, L.E.M. Gevers, and I.F.J. Vankelecom, Solvent resistant nanofiltration: separating on a molecular level, Chem. Soc. Rev., 37 (2008) 365-405.

136. P. Silva, S. Han, A.G. Livingston, Solvent transport in organic solvent nanofiltration membranes, J. Membr. Sci., 262 (2005) 49-59.

137. D. Bhanushali and D. Bhattacharyya, Advances in Solvent-Resistant Nanofiltration Membranes – Experimental Observations and Applications, Ann. N.Y. Acad. Sci., 984 (2003) 159-177.

138. S.S. Köseoglu and D.E. Engelgau, Membrane applications and research in the edible oil industry: An assessment, J. Am. Oil Chem. Soc., 67 (1990) 239-249.

139. K.K. Reddy, T. Kawakatsu, J.B. Snape, and M. Nakajima, Membrane Concentration and Separation of L-Aspartic Acid and L-Phenylalanine Derivatives in Organic Solvents, Sep. Sci. Technol., 31 (1996) 1161-1178.

140. E.M. Tsui and M. Cheryan, Membrane processing of xanthophylls in ethanol extracts of corn, J. Food Eng., 83 (2007) 590-595.

141. X. Cao, X.Y. Wu, T. Wu, K. Jin, and B.K. Hur, Concentration of 6-Aminopenicillanic Acid from Penicillin Bioconversion Solution and Its Mother Liquor by Nanofiltration Membrane, Biotech. Bioprocess Eng., 6 (2001) 200-204.

142. J.P. Sheth, Y. Qin, K.K. Sirkar, and B.C. Baltzis, Nanofiltration-based diafiltration process for solvent exchange in pharmaceutical manufacturing, J. Membr. Sci., 211 (2003) 251-261.

143. F.C. Ferreira, H. Macedo, U. Cocchini, and A. G. Livingston, Development of a Liquid-Phase Process for Recycling Resolving Agents within Diastereomeric Resolutions, Org. Process Res. Dev., 10 (2006) 784-793.

144. P.B. Kosaraju and K.K. Sirkar, Novel solvent-resistant hydrophilic hollow fiber membranes for efficient membrane solvent back extraction, J. Membr. Sci., 288 (2007) 41-50.

145. L.S. White, Development of large-scale applications in organic solvent nanofiltration and pervaporation for chemical and refining processes, J. Membr. Sci., 286 (2006) 26-35.

146. J.T. Scarpello, D. Nair, L.M. Freitas dos Santos, L.S. White, A.G. Livingston, The separation of homogeneous organometallic catalysts using solvent resistant membranes, J. Membr. Sci., 203 (2002) 71-85.

147. S.S. Luthra, X. Yang, L.M. Freitas dos Santos, L.S. White, A.G. Livingston, Homogeneous phase transfer catalyst recovery and re-use using solvent resistant membranes, J. Membr. Sci., 201 (2002) 65-75.

148. A.G. Livingston, L. Peeva, S. Han, D.A. Nair, S.S. Luthra, L.S. White, L.M. Freitas dos Santos, Membrane separation in green chemical processing - solvent nanofiltration in liquid phase organic synthesis reactions, Ann. N.Y. Acad. Sci., 984 (2003) 123-141.

149. A.A. Chin, B.M. Knickerbocker, J.C. Trewella, T.R. Waldron, L.S. White, Recovery of aromatic hydrocarbons using lubricating oil conditioned membranes, US Patent, 6,187,987, 2001.

150. L.S. White, R.F. Wormsbecher, and M. Lesemann, Membrane separation for sulfur reduction, US Patent 7,048,846, 2006.

151. G.L. Wick and W. R. Schmitt, Prospects for renewable energy from the sea, Marine Tech. Soc. J., 11 (1977) 16-21.

152. R.E. Pattle, Production of Electric Power by mixing Fresh and Salt Water in the Hydroelectric Pile, Nature, 174 (1954) 660.

153. S. Loeb, Production of energy from concentrated brines by pressure retarded osmosis, J. Membr. Sci., 1 (1976) 49-63.

154. S. Loeb, Method and apparatus for generating power utilizing pressure-retarded osmosis, US Patent 4,193,267, 1980.

155. M. Olsson, G.L. Wick, and J.D. Isaacs, Salinity gradient power - utilizing vapor-pressure differences, Science, 206 (1979) 452–454.

156. J.W. Post, J. Veerman, H.V.M. Hamelers, G.J.W. Euverink, S.J. Metz, K. Nymeijer, C.J.N. Buisman, Salinity-gradient power: Evaluation of pressure-retarded osmosis and reverse electrodialysis, J. Membr. Sci. 288 (2007) 218-230.

Chapter 9

Advanced Catalytic Materials for the Refining and Petrochemical Industry: TUD-1

Philip J. Angevine[1], Anne M. Gaffney[1], Zhiping Shan[1,2], and Chuen Y. Yeh[1]

[1]Lummus Technology Inc., 1515 Broad Street, Bloomfield, NJ 07003
[2]Current address: Huntsman Corporation, 8600 Gosling Road, The Woodlands, TX 77381

TUD-1, a new family of mesoporous materials, is a three-dimensional amorphous structure of random, interconnecting pores. The original emphasis was on the silica version, which has since been extended to about 20 chemical variants (e.g., Al, Al-Si, Ti-Si, etc.). Multimetallic oxides have been made so the catalytic opportunities are almost endless.

In principle, these "mesoporous" materials (i.e., pore diameters from 2 to 20 nm) should be useful for processing high molecular weight materials, such as petroleum residua, lubricants, etc. If synthesized with high surface areas, the resultant catalysts could have significant benefit for high conversion, fast reactions where mass transfer plays a critical role. In addition to efficiently transporting reactants into the active sites, mesopores enable the products to leave the active sites and thereby reduce unwanted side reactions.

Since discovery of the famous M41s family of crystalline mesoporous materials, a worldwide effort has been made on synthesis, characterization, and catalytic evaluation. Many research programs focused on transforming MCM-41 into active and stable forms, but with medium success. Many scientists since have refocused attention towards other mesoporous materials.

Since mesoporous catalysts held promise for hydrocarbon chemistry such as refining and petrochemical, Lummus initiated a joint research project with the Delft University of Technology. From this collaboration came the discovery of a new mesoporous material, named TUD-1.

The key common properties of TUD-1 are:

- Tunable porosity and pore size
- High surface area
- Excellent stability
- Random, three-dimensional interconnecting pores

Variants of TUD-1 have been shown to be effective for a wide range of reactions, some of which include alkylation, cracking, epoxidation, hydrogenation, hydrogenolysis, etc. One exciting area is the use of zeolites embedded in TUD-1 to give a synergistic performance for various probe reactions.

Introduction – Historical Background

Academic and industrial scientists have long sought to synthesize larger pore materials that bridged the gap between the microporous and macroporous range. In principle, these "mesoporous" materials (i.e., pore diameters from 2 to 20 nm) should be useful for processing high molecular weight materials, such as petroleum residua, lubricants, amino acids, etc. If synthesized with high surface areas, the resultant catalysts could have significant benefit for high conversion, fast reactions where mass transfer plays a role. In addition to efficiently transporting reactants into the active sites, mesopores enable the products to leave the active sites and thereby reduce unwanted side reactions.

Since discovery of the now-famous M41s family (MCM-41, etc.) of crystalline mesoporous materials (1, 2, 3, 4), an enormous worldwide effort has been expended on synthesis, characterization, and catalytic evaluation. Using solvents, the discrete pore size of MCM-41 can be created in a controlled manner. Unfortunately, the M41s materials generally lacked significant catalytic activity and also suffered from subpar structural, thermal, and hydrothermal stability. Many research programs focused on transforming MCM-41 into active and stable forms, but with medium success. Many scientists then refocused attention on other mesoporous materials, but to date no materials have been both catalytically significant and inexpensive to synthesize.

Figure 1 shows a schematic diagram of MCM-41 formation and pore control (5). MCM-41 is a one-dimensional structure whose pore size is determined by the alkyl chain length of the surfactant micelle around which the molecular sieve is formed. A key surfactant employed is cetyltrimethyl

- *Surfactant's alkyl chain length determines pore diameter*

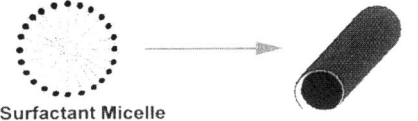

Surfactant Micelle

- *Use of solubilization agents increases pore diameter*

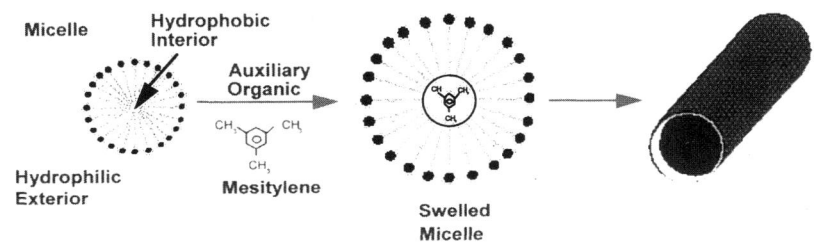

From T.F. Degnan et al, Catalysis Society Meeting, New York, 2003.

Figure 1. MCM-41 formation and pore tuning

ammonium chloride ("CTMACl"). Moreover, addition of a solubilization agent (e.g., mesitylene) can swell the micelle, thus forming a larger pore diameter.

The M41s family has several structures, of which the major ones are shown in Figure 2 (5). A common characteristic of virtually all mesoporous materials (including amorphous ones) is a broad x-ray defraction (XRD) peak at low 2θ. MCM-48 is a cubic structure with two nonintersecting pores. The original structure, MCM-41, is a one-dimensional, hexagonal structure. MCM-50 is a layered ("lamellar") structure with silica sheets between the layers. The utility of these materials is expected to be limited due to the lack of intersecting, three-dimensional pores such as in the major zeolites, i.e., zeolite Y, ZSM-5, and Zeolite Beta.

Table I describes some better-known mesoporous molecular sieves (5). It is interesting to note that various types of directing agents and synthesis mechanisms are employed, resulting in different physical characteristics. The prefixes "HMS" and "MSU" refer to the Michigan State University group (T. Pinnavaia et al), and the "SBA" term refers to the UC Santa Barbara group (G. Stucky et al.). The symbols "S" and "I" denote the surfactant and inorganic oxide precursors, respectively. The symbols "X" and "M" denote counterions (e.g., $X^-=Cl^-$, etc. and $M^+=Na^+$, etc.).

338

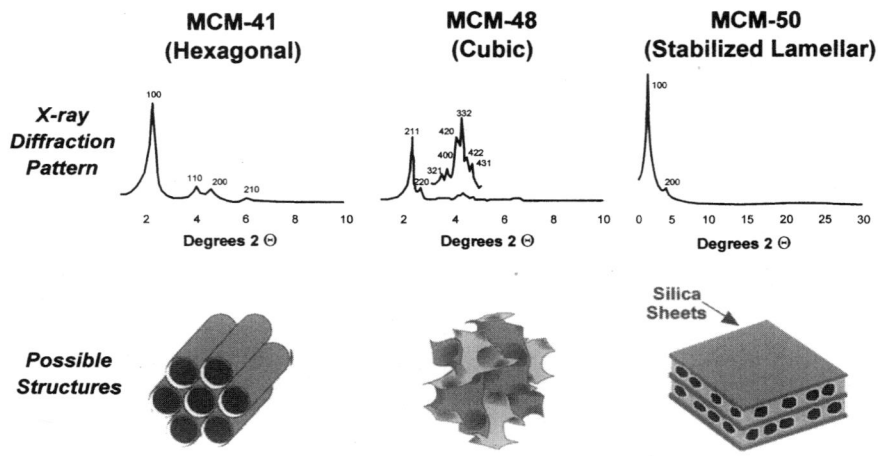

Figure 2. Members of the M41s family

Based on patents and publications, there are many potential applications for MCM-41 catalysts, including:

- Acid catalysis (e.g., metal-substituted material, zeolite/ mesoporous composites, sulfonic acid-functionalized)
- Base catalysis (Na-, Cs-MCM-41)
- Hydrotreatment (Ni/W on Si- or Al-MCM-41)
- Hydrogenation (Pt on Al-MCM-41)
- Redox catalysis (Ti-containing materials)
- Anchored complexes (enantioselective reactions)

Since MCM-41 can be made in a variety of chemical compositions, its high surface area can serve as either a simple substrate or an active component of the catalyst.

TUD-1

Since mesoporous catalysts held promise for hydrocarbon chemistry such as those utilized in refining and petrochemical industry, Lummus initiated a joint research project with the Delft University of Technology. The major product of this collaboration was the discovery of a new mesoporous material, named TUD-1. The composition of matter patent (6) is owned jointly by Lummus and

Table I. Some Mesoporous Molecular Sieves

Mesoporous Material	Directing Agent(s)	Mechanism	Characteristics
M41S	Ionic surfactants	S+I-, S-I+, S+X-I+, S-M+I-	Well controlled, ordered pores, hexagonal, cubic, lamellar structures
HMS, MSU	Non-ionic primary amines, propylene oxide surfactants	Neutral oligomeric silicas, S^0I^0	Less ordered, worm-like structures, thicker walls
SBA	Amphiphilic di-and tri-block copolymers	Neutral oligomeric silicas, S^0I^0	Long range order, monodispersed mesopores (to 30nm), thicker walls

Delft University. TUD-1 is a three-dimensional amorphous structure of random, interconnecting pores. As will be later described, the pore size can be tailored. The original emphasis was on the silica version, which has since been extended to about 20 chemical variants (e.g., Al, Al-Si, Ti-Si, etc.).

TUD-1 is clearly an amorphous material. Unlike crystalline structures, it has no characteristic x-ray diffraction pattern, no planes of symmetry and an associated space group, no specific morphology, no characteristic phase diagram, no heat of crystallization, and no characteristic density, refractive index (R.I.), cleavage, planes, Madelung constant, etc.

Figure 3 illustrates the pore diameter of TUD-1 versus the major molecular sieves, ZSM-5, Zeolite Y, and MCM-41. Of note, the pore diameter can be varied from about 50Å to 250 Å.

General Method of Synthesis

A key enabler often employed in the synthesis of zeolites is the template, often called an organic directing agent. The template type is frequently different for microporous zeolites, mesoporous materials, and macroporous materials. The template can be an individual molecule (e.g., quaternary salts or linear amines), *in-situ* formed micelle clusters, or preformed structures (e.g., polyethylene spheres).

In TUD-1 synthesis, we have used a new approach to template formation and utility. A chemical intermediate is employed with a network of meso-sized organic aggregates that penetrate the inorganic phase. At low temperature, the templates form a homogeneous mixture with the inorganic phase at the

340

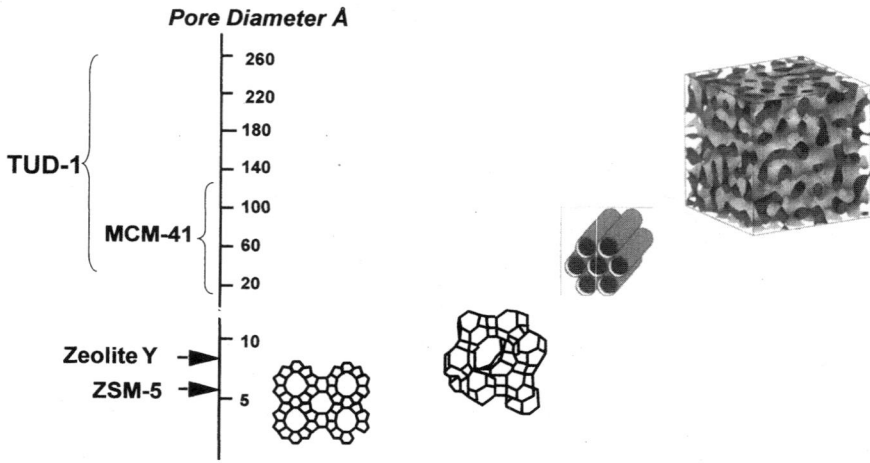

Figure 3. What are the pore diameters?

molecular scale. At high temperature, the templates undergo phase separation with the inorganic phase at the meso-sized level. Desirable properties of the TUD-1 template are: physically stable at elevated temperatures (200-250°C), chemically interactive with the inorganic phase, and inexpensive.

Synthesis Procedures: First Generation

Figure 4 is a schematic visualization of the major chemical states that take place in TUD-1 synthesis (7). The three major steps are: (a) formation of a homogeneous mixture, (b) migration of the template to achieve meso-sized aggregation, and (c) pore generation. Some micropores are formed in addition to the mesopores, which is another key differentiator from many other crystalline mesoporous materials. In terms of unit operations, the six operations include: (a) mixing, (b) hydrolysis, (c) aging, (d) drying, (e) heat treatment [optional], and (f) calcination.

Figure 5 is a simplified diagram of the original synthesis route (i.e., Si-TUD-1). The first step − the formation step − involves a monomeric silica source (here, TEOS), triethanolamine ("TEA"), and optionally TEAOH. The TEA serves as a template for the mesopore formation. The TEAOH serves as both a source of quaternary cation (to generate some micropores if necessary) and a basic environment to accelerate TEOS hydrolysis. The reaction rate increases with pH (i.e., the [OH-]/ [SiO$_2$] ratio), which can also be achieved in part or wholly by increased temperature. The second step involves an aging/drying phase to establish the primary pore structure. The last step −

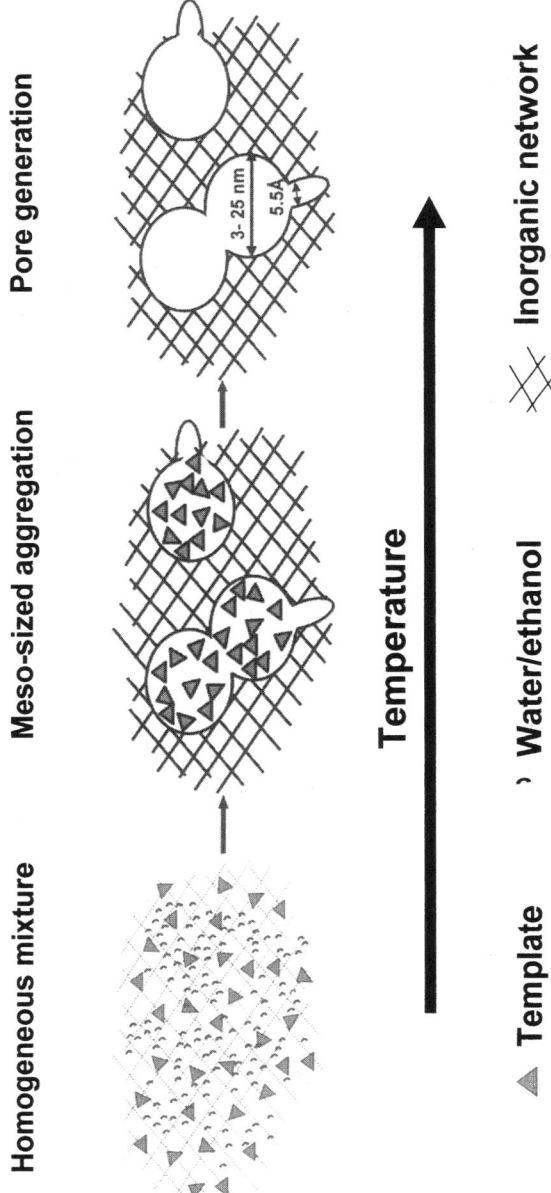

Figure 4. A schematic visualization of TUD-1 function.

calcination – is required to remove the large quantities of organics. An optional step, between drying and calcination, is a pore modification step employing elevated temperature (e.g., 150-190°C for Si), which we call "heat treatment" ("HT").

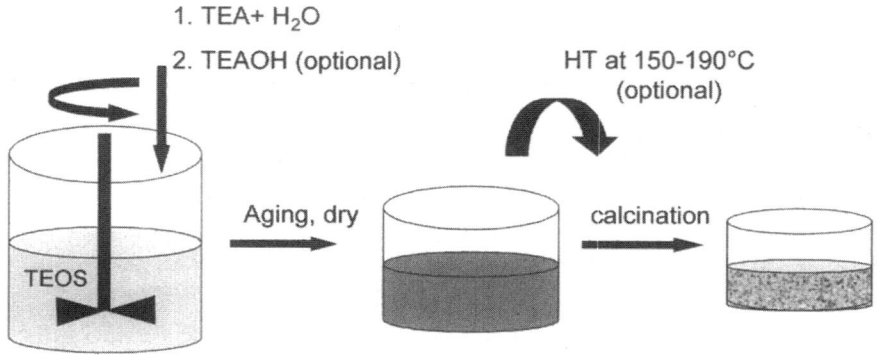

TEOS – **tetraethyl orthosilicate**
TEA – **triethanolamine**
TEAOH – **tetraethylammonium hydroxide**

Figure 5. Original TUD-1 synthesis route

For an Al-Si-TUD-1, the aluminum source can be aluminum isopropoxide, aluminum tri-sec-butoxide, or another organoalumina species that forms a monomeric Al as an intermediate component.

Figure 6 shows a typical XRD pattern of TUD-1 (6). As with other mesoporous materials, TUD-1 has a broad peak at low 2θ, but it also has a broad background peak, commonly called an "amorphous halo".

Figure 7 illustrates the typical capabilities in tailoring TUD-1 pore size via post-synthesis heat treatment (7). Here, the mesopore diameter and surface area are plotted versus heating time. For example, the starting sample had a pore diameter of about 4 nm and a surface area of about 800 m^2/g. After a heating time of 48 hours, the sample had a pore diameter of 19 nm while retaining a surface area of 400 m^2/g. This specific example was a Si-TUD-1, but similar curves can be generated for other chemical variants. In general, the final surface area approaches an asymptote to 50% of the original surface area.

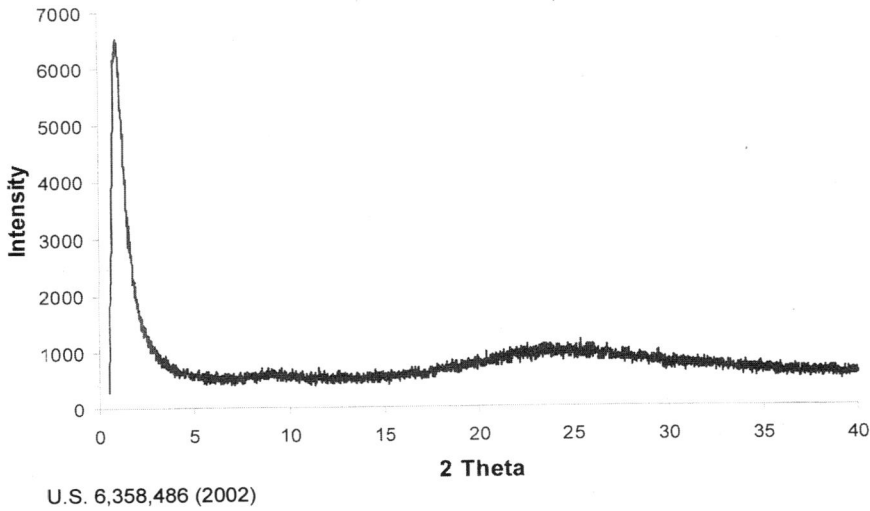

U.S. 6,358,486 (2002)

Figure 6. TUD-1 XRD

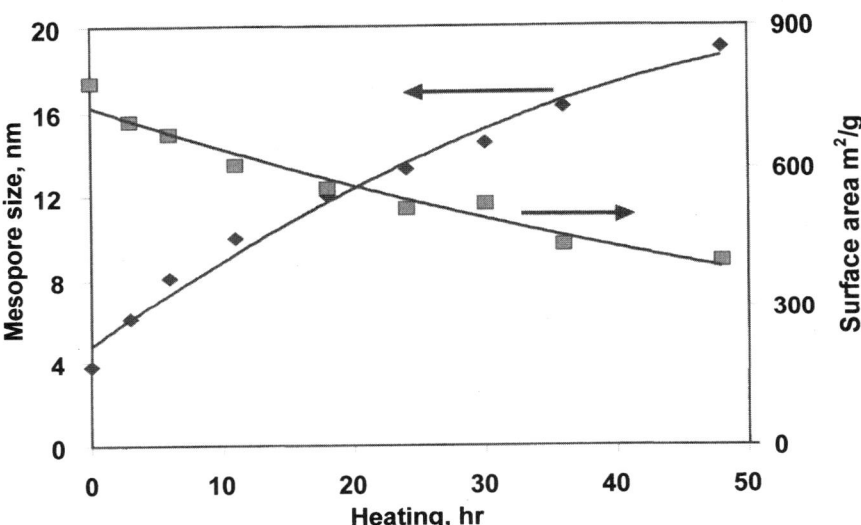

Figure 7. Heat treatment tunes porosity (pore diameter versus surface area)

Characterization (TEM, XRD, Physicals, Inverse Carbon Skeleton)

The key common properties of TUD-1 are:

- Tunable porosity
- Pore volume of 0.3-2.5 cc/g
- Pore diameter of 2.5-25 nm
- High surface area: 400-1000 m^2/g
- Excellent thermal, hydrothermal, and mechanical stability
- Random, three-dimensional interconnecting pores

Most TUD-1 variants are either Si-TUD-1 or an M-Si version where M is another element (e.g., Ti, Al, Cr, Fe, Zr, Ga, Sn, Co, Mo, V, etc.). Non-siliceous versions of Al- and Ti-TUD-1 have also been made. The typical SiO_2/ M_xO_y molar ratio is normally 20-∞.

The structural determination of TUD-1 was an early challenge. Figure 8 is a TEM comparison of TUD-1 versus MCM-41 and MCM-48. The TEM shows that the TUD-1 pores have no periodicity. One can also see the various depths of the pores. In this example the pore diameter is about 5 nm. From simple inspection, one can see the clear-cut periodicity of MCM-41 and -48; however, TUD-1 appears more like a sponge topology. This can be seen clearly in Figure 9, where ceramic foam is shown versus TUD-1, albeit at two very different scales.

One of the early issues with TUD-1 dealt with its pore structure: did it have intersecting or nonintersecting pores? Z. Shan executed a simple, but conclusive study. He generated a carbon replica of TUD-1 by filling its pores with sugar solution, carburizing it, and dissolving the silica, then compared the original Si-TUD-1 with its carbon replica. As shown in Figure 10, the TEM of the carbon replica looked very much like its silica counterpart in Figure 9. Also, XRD patterns of the Si-TUD-1, the combined Si-TUD-1/carbon replica, and the carbon replica looked very similar. If the pores had been nonintersecting, the carbon replica, like a pile of sticks, would have collapsed. Instead, they maintained the gross structure of the parent. Also shown is a scanning electron micrograph (SEM) of the carbon crystals.

Another conclusive characterization was carried out with a silica TUD-1 with Pt inserted, which was analyzed by 3-D TEM (9). The Pt anchors were used as a focal point for maintaining the x, y, z orientation. As shown in Figure 11, the TUD-1 is clearly amorphous. While not quantitatively measured, the ores appear rather uniform, consistent with previous porosimetry measurements.

Synthesis Procedures: Second Generation

While the original synthesis route was universally applicable to making many chemical variants with a spectrum of physical properties, the raw material

TUD-1

MCM-41

MCM-48

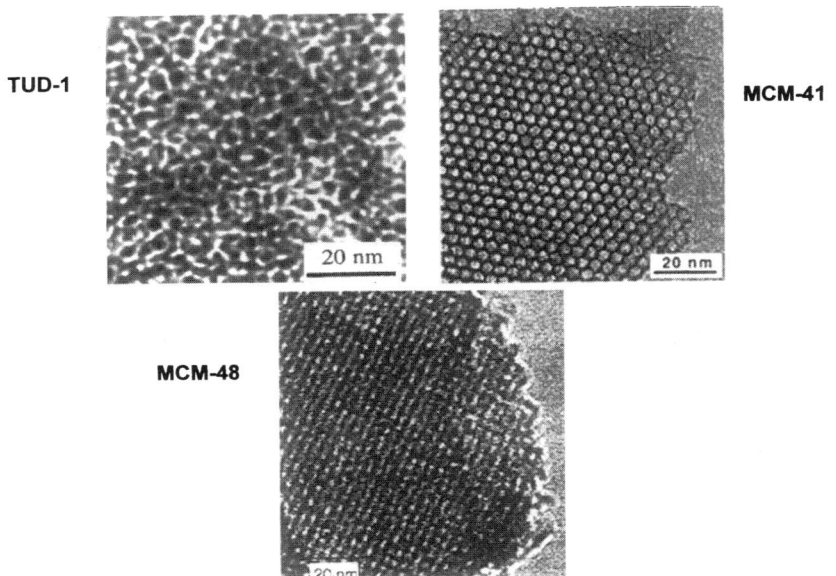

Figure 8. TEM comparison of TUD-1 versus MCM-41 and -48

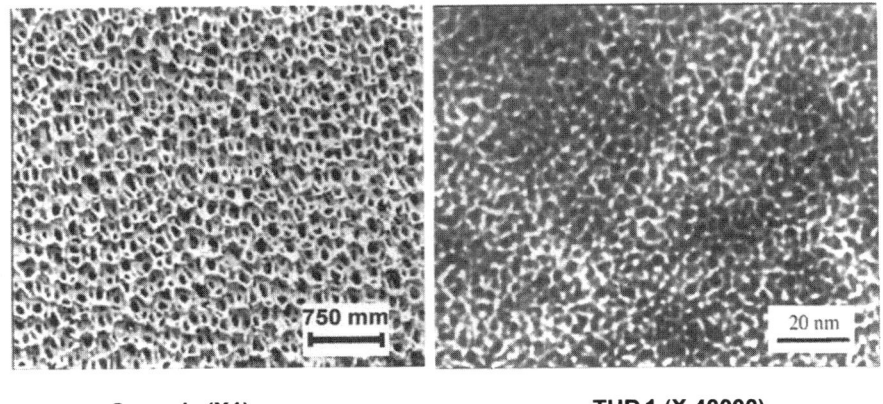

Ceramic (X1)

TUD-1 (X 40000)

Figure 9. Ceramic foam versus TUD-1 – an analogy

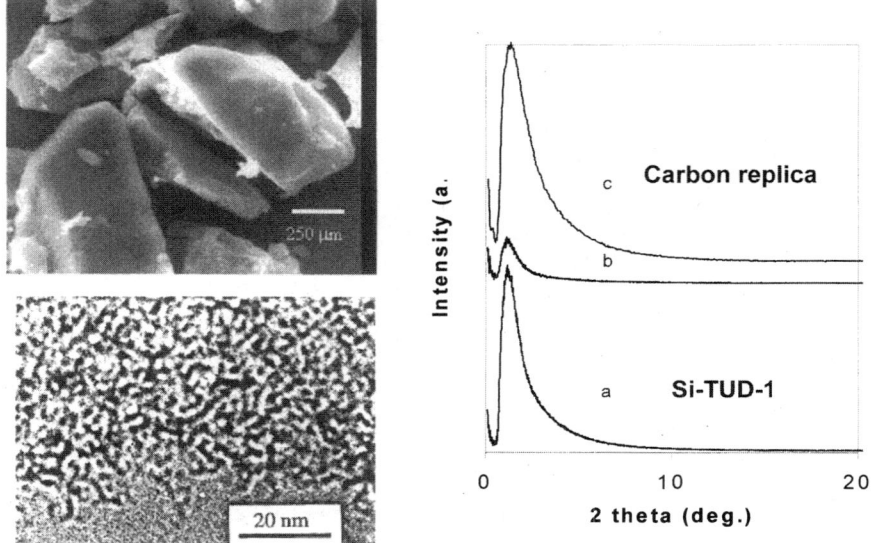

Figure 10. Carbon replica of TUD-1

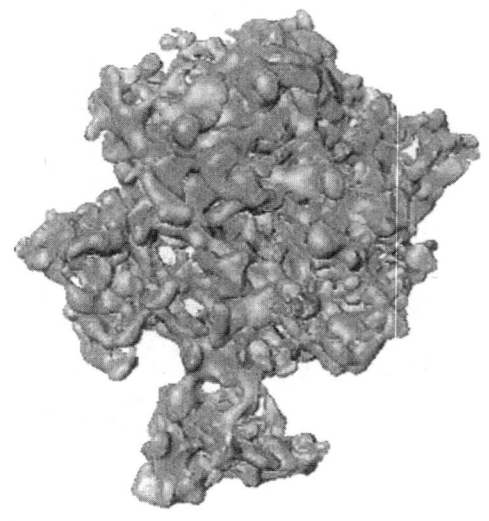

Figure 11. 3D TEM of Pt/Si-TUD-1

costs were quite high. As such, a second approach was developed. This procedure is called the "complexation" route (10). Typical raw materials are silica gel, TEA acting as both a complexing agent and a templating agent, ethylene glycol ("EG") used mainly as the solvent during complexation, and water for a hydrolysis agent. For an Al-Si-TUD-1, aluminum hydroxide can serve as an effective yet inexpensive Al source.

Figure 12 shows the major steps of this synthesis: (a) complexation (addition of the TEA and EG, then silica gel), (b) hydrolysis via water addition and condensation, (c) aging, (d) drying, (e) heat treatment, (f) extraction (optional) and (g) calcination. In this diagram, most of the TEA and EG are recycled.

The complexation route has several benefits: it yields an equivalent TUD-1 product, but it employs significantly cheaper raw materials (a low cost silica source and no alkali); chemicals are used efficiently, so less chemicals are involved; and since most of two major raw materials (EG and TEA) can be recycled, the procedure is environmentally friendly.

Catalytic Performance

Al-TUD-1 and Al-Si-TUD-1

While siliceous TUD-1 provides a large-pore/high–surface-area material, the Si does not provide a strong anchor for metals. Hence, metals dispersion is often low on silica-type supports. With this in mind, an early endeavor was to employ other elements in place of the Si so that various catalytic properties could be achieved. One of the first elements added was Al (11).

While several specific synthesis routes have been proven, one new synthesis process for mesoporous aluminum oxide comprises the following: (a) dissolving an organic aluminum source alone or together with a framework-substituted element in a solvent, (b) adding a pore-forming agent to the mixture, followed by solvent addition, (c) optionally aging the mixture at a temperature ranging from about 10 to 90°C; (d) drying, and (e) removing the pore-forming agent, preferably by conventional methods such as calcination or solvent extraction.

This general synthesis route represents several improvements: use of inexpensive, small organic chemicals as pore-forming agents instead of surfactants; no micelles formed in the mesopore templating process, whereas most other mesoporous materials are synthesized based on micelle formation; mesoporosity of the aluminum oxide can be easily and continuously tuned; and use of inexpensive inorganic aluminum sources. While the original alumina synthesis utilized aluminum isopropoxide ("AIP"), the more recent efforts successfully used aluminum sulfate or nitrate.

348

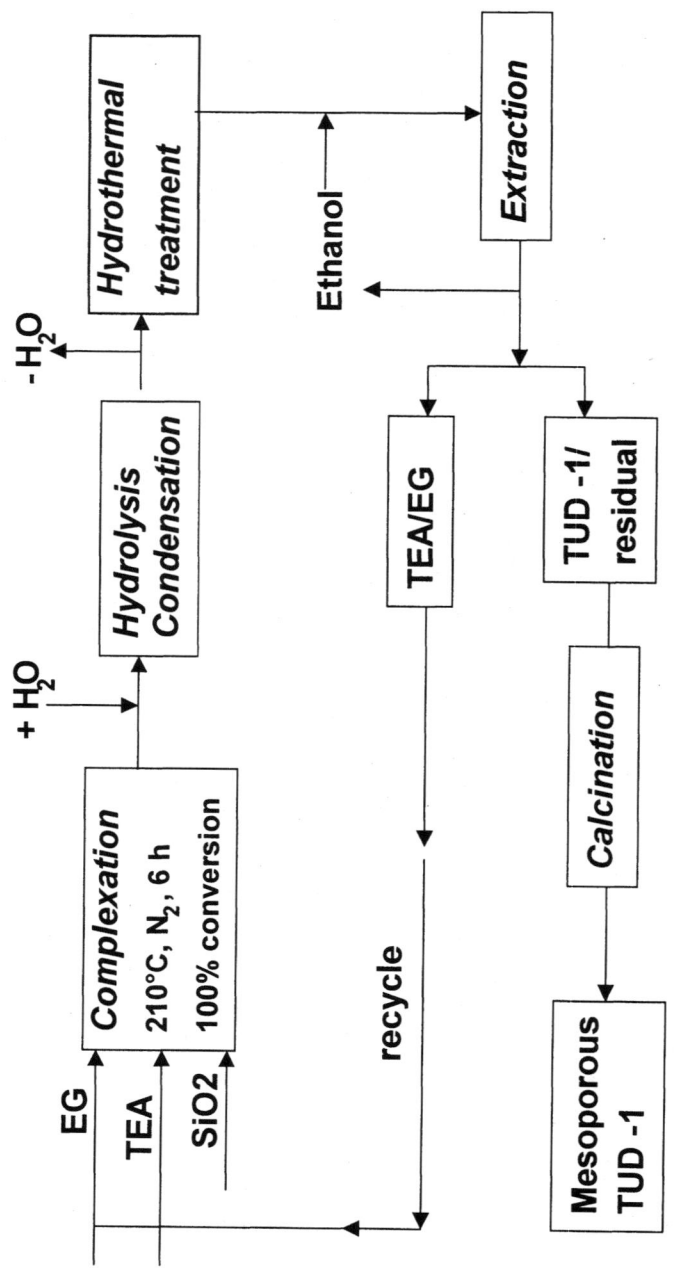

US Pat Appl 20050164870 (filed 2004)

Figure 12. Complexation route for TUD-1 synthesis

A 2007 patent (11) and subsequent divisionals describe the synthesis (also given in ref. 12) and catalytic use of Al-containing TUD-1 materials. Some of the reactions demonstrated include hydrogenation of mesitylene (Pt as active metal) and dehydration of 1-phenyl-ethanol to styrene. Conceptual reactions include:

- Dehydrogenation of propane to propylene (Pt/Sn as active metal with K promoter)
- Demetalation (NiMo or CoMo as active metal in oxide form)
- Steam reforming of methane (Ni form)
- Diels-Alder reaction of crotonaldehyde and dicyclopentadiene
- Amination of phenol with ammonia
- Hydrotreating of FCC clarified slurry oil

Al-Si-TUD-1 can be a stand-alone cracking catalyst, much like traditional amorphous silica-alumina catalysts, albeit with higher surface area/porosity. Since molecular intimacy is required for the silica and alumina to be effective, the most successful synthesis approach here used TEOS and AIP – critical in forming monomeric species. A simple indicator for nonuniformity is a bimodal pore size distribution, indicative of two separate phases.

Ti-Si-TUD-1

The synthesis of Ti-Si-TUD-1 is similar to the silica version, but a portion of the reactant is a titanium alkoxide, such as titanium (IV) n-butoxide. The SiO_2/TiO_2 ratio can be 10-∞, but a preferred ratio is about 50. The surface area can range from 400-1000 m^2/g, and the pore volume is 0.4-2.0 cc/g – slightly lower than the silica version. One of the early head-to-head catalytic tests of TUD-1 versus MCM-41 was for epoxidation using the titanosilica versions. The genesis of this concept was the extrapolation of epoxidation via TS-1, the titanosilicate isostructure of ZSM-5.

There are two major paths to generate a mesoporous titanosilica: framework substitution by one-pot synthesis and post-synthesis grafting. Framework substitution, as used with MCM-41 and MCM-48, can result in low utilization of the available titanium centers. Post-synthesis grafting is more effective at generating a high concentration of accessible Ti centers; however, it can be a complicated and, hence, costly synthesis method.

Table II summarizes some early work comparing various Ti-TUD-1 and Ti-MCM-41 for cyclohexene epoxidation (13, 14, 15). The as-synthesized Ti-TUD-1 is five times more active than Ti-MCM-41, even though they have equivalent surface area. The grafted MCM-41 is more active than its as-synthesized counterpart.

There are at least two possible explanations for the higher activity of the Ti-TUD-1: its three-dimensional pore structure and active site distribution. Figure 13

Table II. Epoxidation of Cyclohexene

Catalyst	S_{BET} (m^2/g)	D (Å)	TOF (hr^{-1})
Ti-TUD-1	870	50	20
Ti-MCM-41	920	30	4
Ti grafted TUD-1	560	100	28
Ti grafted MCM-41	1020	30	23

* Ti loading: 1.5-1.8 wt%   Selectivity: ~ 100%

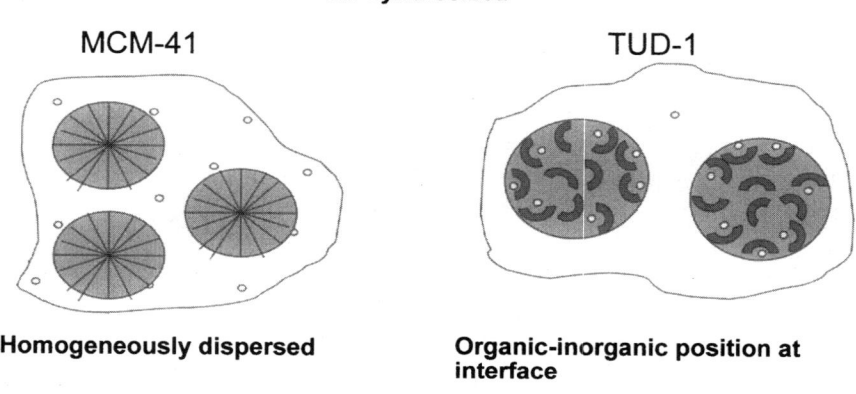

As-synthesized

MCM-41   **TUD-1**

Homogeneously dispersed   **Organic-inorganic position at interface**

——— Surfactant   ○ Ti site   ❨ Triethanolamine

Figure 13. Why is Ti-TUD-1 more active than Ti-MCM-41?

illustrates the synthesis and conceptual Ti site distribution for MCM-41 and TUD-1. The TUD-1 synthesis employs a bifunctional templating agent – triethanolamine – resulting in most of the Ti being accessible via the pores. After calcination, the Ti is homogeneously dispersed within the walls of MCM-41, whereas the Ti in the TUD-1 is preferentially dispersed on the surface of the pore walls.

Figure 14 shows a schematic of the bifunctional templating for forming Ti-TUD-1. Here, the TEA acts both as a traditional template as well as a complexing agent. Ti-TEA complexes are more stable than their Si-TEA counterparts. During the mesopore formation process or meso-sized TEA aggregate formation process, Ti-TEA preferably stays with the TEA aggregate due to being similar organic entities, whereas Si-TEA preferably undergoes hydrolysis and condensation and forms the inorganic phase. After calcination, TEA template is removed and consequently, the Ti-TEA decomposes and the Ti centers are then grafted onto the pore walls.

Other TUD-1 catalysts proven for selective oxidation include Au/Ti-Si-TUD-1 for converting propylene to propylene oxide (96% selectivity at 3.5% conversion), Ag/Ti-Si-TUD-1 for oxidizing ethylene to ethylene oxide (29% selectivity at 19.8% conversion), and Cr-Si-TUD-1 for cyclohexene to cyclohexene epoxide (94% selectivity at 46% conversion).

Hydrogenation

While the high-surface-area TUD-1 can serve as an anchor for many catalysts, one application deals with the hydrogenation of olefins and aromatics. In the refining industry one use is the hydrogenation of polynuclear aromatics ("PNAs") in diesel fuel, which can impact the fuels' carcinogenic properties. Also, jet fuel has an aromatics constraint, which relates to lessened smoke formation. Cracked stocks (e.g., coker or visbreaker liquids) generally have undesirable olefins that also need to be saturated prior to final processing.

In the referenced cases (17, 18), Si- or Al-Si-TUD-1 can be used as a support for various noble metals (Pt, PtPd, Ir, etc.). These noble metal catalysts are sensitive to sulfur so quite often the feeds are desulfurized by hydroprocessing before the hydrogenation step.

Table III shows a performance comparison of Pt/Pd TUD-1 with a similar commercial catalyst (17). The feedstock is a typical straight run gasoil ("SRGO"), a distillate precursor to diesel fuel. Under identical test conditions, the TUD-1 achieved 75% aromatics saturation versus 50% for the commercial catalyst. This superior result is especially interesting because the TUD-1 catalyst has a much lower density so that less catalyst by weight was used in the experiment.

352

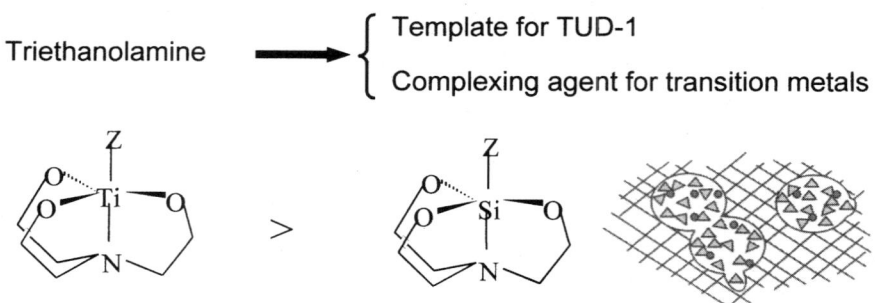

Figure 14. TEA Bi-functional templating for Ti-TUD-1: in-situ grafting

Table III. Hydrogenation via Pt/Pd TUD-1

	Feed	*Commercial Pt/Pd Catalyst*	*Pt/Pd Si-TUD-1 Catalyst*
Aromatics, vol%	21.2	10.1	5.1
Specific Gravity	0.8344	0.8241	0.8220
Cetane Index (D976)	44.7	46.7	47.8

From US Patent Application 2006009665 (filed 7/8/2004)

Hydrogenolysis

The need for improved hydrocarbon fuel efficiency has focused more attention on diesel fuels. Due to engine fundamentals, the diesel engine is about 15% more efficient than the gasoline engine. As we move toward "dieselization", the diesel-to-gasoline ratio will increase and refineries will need to produce more high quality diesel fuel. In many refineries, the major upgrading unit is the fluid catalytic cracker (FCC), which is designed to maximize gasoline yield. The distillate product, often known as light cycle oil ("LCO"), is a low-quality component for diesel fuel use.

Just as octane number is important for gasoline, cetane number generally defines the combustion properties of diesel fuel. The cetane number correlates to the ability to generate free radicals, essential for diesel combustion. The best diesel components (high cetane) are n-paraffins, followed by slightly branched paraffins, alkylcyclohexenes, and alkylbenzenes. LCO is primarily comprised of the poorest cetane components – dicyclics (naphthalenes, tetralins, and decalins). All have cetane numbers in the 10-30 range, significantly below the 40+ specs in the U.S. and 50+ specs in Europe.

Aromatics saturation and hydrocracking have been used to improve diesel fuel cetane quality. Unfortunately, aromatics saturation brings about only a marginal improvement in cetane number and that at a high hydrogen consumption per cetane barrel improvement. Hydrocracking naphthalenes and their alkyl homologues into the jet fuel and naphtha boiling ranges achieves a net increase in high-cetane-value distillate components (e.g., alkyl cyclohexanes, alkyl benzenes, paraffins, and slightly branched paraffins). One negative of conventional hydrocracking is its poor selectivity for distillate and high C_3/C_4 production.

An alternative to hydrogenation is selective ring opening via hydrogenolysis. For example, conversion of decalins to alkylcyclohexenes raises that component's cetane number by 20-30 points. Hydrogenolysis here is a carbon-carbon bond breaking via a free radical mechanism. The most effective metals are noble metals with virtually no acidity from the binder. Any residual acidity can cause unwanted hydrocracking, thereby converting much of the hydrocarbon stream into lighter products including C_3-C_4s, outside of the preferred diesel range,. Moreover, the feedstock must be very low in S content (e.g., less than 50 ppm) so as not to poison the noble metal.

Earlier work on selective ring opening (19, 20, 21) employed Pt on low acidity Zeolite Y. The choice of Zeolite Y was undoubtedly because of its high surface area and reasonably large aperture. TUD-1's pore size is several times that of Zeolite Y, and it normally has a significantly higher surface area. However, when corrected for density, its volumetric surface area is comparable to Zeolite Y. As such, TUD-1's *a priori* advantage over zeolites is its improved accessibility.

In recent work (22), it was shown that certain TUD-1 catalysts had effective hydrogenolysis activity. For example, a silica TUD-1 catalyst containing 0.9% iridium was tested for the selective ring opening of decalin. Decalin conversion was 76% while the total ring-opening yield was 61%. The reaction was carried out at 300°C, a pressure of 31 bars, and WHSV of 0.5 h^{-1}. In this process, hydrogen partial pressure must be relatively moderate because higher pressures favor hydrocracking versus hydrogenolysis.

Other Active Metals

One recent patent (23) and related patent application (24) cover incorporation and use of many active materials into Si-TUD-1, including Al, Ti, V, Cr, Zn, Fe, Sn, Mo, Ga, Ni, Co, In, Zr, Mn, Cu, Mg, Pd, Pt and W. Some active materials were incorporated simultaneously (e.g., NiW, NiMo, and Ga/Zn/Sn). The various catalysts have been used for many organic reactions [TUD-1 variants are shown in brackets]:

- Alkylation of naphthalene with 1-hexadecene [Al-Si]
- Friedel-Crafts alkylation of benzene with chlorobenzene [Fe-Si, Ga-Si, and Sn-Si]
- Oligomerization of 1-decene [Al-Si]
- Selective oxidation of ethylbenzene to acetophenone [Cr-Si, Mo-Si]
- Selective oxidation of cyclohexanol to cyclohexanone [Mo-Si]

One example (36) describes Co-TUD-1 for liquid-phase oxidation of cyclohexane. Another example (37) describes the synthesis, characterization, and catalytic performance of Fe-TUD-1 for Friedel–Crafts benzylation of benzene. Other reactions were described:

- Acylation (e.g., 2-methoxynaphthalene to 2-acetyl-6-methoxynaphthalene)
- Hydrotreating (e.g., S, N, and CCR reduction)
- Paraffin isomerization
- Resid demetalation
- Catalytic dewaxing via hydroisomerization
- Hydroxylation
- Hydrogenation
- Lube hydrocracking
- Ammoximation
- Dehydrogenation
- Cracking (e.g., FCC)

Immobilized Catalyst

An immobilized catalyst concept has been described in a recent patent application (25). The catalyst, used for the dehydrogenation of organic compounds, comprises an organometallic pincer complex bonded to a TUD-1 support. The pincer complex possesses catalytic activity for alkyl group dehydrogenation. This catalyst can be used for various refining and petrochemical processes, such as paraffin dehydrogenation.

Zeolites in TUD-1

Perhaps one of the most interesting concepts in mesoporous catalysts is its combination with embedded zeolites. With the zeolite distributed throughout the TUD-1, the synergistic benefits are (a) high accessibility to the internal zeolite crystal (achieving higher effective activity) and (b) easy egress from the zeolite surface (potentially achieving less secondary reactions to form unwanted byproducts, also reduced coking, pore-mouth plugging, and associated aging). The ability to tune both zeolite and TUD-1 physicochemical properties gives the catalyst many added degrees of freedom.

Figure 15 is a schematic of the zeolite/TUD-1 composite (26). As shown, the ideal composite is the dispersion of small crystals ("nanozeolites") in the highly porous TUD-1 matrix.

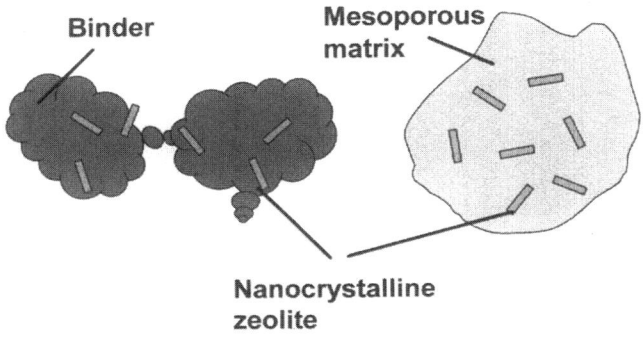

Ideal composite: Dispersed Nano-zeolite in porous matrix

Figure 15. Zeolite/TUD-1 composite: catalyst tailoring

Figure 16 illustrates the many scales involved in a catalytic process: the reactor is often several meters in length, the catalyst particle is typically several millimeters (up to 30) in cross-section, the zeolite/TUD-1 clusters can be microns in size, and (not shown) the ultimate crystals can be as small as 10-50 nm.

One lab-scale synthesis route is shown in Figure 17 (26). The zeolite embedding method involves suspending the zeolite in water, adding an inorganic oxide precursor to the water and mixing. The inorganic oxide precursor can be a silicon-containing compound such as tetraethyl orthosilicate (TEOS) or a source of aluminum such as aluminum isopropoxide, which reacts with water to form the inorganic oxide.

The pH of the mixture is preferably kept above 7.0. Optionally, the aqueous mixture can contain other metal ions such as those indicated earlier. After stirring, an organic templating agent is added into the mixture to help form the mesopores during the pore-forming step, as discussed below. The organic templating agent, which should not be so hydrophobic so as to form a separate phase in the mixture, can be one or more compounds. It is preferably added by dropwise addition with stirring to the aqueous inorganic oxide solution. After

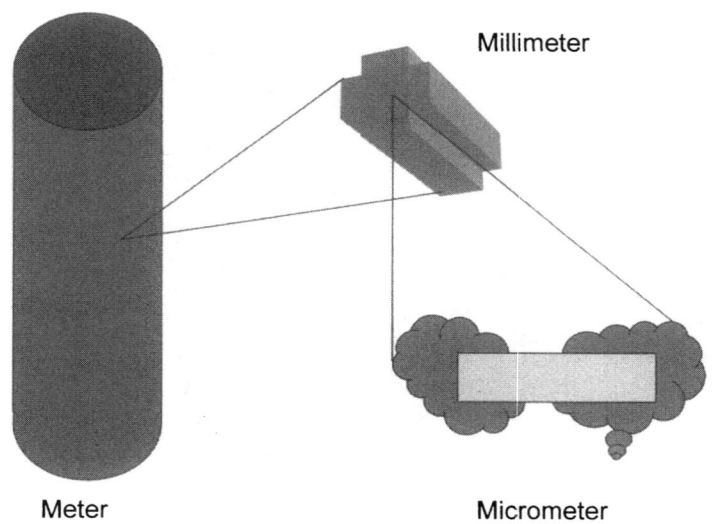

Meter   Micrometer

Figure 16. Zeolite/TUD-1 composite – scale

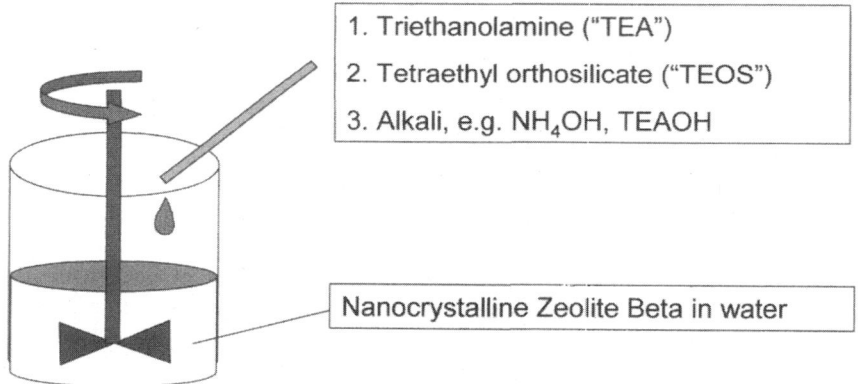

1. Triethanolamine ("TEA")
2. Tetraethyl orthosilicate ("TEOS")
3. Alkali, e.g. NH_4OH, TEAOH

Nanocrystalline Zeolite Beta in water

Objective:
Uniform dispersion of zeolite crystals in the finished composite

Figure 17. Synthesis of Zeolite Beta/TUD-1

the reaction period, the mixture forms a thick gel, which avoids zeolite particle precipitation or segregation and thereby ensures zeolite particles are homogeneously dispersed throughout the gel. The solution should include an alcohol, which can be added to the mixture and/or formed *in situ* by the decomposition of the inorganic oxide precursor. (For example, heating TEOS produces ethanol; propanol can be produced by the decomposition of aluminum isopropoxide.) Following drying, optional hydrothermal treatment, and calcination, a zeolite composite is obtained within a homogenously synthesized mesoporous material.

Figure 18 shows a high resolution, transmission electron microscopy image of the mesoporous inorganic oxide support with embedded Zeolite Beta, and an inset showing an electron diffraction ("ED") pattern of the zeolite domain (27). The ED pattern clearly shows that the Zeolite Beta has retained its crystallinity.

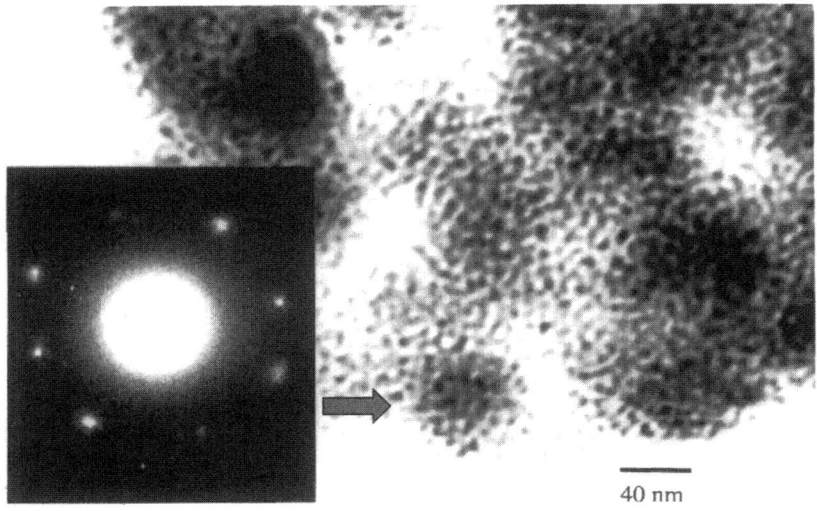

40 nm

Figure 18. Zeolite Beta in a mesoporous matrix

Figure 19 shows a series of syntheses, starting from TUD-1 (0% Zeolite Beta), then 10%, 20%, 40%, 60%, and 100% Zeolite Beta (27). The key pattern to note is that the Zeolite Beta maintains its crystallinity throughout all of these preparations. As such, one would expect that the resultant catalysts will have catalytic performance similar to Zeolite Beta, but impacted in some way by the TUD-1 mesopores.

Another zeolite TUD-1 study was carried out where a commercial Zeolite Y was compared to a Zeolite Y embedded in TUD-1. In this study the two

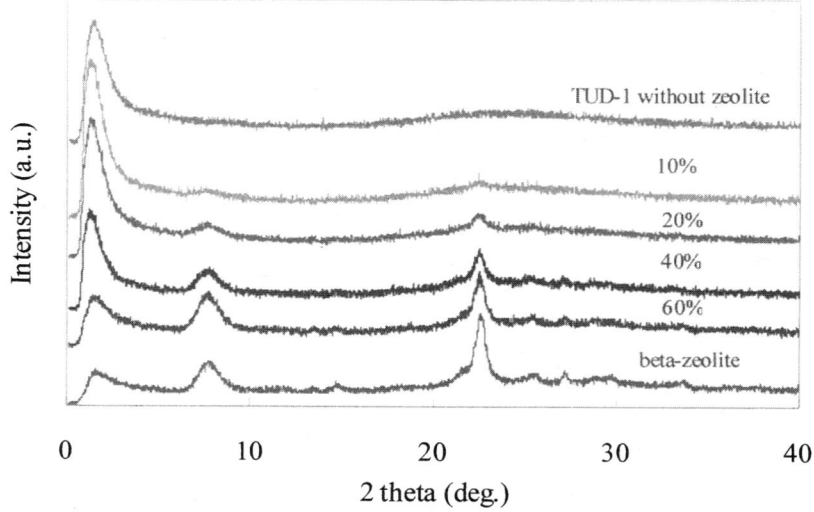

Figure 19. XRD pattern of Zeolite Beta-TUD-1

catalysts were tested using ethylbenzene synthesis as a probe reaction. Two different particle sizes (0.3 and 1.3 mm) were used for each catalyst. In Figure 20, the first-order rate constants were plotted versus particle size, analogous to a linear plot of effectiveness versus Thiele modulus. Using the mathematical model for effectiveness factor versus Thiele modulus, the rate constants were fitted for both catalysts. Interestingly, the Y/TUD-1 catalyst was twice as active as the commercial Y catalyst, primarily due to its very high diffusivity – more than 10 times higher than that of the commercial Zeolite Y. Extrapolating to zero particle diameter, one can see that the commercial Y is intrinsically more active than the Y embedded in the TUD-1. If the Y in TUD-1 had been optimized like the commercial Y catalyst, one should expect an even greater boost in performance.

Another example of zeolite /TUD-1 synergism is with a series of Zeolite Beta/ TUD-1 catalysts (29):

- 20 wt% Zeolite Beta in Al-Si-TUD-1 (Si/Al = 150)
- 40 wt% Zeolite Beta in Al-Si-TUD-1 (Si/Al = 150)
- 60 wt% Zeolite Beta in Al-Si-TUD-1 (Si/Al = 150)
- Zeolite Beta only

These four catalysts were tested in a fixed bed reactor, at atmospheric pressure, with constant residence time, and at 500-600°C. The first-order rate constants for n-hexane are shown in Figure 21. Note that the 40% and 60% zeolite-loaded catalysts were clearly superior to both the lower loading (20%)

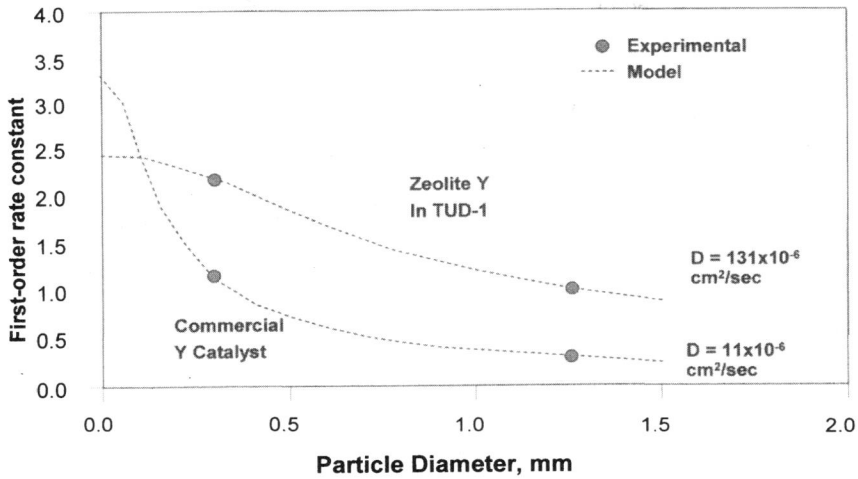

Figure 20. EB alkylation activity versus particle diameter

and pure Zeolite Beta catalyst. Again, this is evidence that the TUD-1 enhances the performance of the zeolite. While this study used a simple model compound, one can infer that a similar relationship would occur in catalytic cracking.

One patent application (30) describes the use of zeolite/TUD-1/ metal function for various refining processes, including hydrotreating, middle distillate hydrocracking, hydrocracking for lubes (commonly called "lubes hydrocracking"), and hydroisomerization of heavy distillate.

Another patent application (31) describes the use of zeolite/TUD-1/ (optionally) metal function for acylation, alkylation, dimerization, oligomerization, polymerization, dewaxing, hydration, dehydration, disproportionation, hydrogenation, dehydrogenation, aromatization, selective oxidation, isomerization, hydrotreating, catalytic cracking and hydrocracking. In one example, a commercially available, alumina-bound Zeolite Beta (80 wt%) was tested for ethylene/benzene alkylation to produce ethylbenzene using the same methodology and apparatus described earlier. A first-order rate constant of 0.29 cm^3/g-sec was obtained. A 16% Zeolite Beta in TUD-1 catalyst also had a rate constant of 0.30 cm^3/g-sec for the same reaction. These results indicate that: (a) the integrity of the zeolite crystals in the mesoporous catalyst support is maintained during the synthesis; (b) the microporous Zeolite Beta in the mesoporous support was still accessible after the synthesis of the catalyst; and (c) the mesopores of the support facilitate mass transfer in aromatic alkylation reactions.

In another example, as-synthesized swollen MCM-22 / TUD-1 was tested (31) for acylation of 2-methoxynaphthalene with acetic anhydride to 2-acetyl-6-

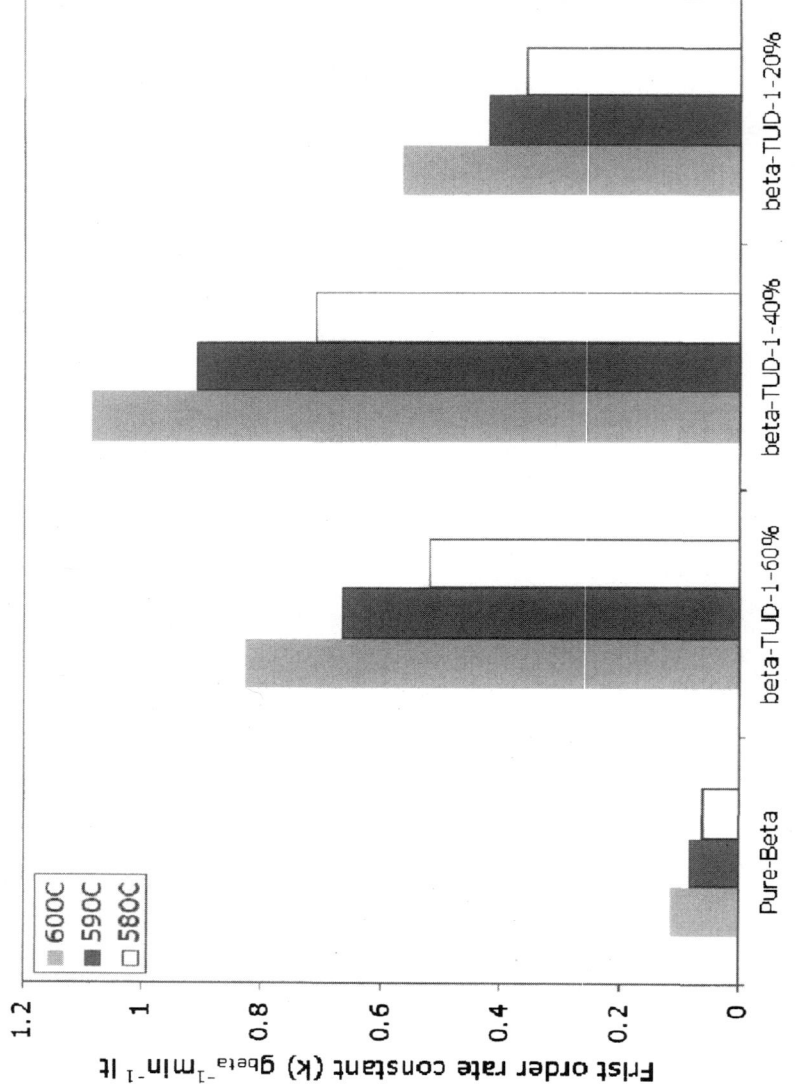

Figure 21. n-Hexane cracking over Zeolite Beta / TUD-1

methoxynaphthalene at 240°C. After reaction for six hours, conversion of 2-methoxynaphthalene reached 56% with 100% selectivity to 2-acetyl-6-methoxynaphthalene. Other zeolite catalysts were similarly tested, but none were nearly as effective.

Other references (32, 33, 34, 35) describe the synthesis and performance benefits of zeolites embedded in TUD-1. This concept has been reported in the literature with other mesoporous materials, but the three-dimensional nature of TUD-1 should make these zeolite combinations of special benefit.

General Applicability

This chapter has described some of the catalytic applications for TUD-1. Since there are nearly 20 known chemical variants of TUD-1, these clearly are not all possible applications. We have focused on major hydrocarbon processes, but one can easily envision many reactions to form fine chemicals and pharmaceuticals. In refining, the areas of attention should be the upgrading of high molecular weight streams (e.g., resid and lubes). Resid upgrading could take the form of FCC feed hydrotreating, resid hydrocracking (e.g., LC-Fining type), or RFCC (resid fluid catalytic cracking). Lubes upgrading could include lube hydrocracking or hydrofinishing.

Issues

As with many new catalysts, there is a "chicken and egg" syndrome. Which comes first: the catalyst application or the catalyst development? Some catalysts may require a simple change in recipe (e.g., Zeolite X versus Y, Zeolite Beta versus ZSM-12, etc.). For that situation, the needed performance improvement may be relatively small to justify a development effort. However, the synthesis of TUD-1 is significantly different than many other new materials and as such, a major economic incentive will be needed to justify catalyst commercialization.

Even with all of its resources and renowned catalyst expertise, ExxonMobil required a substained 10+ years of active effort with MCM-41 before it was first commercialized. The good news is that there are clearly niche applications where a mesoporous material will be competitive.

The Future

Fossil fuel energy supplies (mostly natural gas) are growing at 1%/ year, and overall demand is growing at 2+%/ year with no major relief in sight. Except for some recent oil discoveries (e.g., Cambodia and off-shore China), most oil-producing countries have peaked in production and rates are declining. As crude oil prices continue to climb, there will be a significant boost in price spreads (resid to gasoline and diesel, resid to lubes, etc.) and mesoporous catalysts will be further studied for heavy oil and resid upgrading.

We project that, with continued effort, many uses will be discovered for TUD-1 and other ultra-large pore catalysts.

On the cost side of mesoporous materials, our second-generation synthesis route has dramatically lowered the manufacturing costs. Assuming an 80% recycle of EG and TEA, the raw material costs are about US$2-3/lb catalyst.

Conclusion

In catalysis, the major zeolites of commercial interest (zeolite Y, ZSM-5, and Zeolite Beta) all have three-dimensional, intersecting pores, a property important for aging stability and selectivity. TUD-1 is one of the few mesoporous materials also having three-dimensional, intersecting pores. As such, its catalytic potential appears quite promising. Since it can be made with many chemical versions, its utility is almost endless.

Acknowledgment

The technical leadership by Professors J.C. Jansen and Th. Maschmeyer is greatly appreciated. They each played a visionary role in the discovery, synthesis, and characterization of the TUD-1 family.

References

1. Kresge, C.T. et al. US 5,098,684 (1992)
2. Kresge, C.T. et al. US 5,102,643 (1992)
3. Beck, J.S. et al. US 5,145,816 (1992)
4. Kresge, C.T. et al. US 5,198,203 (1993)
5. Degnan, T.F. et al. Catalysis Society of New York Meeting, 2003
6. Shan, Z. et al. US 6,358,486 (2002)
7. Shan, Z. et al. International Symposium on Silica, 2000, France
8. Shan, Z. et al. NCCC2000, The Netherlands
9. Dautzenberg, F.M. 13th International Congress on Catalysis, 2004, Paris
10. Shan, Z. et al. US Pat. Appl. 20050164870 (filed 1/26/04)
11. Shan, Z. et al. US 7,211,238 (2007)

12. Shan, Z.; Jansen, J.C.; Zhou, W.; Maschmeyer, Th. "Al-TUD-1, Stable Mesoporous Aluminas with High Surface Areas", Applied Catalysis A: General 254 (2003) 339-343

13. Shan, Z.; Jansen, J.C.; Marchese, L.; Maschmeyer, Th. "Synthesis, Characterization, and Catalytic Testing of a 3-D Mesoporous Titanosilica, Ti-TUD-1", Mesoporous and Mesoporous Materials 48 (2001) 181-187.

14. Shan, Z.; Gianotti, E.; Jansen, J.C.; Peters, J.A.; Marchese, L.; Maschmeyer, Th. "One-Step Synthesis of a Highly Active, Mesoporous, Titanium-Containing Silica by Using Bifunctional Templating", Chem. Eur. J. 2001,7, no.7, 1437-1443

15. Shan, Z. et al. US 6,906,208 (2005)

16. Shan, Z. et al. US 7,091,365 (2006)

17. Ramachandran, B. et al. US Pat. Appl. 20060009665 (filed 7/8/04)

18. Ramachandran, B. et al. US Pat. Appl. 20060009666 (filed 1/10/05)

19. Tsao, Y.P. et al. US 6,210,563, 4/3/01

20. Tsao, Y.P. et al. US 6,241,876, 6/5/01

21. Tsao, Y.P. et al. US 6,500,329, 12/31/02

22. Ramachandran, B. et al. US Pat. Appl. 20060014995 (filed 9/23/05)

23. Shan, Z. et al. US 6,930,219 (2005)

24. Shan, Z. et al. US Pat. Appl. 20030188991(filed 12/6/02)

25. Yeh, C.Y. et al. US Pat. Appl. 20040181104 (filed 3/2/04)

26. Shan, Z. et al. NCCCIII, 2002, The Netherlands

27. Shan, Z. et al. US 6,762,143 (2004)

28. Dautzenberg, F.M. 13th International Congress on Catalysis, 2004, Paris

29. Shan, Z. et al. US Pat. Appl. 20040138051 (filed 10/22/03)

30. Angevine, P.J. et al. US Pat. Appl. 20060052236 (filed 9/7/05)

31. Shan, Z. et al. US Pat. Appl. 20060128555 (filed 2/8/06)

32. Shan, Z. et al. US 6,814,950 (2004)

33. Shan, Z. et al. US 6,930,217 (2005)

34. Shan, Z. et al. US 7,084,087 (2006)

35. Angevine, P.J.; Shan, Z. "Mesoporous Materials", USA-Netherlands Catalysis Conference, August 19-20, 2004, Philadelphia, PA

36. Anand, R. et al. Catalysis Letters Vol. 95, No. 3-4, June 2004

37. Hamdy, M.S. et al. Catalysis Today 100 (2005) 255–260

Chapter 10

Evolving Production of the Acetyls (Acetic Acid, Acetaldehyde, Acetic Anhydride, and Vinyl Acetate): A Mirror for the Evolution of the Chemical Industry

Joseph R. Zoeller

Research Laboratories, Eastman Chemical Company, P.O. Box 1972, Kingsport, TN 37660

Introduction

Acetic acid is one of the earliest chemical entities produced by mankind, with intentional production evident as early as 3,000 BC,[1a] although it was likely produced concurrently with the earliest attempts at wine making (ca. 10,000 BC.) The earliest applications were to preserve food by pickling, but its use in metal refining and some studies on its properties appear from ancient Greece and the period of alchemy.[1a] Given its antiquity and long history of utility to mankind, acetic acid would appear to be a natural early choice for production and utilization by the chemical industry. As a consequence, the production of acetic acid and its related products and starting materials have mirrored the development of the chemical industry, reflecting the availability of new resources, the development of new technologies, and the changing needs of the various industries dependent upon chemicals. Indeed, changes in its production methods and those for its key derivatives have often been harbingers of changes in the industry.

In this chapter, we will discuss the evolution of processes for the production of the acetyl chemicals - acetic acid (AcOH), acetic anhydride (Ac$_2$O), acetaldehyde (AcH), and vinyl acetate (VA). These four materials coevolved in a series of synergistic relationships which ultimately led to the modern acetyl stream which now exceeds 6 X 10^9 kg/yr of acetyl (as acetic acid equivalents) per year and will be discussed in evolutionary context with each other. Opportunities and resultant innovations in the chemical industry are normally created when lower cost raw materials become available or existing products fail to meet the needs and requirements generated by new industries and products. The production and interplay of the acetyl products is a mirror of the evolution of the chemical industry as a whole and we shall discuss how each process evolved over time. (There are numerous good sources for more detailed overviews of these processes provided in the references.[1-5])

Acetyl Processes 1700-1910.

Using What Nature Provided - Acetic Acid from Natural Sources.

Like the chemical industry, the roots of acetyl production start with the production of acetic acid by natural processes. While we already discussed the ancient history of acetic acid production, larger scale production and isolation began in the 1700's with the "Orleans" (or "Slow") process which consisted of intentional fermentation of wines in a semi-batch process in which the wine was added in increments until the barrel was full. However, this was displaced by the "German" (or "Fast") process in which wine was continuously fed over charcoal or wood chips and oxidized with a continuous flow of air to increase the activity of the bacteria. (As a continuous process, the "German" process likely represents one of the earliest examples of a continuous reactor system.) These fermentation processes were reportedly capable of producing up to 10 tons/day of product.[1,6]

The fermentation processes were ultimately displaced in the late 1800's when destructive (pyrolytic) wood distillation, which co-produced methanol, acetic acid, methyl acetate, acetone, methyl ethyl ketone, and creosote, became the common method of production for acetic acid. While these natural sources served mankind for several millennia, and research continues on fermentation processes,[6] they were inefficient processes which generated large amounts of waste and consumed large amount of energy. Further, none of these processes produced glacial acetic acid which had to be produced by precipitation as the calcium salt and subsequent acidification with, and distillation from, sulfuric acid. However, demand for glacial acetic acid was relatively small in this period and this process suffced until the beginning of the 20th century.

The Acetylene Period (1910-1950).

Genesis of Acetyls as a Chemical Enterprise (1910-1920)

New enterprises and rapid innovation in the chemical industry are normally created when existing products fail to adequately meet the material needs and requirements generated by new industries or products. The newly introduced product then finds numerous new, often unanticipated, applications. Slower, more evolutionary, innovations occur when lower cost raw materials become available, or new, more efficient, technologies are developed. In the case of acetyls, all of these events occurred nearly simultaneously in the period 1910-1920 and led to the parallel development of an array of successful processes.

Let's begin with the advent of the new raw material and a new technology. At the dawn of the 20th century, acetylene, which was generated by heating coal and lime and subsequently hydrolyzing the calcium acetylide (also called calcium carbide), became available as the first simple building block for the chemical industry. The first fully synthetic process for generating acetic acid involved the addition of water to acetylene in the presence of a mercuric salt and sulfuric acid and subsequent air oxidation of the resultant acetaldehyde. (See equations [1] and [2].) This process was first commercialized in 1911 and provided the first readily available source of glacial acetic acid. By 1916, the acetaldehyde portion of the process was improved by the addition of iron which aided the reoxidation of mercury. The oxidation of acetaldehyde would remain the predominant means of generating acetic acid for the next 60 years, although the raw materials leading to acetaldehyde would change over time. While this potentially made large amounts of glacial acetic acid available, there was limited demand for the glacial acetic acid product and its primary outlet was to make acetone by the pyrolysis of calcium acetate.

$$HC\equiv CH + H_2O \xrightarrow{\text{HgSO4, FeSO4, H2SO4}} AcH \qquad [1]$$
$$AcH + \tfrac{1}{2}O_2 \longrightarrow AcOH \qquad [2]$$

The commercial driving force which created the acetyl industry was provided by the onset of World War I and the fledgling aviation industry. While World War I temporarily increased the requirements for acetic acid to make acetone (needed to make "cordite"), the permanent change would result from requirements in aviation. At the dawn of the 20th century, outside of a few natural polymers and monomers which had limited utility, the only plastic available for films, coatings, and plastics was cellulose nitrate. While the first planes used fabric wings, they were not strong enough and were too porous, so cellulose nitrate was applied to form a lightweight wing, which today we would recognize as a fiber reinforced plastic. Given the highly flammable (sometimes explosive) nature of cellulose nitrates, there were obvious problems with their use and the early airplanes were firetraps which could ignite in midair. The advent of World War I, and the obvious unsuitability of cellulose nitrate in a war environment, led to a sudden demand for a new chemical entity, not yet commercially available, to replace cellulose nitrate.

While of little use at the time of its discovery, the solution to this problem was provided by a discovery by Miles and Eichengrun[7], who demonstrated in 1904 that cellulose triacetate (made by sulfuric acid catalyzed acetylation of cellulose with acetic anhydride) could be made soluble (in solvents like acetone and acetic acid) by partial hydrolysis of a portion of the acetate functionality.

Cellulose acetate was stable, not readily flammable, and water repellent, so it was a suitable substitute for cellulose nitrate. Further, in its soluble form, cellulose acetate could be applied by simply painting the fabric as had been done with cellulose nitrate. However, cellulose acetate manufacture would require large quantities of acetic anhydride, and there was no good way to produce acetic anhydride other than the existing neutralization of calcium acetate with sulfuric acid. What ensued would be one of those creative bursts that occur sporadically in the chemical industry.

Expanding on the mercuric sulfate – sulfuric acid technology, it was known as early as 1912 that addition of a single equivalent of acetic acid to acetylene in the presence of mercuric sulfate and sulfuric acid resulted in the generation of vinyl acetate (a mere laboratory curiosity at the time) whereas addition of two or more equivalents of acetic acid provided ethylidene diacetate (1,1-diacetoxyethane, EDA). EDA was known to evolve acetaldehyde upon heating in the presence of acids. Upon complete removal of a mole of acetaldehyde, an equimolar amount of acetic anhydride could then be distilled from the residue. If one balances the sequential processes as shown in equations [3] through [6], one finds that the process is in stoichiometric balance for conversion of acetylene and air to cellulose acetate without by-products. (In practice there is some net acetic acid production since cellulose is never completely dry.)

$$HC\equiv CH \ + \ 2 \ AcOH \ \longrightarrow \ (AcO)_2CHCH_3 \qquad\qquad [3]$$

$$(AcO)_2CHCH_3 \ \longrightarrow \ Ac_2O \ + AcH \qquad\qquad [4]$$

$$AcH + \tfrac{1}{2} O_2 \ \longrightarrow \ AcOH \qquad\qquad [5]$$

$$Cellulose + Ac_2O \ \longrightarrow Cellulose\text{-}Acetate + AcOH \qquad\qquad [6]$$

Net Reaction:
$$Cellulose + HC\equiv CH + \tfrac{1}{2} O_2 \ \longrightarrow Cellulose\text{-}Acetate \qquad\qquad [7]$$

The chemistry is more complex than would appear on the surface. When acetic anhydride reacts with acetaldehyde, it forms EDA in an equilibrium that favors EDA (Equation [8]), but a subsequent disfavorable equilibrium follows which forms acetic acid and vinyl acetate (Equation [9]). These equilibrium constants indicate that EDA is the most thermodynamically favored product. In the presence of acids, these equilibria occur rapidly. Using LeChatelier's principle, if the most volatile product, acetaldehyde, is continuously removed you can shift the equilibrium until only acetic anhydride remains which can then be distilled independently. (The equilibrium between EDA and vinyl acetate would prove to be pivotal in the later development of a vinyl acetate process.)

$$AcH \ + \ Ac_2O \ \rightleftharpoons \ (AcO)_2CHCH_3 \qquad K_{eq\,(140°C)} = 25 \qquad\qquad [8]$$

$$(AcO)_2CHCH_3 \rightleftharpoons AcOH +(AcO)HC=CH_2 \quad K_{eq(140°C)} = 0.01 \quad [9]$$

While the acetylene based route to acetic anhydride would remain in practice in some locations until the 1940's, a competitive technology was developed within the same decade that would dominate production even today. In this period, chemists discovered that the highly endothermic generation of ketene could be accomplished by heating either acetone (equation [10]) or acetic acid in the presence of a phosphate catalyst (equation [11]) at very high temperature (>700°C).[4] Subsequent reaction of ketene with glacial acetic acid gave acetic anhydride (equation [12]).

Acetone to ketene:
$$CH_3C(=O)CH_3 \longrightarrow H_2C=C=O + CH_4 \qquad [10]$$

Acetic acid to ketene:
$$CH_3C(=O)OH \longrightarrow H_2C=C=O + H_2O \qquad [11]$$

Acetic anhydride from ketene:
$$H_2C=C=O + AcOH \longrightarrow Ac_2O \qquad [12]$$

While acetone would serve as a source of ketene in a number of locations, the acetic acid dehydration would predominate and the acetic acid based ketene process is still widely practiced at the start of the 21st century. While acetone had some attractive features, particularly the generation of an inert co-product (methane) rather than reactive water which can destroy ketene, there are some sound reasons the acetic acid process predominated. First, let's recall that during the period 1910-1920, representing the time these processes were introduced, acetone was still made from calcium acetate, so acetone was obtained, at the time, in a two step process from acetic acid. The choice of acetic acid skipped two steps and eliminated wastes.

Further, the acetic acid based ketene process was also aided by a chemical engineering advance. Chemical engineers demonstrated that glacial acetic acid could be obtained by dehydration with azeotroping agents and subsequent distillation. Glacial acetic acid was now available even if the source was very wet. This had multiple impacts. First, the advent of azeotropic drying enabled facile dehydration of the wet acetic acid co-product obtained in the ketene generation. Second, recall that the useful form of cellulose acetate required partial hydrolysis of the cellulose triacetate. The addition of water to the cellulose acetate product mixture simultaneously quenched any excess acetic anhydride used in the reaction, performed the necessary hydrolysis, and removed the acetic acid by-product obtained during the initial acetylation. However, this hydrolytic step resulted in a very wet acetic acid stream that required recovery and recycle. Azeotropic distillation permitted the manufacturer to recycle this acetic acid co-product for use in acetic anhydride

370

production rather than discarding it or selling it at a loss. (Until the 1950's, the market for acetic anhydride and cellulose acetate was much larger than that for acetic acid. Therefore, any process that generated a glut of acetic acid could not be absorbed by the market.) Lastly, this technology also reinvigorated the wood distillation sources since wood distillation could now provide glacial acetic acid and diversify available feedstocks for creating acetic anhydride.

The feedstock picture further diversified in 1920 with the commercialization of ethanol dehydrogenation to generate acetaldehyde. (The process was conducted in the vapor phase at 260-290°C using copper-chromite catalysts.) While the process was known as early as 1886[3b], the development of adequate catalysts for the endothermic process would take nearly 35 years. Subsequent oxidation to acetic acid provided an additional source of acetic acid. These technologies would largely stay in place with only minor modification until the 1950's. A summary of the chemical routes to the various acetyls in 1920 is shown in Figure 1.

$$EtOH \longrightarrow AcH + H_2 \qquad \Delta H= +19.7kcal/mol \qquad [13]$$

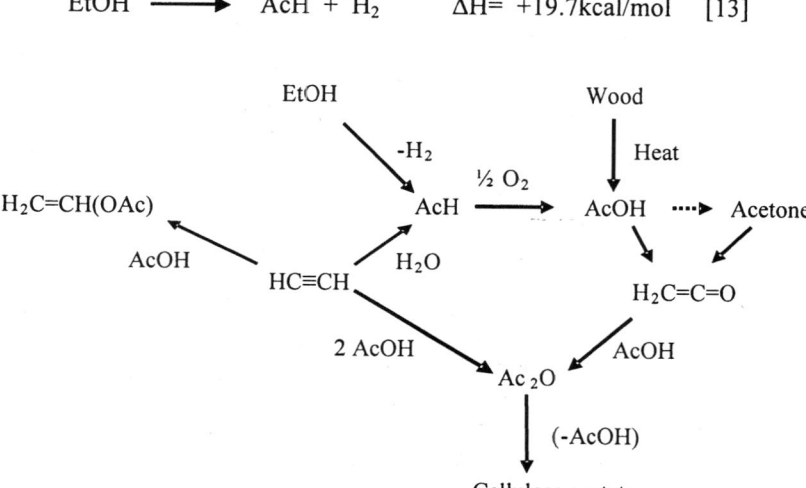

Figure 1. Acetyl Production Circa 1920.

Maturation of Cellulose Acetate and the Introduction of Vinyl Acetate. (1920-1950)

As often happens when a new material is introduced, there was a rapid expansion into new, often unanticipated, applications. Cellulose acetate was no exception. In the period 1920-1935, cellulose acetate and related esters helped usher in the era of plastics when cellulose acetate became the first thermoplastic available for molded parts. Cellulose acetate was spun into textile fibers, cast into films for packaging and photography, used to generate molded parts, and saw expanded applications in coatings and reinforced plastics.

Over the next 30 years, wood distillation declined relative to ethanol dehydrogenation and acetylene based processes as a source of acetyls, but all three would contribute to the acetyl supply chain required to generate the acetic anhydride needed to meet the market demands for cellulose acetate as the product grew and matured over the period 1920-1940. However, while cellulose acetate and cellulose ester markets would grow, improvements in technology for the basic acetyl products were limited during this time period.

Long term, the most significant new development in the acetyl market in this period was the introduction of vinyl acetate. While the liquid phase addition of acetic acid to acetylene in the presence of a mercuric sulfate catalyst at 60-100°C to generate vinyl acetate was first discovered in 1912, it was introduced as a commercial scale process in 1925 and introduced vinyl acetate to the market as a commodity scale product. However, the mercury process was inefficient and toxic and did not last long as a commercial process. In 1921, it was discovered that Zn acetate on activated charcoal could catalyze the addition of acetic acid to acetylene in the vapor phase. By 1940, sequential improvements in the stability of activated charcoal provided Zn on carbon catalysts that were sufficiently stable to render any remaining Hg based processes untenable. By about 1950 the mercury based process was extinct and completely replaced with the vapor phase Zn on activated charcoal process which was operated at 170-210°C and pressures just exceeding 1 atmosphere with an excess of acetylene. However, vinyl acetate still represented a relatively small portion of the market in acetyl related products.

There were also improvements in acetaldehyde and acetic anhydride manufacture. Ag based catalysts for the partial oxidation of ethanol became available around 1940. When used to oxidatively dehydrogenate ethanol [14], the conversion of ethanol to acetaldehyde was no longer equilibrium limited since the reaction was now very exothermic. Fortunately, the process still displayed excellent selectivity (ca. 93-97%) for acetaldehyde. This technology replaced the older Cu-Cr processes over the period of the 1940-1950 and made ethanol a much more attractive resource for acetaldehyde. When ethylene became available as a feedstock in the 1940's through 1950's, ethanol became cheaply available via ethylene hydration (as opposed to traditional fermentation). With ethanol now cheaply available from ethylene, the advent of the Ag catalyzed oxidative dehydration to acetaldehyde rapidly accelerated the shutdown of the last remaining wood distillation units.

$$\text{EtOH} + \tfrac{1}{2}\,O_2 \xrightarrow{\;400°C,\;Ag\;} \text{AcH} \qquad \Delta H = -58 \text{ kcal/mol} \qquad [14]$$

Acetaldehyde oxidation was also marginally improved, especially for the manufacture of acetic anhydride. In 1935, workers at Shawingan Chemicals discovered that the oxidation of acetaldehyde, if conducted in the presence of cobalt, copper, or better yet, a mixture of the two catalysts, yielded a mixture of acetic anhydride and acetic acid providing the water co-product was rapidly separated by azeotropic distillation, normally with a compatible material such as ethyl acetate. It would not be until the 1940's that this became widely practiced, but the process was eventually widely adopted. While experimental units produced ratios of acetic anhydride: acetic acid as high as 4:1, it appears that the commercial process normally gave a 5:4 mixture of acetic anhydride: acetic acid under normal operating conditions.[3b] The poor selectivity reflects both an inefficiency in the catalytic process and the difficulty in separating water rapidly enough to prevent hydrolysis of acetic anhydride to acetic acid on a commercial basis. However, this still represented a significant improvement since it reduced the amount of ketene that needed to be generated via the dehydration of acetic acid to ketene during the production of cellulose acetate.

The Petrochemical Era (1945-1970).

The dawn of the petrochemical industry is normally marked as 1940 in the United States and 1950 in Europe and Japan. Throughout this period, the chemical industry rapidly phased out coal and acetylene as its raw material base and shifted to using products obtained from petroleum cracking and fractionation. The new key building blocks would be hydrocarbons, olefins (especially ethylene and propylene) and the aromatics (benzene, toluene, and xylene) which were cheaper and cleaner to obtain. As expected, this change affected the methodology for acetyl production and a new set of processes began to emerge. However, the emergence of two technologies had as profound an affect on the production methods for acetyls as the change in feedstocks.

First, the pioneering work of Rölen in cobalt catalyzed hydroformylation (the addition of CO and H_2 to olefins to generate aldehydes discovered in 1938) and Reppe (the nickel and cobalt catalyzed carbonylation of alcohols, olefins, and acetylene to carboxylic acids during the period ca. 1940) ushered in the era of organometallic chemistry and modern transition metal based homogeneous catalysis. The advent of organometallic chemistry and transition metal homogeneous catalysis would profoundly affect the methods for the production of all acetyls. The second major event was the development of new materials of construction for reaction vessels, especially those resistant to halide induced corrosion. The carbonylation work of Reppe, while promising, used iodine and very high temperatures and pressures. In the original schemes this required silver lined reactors, but the advent of Ti and a selection Ni-Cr alloys as materials of construction in the mid 1950's enabled the economic construction of reactors using halides under high temperature and pressure conditions. While often overlooked, this was one of the most significant events in the history of acetyl chemicals. These two major events would significantly impact methods of production for all the acetyls.

Changes in Acetic Acid Manufacture

While reports of the carbonylation of methanol to acetic acid (equation [15]) date back to 1913 (at BASF)[1b] the first full descriptions appeared in 1926 when workers at Celanese described the carbonylation of methanol over acidic heterogeneous catalysts.[8] None of these processes looked promising at the time. However, in the 1950's, Reppe (BASF) reported a successful homogeneous cobalt catalyzed process operated in the presence of a methyl iodide co-catalyst. The process was commercialized in 1960 and leveraged cheaply available natural gas to generate synthesis gas (a mixture of CO and hydrogen sometimes referred to simply as syngas) via steam reforming or partial oxidation of methane. The synthesis gas was used to synthesize both the methanol feedstock and as a source of CO in the carbonylation.

$$MeOH + CO \longrightarrow AcOH \qquad [15]$$

Unfortunately, the process required very high temperatures and pressures (250°C, 680 atm.) and selectivity was not good, providing yields of 90% based on methanol and 70% based on CO. Several plants would be built using this technology between 1960 and 1966, but, at these extreme pressures and temperatures, the plants were *very* capital intensive and the yields were mediocre. Corresponding improvements in acetaldehyde production (described later in this section) which were also attributable to advances in homogeneous catalysis and the advent of corrosion resistant materials of construction, prevented the cobalt based methanol carbonylation process from displacing acetaldehyde oxidation as a means of producing acetic acid.

Another competing acetic acid process also arose in the same time period. About 1957, Celanese began to operate a facility for the oxidation of butane to acetic acid to take advantage of low cost hydrocarbons. The oxidation, which is conducted in the presence of a Co catalyst at 180°C and 15-20 atm. of oxygen at a butane conversion of 10-20% yielded a mixture of products in the following portions:

acetic acid	12.5
formic acid	1.25
2-butanone	2
propionic acid	1

This process still operates today, but is only viable due to the added value of the by-products and can not compete with modern methanol carbonylation for acetic acid production. There is continued interested even today in the oxidation of hydrocarbons, especially ethane, but these technologies are not competitive with modern methanol carbonylation for the generation of acetic acid.

Changes in Acetaldehyde Manufacture

Developments in acetaldehyde manufacture allowed acetaldehyde oxidation to remain competitive with, or superior to, Co catalyzed methanol carbonylation and butane oxidation throughout the period 1945-1970 as a means of generating acetic acid. First, as mentioned earlier, ethanol was now available from ethylene via hydration, which lowered the cost of ethanol. Therefore, the oxidative dehydration of ethanol (equation [14]) was now more attractive than when ethanol was derived from fermentation.

However, more significant was the introduction of a new one step process for the oxidation of ethylene to acetaldehyde using a $PdCl_2$-$CuCl_2$ catalyst (equation [16]) sometime between 1957-1959. The process, now known as the Wacker Process, is operated at ca. 10 atm and 100-110°C, requires a large excess of chloride, and produces AcH in about 95% yield. The mechanism for this reaction is shown in figure 2.

$$H_2C=CH_2 + \tfrac{1}{2} O_2 \longrightarrow \quad AcH \quad [16]$$

The process displaced ethanol oxidation in new installations and is the predominant route to acetaldehyde today. The role of the various components was well understood. To understand the role of copper, we must first understand that, while thermodynamically feasible, Pd(0) did not readily react with oxygen and required a catalyst. Copper filled the catalyst role, since Cu^{2+} readily oxidized Pd(0) to Pd^{2+} in the presence of excess chloride anion (as HCl) and the resultant Cu^+ was readily oxidized by air. (See equations [17] and [18].) The chloride fulfilled two other critical roles. First, it was critical to the oxidation of Pd since the oxidation appeared to occur via a chloride bridge. Second, a large excess of chloride was necessary to maintain Cu^+ in a soluble form. (CuCl is insoluble but in the presence of excess chloride, CuCl formed soluble $CuCl_2^-$, $CuCl_3^{2-}$, and $CuCl_4^{3-}$ salts.) The reaction required access to corrosion resistant alloys due to the high chloride concentration and tapped into the then newly emerging understanding of homogeneous transition metal catalysis.

In the best practice, the reaction was conducted in two stages in which the Pd catalyzed addition of water to ethylene is carried out in the presence of a large excess Cu^{2+} and the reoxidation of the resulting Cu^+ is conducted in a separate vessel. However, the reaction was reported to be carried out as a one stage process as well.

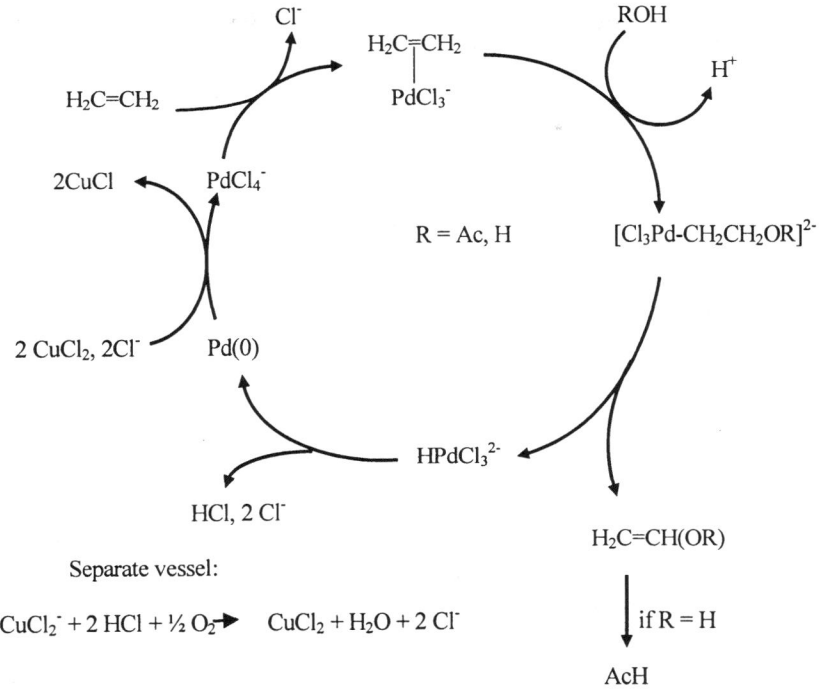

Figure 2. Mechanism for Wacker-Chemie Liquid Phase Acetaldehyde and Vinyl Acetate Process.

$$2\ CuCl_2 + Pd(0) + 2\ Cl^- \longrightarrow 2\ CuCl_2^- + PdCl_2 \qquad [17]$$

$$2\ CuCl_2^- + \tfrac{1}{2}\ O_2 + HCl \longrightarrow 2\ CuCl_2 + 2\ Cl^- + 2\ H_2O \quad [18]$$

(For clarity, only the $CuCl_2^-$ is shown in Equations [17] and [18] and Figure 2. Cu(I) may also be present as $CuCl_3^{2-}$, and $CuCl_4^{3-}$ as indicated above.)

Changes in Vinyl Acetate Manufacture

Until the mid-1950's, vinyl acetate was a moderate volume product used to produce specialty polyolefins, such as cling films. However, in the 1950's, emulsion polymerization ("latex") began to emerge as a major new product line in the chemical industry. (Emulsion polymerization uses surfactants to generate water based emulsions of polyolefin polymers. These products are most familiar to the consumer as water borne paints, but are also found in a wide variety of adhesives, textile sizings, and coatings.[5]) Three major monomers constituted the preferred feedstocks for this new technology: vinyl acetate, acrylic acid, and styrene. Further spurring demand for vinyl acetate was the advent of polyvinyl alcohol applications (obtained by hydrolysis of polyvinyl acetate) and polyvinyl butyral (made by adding butyraldehyde to polyvinyl alcohol), an integral component of safety glass. These new outlets would ultimately drive US production of vinyl acetate from only 59 KMT in 1956, to 356 KMT in 1970, and 850 KMT in 1980. However, the existing acetylene routes were inadequate to meet demand in the petrochemical era. As normally happens in the chemical industry, in the face of rapid changes in demand led to a proliferation of new processes as producers compete to fulfill the new market.

The first new process was introduced by Celanese in the late 1950's. The Celanese vinyl acetate process took advantage of the equilibria:

$$AcH + Ac_2O \rightleftharpoons (AcO)_2CHCH_3 \quad K_{eq\,(140°C)} = 25 \qquad [8]$$

$$(AcO)_2CHCH_3 \rightleftharpoons AcOH + (AcO)HC=CH_2 \quad K_{eq\,(140°C)} = 0.01 \quad [9]$$

which were described earlier in this chapter. Celanese generated the 1,1-diacetoxyethane (EDA) intermediate, by adding excess acetic anhydride and then cracked the EDA to vinyl acetate and acetic acid in the presence of an acid (e.g. benzene sulfonic acid) to generate vinyl acetate. Any acetaldehyde or acetic anhydride coproduct from the cracking step was recycled. Several plants were constructed using this technology, but the technology has not survived into the modern era as the technology was inferior to processes introduced by others quickly after its introduction. The last plant was closed in the 1980's but complete descriptions of the plant are available.[9]

In 1960, quickly after the introduction of the Celanese process, Wacker-Chemie commercialized a liquid phase vinyl acetate process which represented and extension of its earlier acetaldehyde process wherein acetic acid was simply substituted for water. (See equation [19]. This chemical transformation is also referred to as oxidative acetoxylation.) As shown in Figure 2, wherein R=Ac, the liquid phase oxidative acetoxylation of ethylene utilized the same catalytic cycle as the Wacker-Chemie acetaldehyde process.

$$H_2C=CH_2 + AcOH + \tfrac{1}{2}O_2 \longrightarrow (AcO)HC=CH_2 + H_2O \qquad [19]$$

The liquid phase vinyl acetate process dominated new installations until 1970 when Hoechst and Bayer commercialized a jointly developed heterogeneous vapor phase process for the addition of acetic acid to ethylene in the presence of oxygen (Equation [19]) in 1970. National Distillers (now Quantum Chemical) developed a similar process about the same time and commercialized the process rapidly thereafter. The processes (which were developed during the late 1960's) used a Pd-Au or Pd/Cd catalyst with an alkali metal on an inert inorganic oxide support such as alumina, but Pd/Au has dominated as the preferred catalyst. The process, which has recently been reviewed[10], was a heterogeneous analog of the liquid phase process where the Au component replaced Cu as a catalyst for the reoxidation of Pd. The yields with respect to ethylene ranged between 87-94% and 98-99% with respect to acetic acid, although this is a little misleading since a portion of the ethylene was oxidized to acetic acid. The yields were a function of time on line and decreased as the catalyst ages.

There are some operating limitations. First, the reaction conversion has been limited to about 10-15% per pass since the process must be operated below the explosion limits of ethylene and oxygen mixtures (in the presence of excess acetic acid.) It is also critical that the CO_2 formed by over oxidation be removed since it is an inhibitor for the catalyst. Lastly, the catalyst is very sensitive to halide poisons. (This issue would become important in future developments of acetic acid processes.)

The jointly developed Bayer and Hoechst process has dominated and has proliferated across the world due to an aggressive licensing effort. The process has been so successful that vapor phase oxidative acetoxylation rapidly shut down other operations and monopolized production of vinyl acetate by the 1980's.

The acetyl production options commercially practiced in 1970 are shown in Figure 3. However, the new demand for acetic acid would lead to a significant change in acetic acid production soon thereafter.

Radical Changes - Synthesis Gas Based Processes for Acetic Acid and Acetic Anhydride (1970-1985)[1, 2, 11]

The period from 1970 to 1985 saw radical changes in the production of acetic acid and acetic anhydride. By 1985, both products would be generated not from ethylene, but from synthesis gas which in turn could be generated from abundant resources such as coal, natural gas, and in the future, biomass. At the end of this period, acetaldehyde became a very small contributor to the total acetyl product stream since it was no longer required to make acetic acid or acetic anhydride and ethylene would only be required to produce vinyl acetate and to meet a much diminished acetaldehyde market. These advances were the result of two significant process breakthroughs – the Monsanto Acetic Acid Process and the Eastman Chemical Company Acetic Anhydride Process which will be discussed below.

The Monsanto Acetic Acid Process.[1, 11]

Whereas BASF had developed a Co based carbonylation, Monsanto discovered that Rh was a much more active catalyst and could be operated at

378

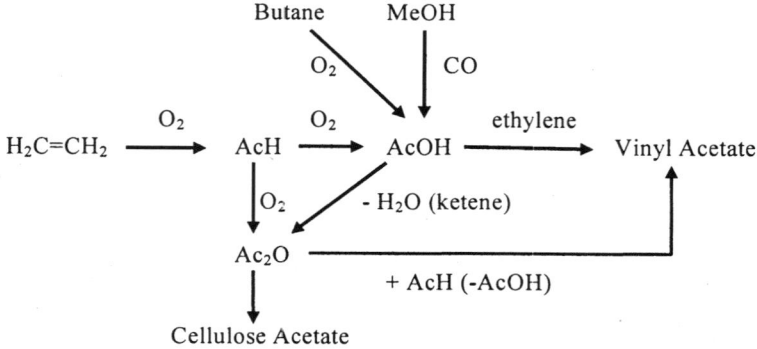

Figure 3. The Acetyl Production Options circa 1970.

much milder conditions (190-195°C, 30 atm vs. 250°C, 680 atm. for the Co process.) The process required 15-20% methyl iodide as a co-catalyst and kinetically, the reaction was 1st order in methyl iodide and Rh and independent of CO pressure and methanol concentration. The mechanism has been well established and is shown in figure 4. (The intermediates, $Rh(CO)_2I_2^-$, $MeRh(CO)_2I_3^-$, and $AcRh(CO)I_3^-$ have all been observed spectroscopically.)

The process, first disclosed in 1968, was commercialized in 1973. Yields in this process were very high (99%) and ease of operation was excellent. The major difficulty encountered was with catalyst precipitation during product removal. To minimize the problematic catalyst precipitation and to stabilize the catalyst, 10-15% water was included in the reaction mixture and the catalyst – product separation was conducted as an adiabatic flash. The inclusion of large amounts of water and the restriction to an adiabatic flash meant that the conversion was limited by product removal, not the reaction rate, and that there were large recycle streams of acetic acid and water. Additional minor difficulties were the cogeneration of traces of acetaldehyde which ultimately lead to propionic acid and iodine containing impurities. While the propionic acid was removable by distillation (with a dedicated unit of operation), the iodine has proven more problematic. It was important to remove essentially all the iodine (to < 40 ppb) during purification since iodine is a poison for the Pd/Au catalyst used in vinyl acetate production.

With the advent of a moderate pressure process, methanol carbonylation was now commercially viable and methanol carbonylation would be the preferred process from this point forward, although there would be significant improvements in the catalyst system as will be described in the next section of this chapter. While often overlooked, this process would not have been feasible without the earlier development of corrosion resistant alloys in the late 1950's.

The Monsanto process was a significant breakthrough for multiple reasons. First, the carbonylation of methanol and its derivatives now represents the one

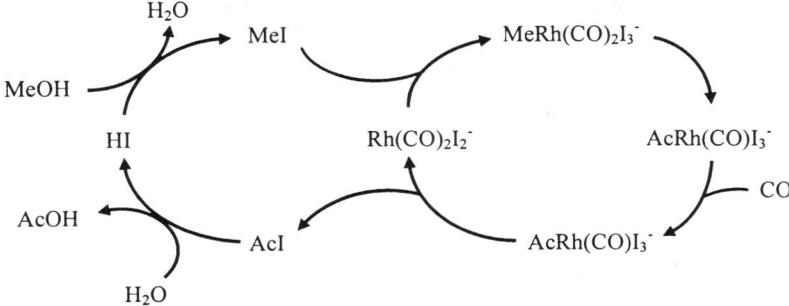

Figure 4. Mechanism for the Monsanto Rh Catalyzed Carbonylation of Methanol. (Reprinted by permission from Reference 11b. Figure 1.)

of largest applications of homogeneous catalysis for the generation of carbon-carbon bonds, possibly surpassing (or soon to surpass) hydroformylation. Second, we must recall that the methanol feedstock is generated from synthesis gas (syngas, a mixture of carbon monoxide and hydrogen) using a CuO-ZnO catalyst. Therefore, the advent of a successful, economically advantaged methanol carbonylation represented the first large scale displacement of a petroleum based process by a purely synthesis gas based process. While the syngas used to feed the methanol plant and CO to feed the acetic acid facility was normally generated from methane (an abundant resource) using steam reforming or partial oxidation, syngas can also be generated from a variety of resources including coal, petcoke, and biological resources. As petroleum resources become scarce, this process may well represent for the first of many migrations from petroleum in the future.

The Eastman Chemical Company Acetic Anhydride Process.[4, 11]

In 1983 Eastman Chemical Company began operating a facility to convert coal to acetic anhydride. This represented the first modern chemicals from coal facility and used coal based synthesis gas to generate acetic anhydride. The lynchpin technologies were a unique, first of its kind, reactive distillation of acetic acid and methanol to generate methyl acetate[12] and the subsequent carbonylation of methyl acetate to generate acetic anhydride.

While the process was commercialized in 1983, Eastman Chemical Company and Halcon began a joint development of the Rh catalyzed carbonylation of methyl acetate to acetic anhydride in the late 1970's. While the catalyst was similar to the Rh catalyst used in the Monsanto process, there were some unique challenges since (i) under anhydrous conditions the active catalyst, $Rh(CO)_2I_2^-$, failed to form, (ii) unlike water, acetic acid did not react with acetyl iodide, and no HI was formed, and (iii) acetyl iodide was only very slowly reactive with methyl acetate. Therefore, the catalyst not only failed to form, but the organic/iodine cycle was inoperative even if it did form.

To overcome these problems, Eastman Chemical added Li^+ which served two key purposes. First, it provided a counter ion for the $Rh(CO)_2I_2^-$ and the catalyst formed with as little as two equivalents of cation. However, any cation served this purpose. The real advantage to Li^+ is that it accelerated the organic cycle. Whereas acetic acid failed to react with AcI, LiOAc reacted very rapidly, generating LiI. However, the subsequent reaction of LiI with methyl acetate to regenerate the critical MeI co-catalyst (see equation [20]) was both reversible (with a slightly disfavorable equilibrium) and significantly slower than the reaction of Rh with MeI. As a consequence, the regeneration of MeI via Equation [20] could have been rate limiting. However, to overcome this limitation, the reaction was normally operated in the presence of excess Li^+ so that the rate began to approximate first order in Rh and MeI. An extensive examination of other cations revealed that only Li and Cr catalyzed the reaction of iodide with methyl acetate fast enough to enter a regime that approximated first order kinetic behavior with respect to MeI and Rh. Otherwise (or in the absence of sufficient Li or Cr), the reaction gives the expected complex kinetics associated with consecutive reactions involving a pre-equilibrium.[12] The mechanism for the Rh-Li-MeI catalyzed carbonylation of methyl acetate is shown in Figure 5.

$$LiI + MeOAc \rightleftharpoons LiOAc + MeI \qquad K_{eq\,(190°C)} = 0.388 \qquad [20]$$

There was an additional requirement in this process. The catalyst was subject to spontaneous oxidation and hydrogen was needed to reduce the catalyst back to the active form. Unfortunately, this led to one of the key by-

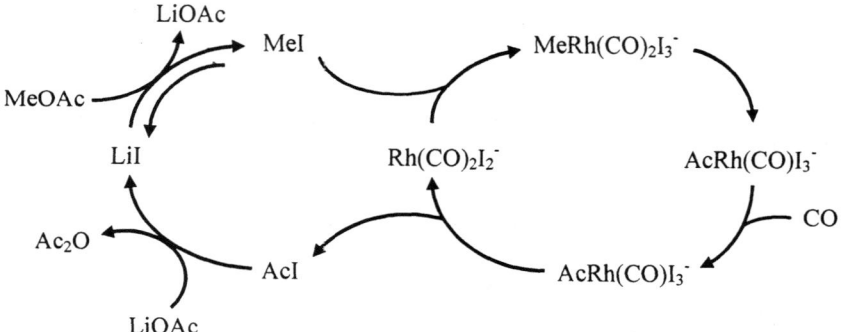

Figure 5. Mechanism of the Eastman Chemical Company Rh-Li Catalyzed Carbonylation of Methyl Acetate.

products from the process, ethylidene diacetate, which must be removed in a separate unit of operation. In the earlier Monsanto acetic acid process, this was not necessary as the presence of water and CO led to spontaneous reduction via water gas shift.

While the generation of ethylidene diacetate was problematic, it was not the biggest issue with this process. While catalyst precipitation was not normally a significant problem, the process produced a high boiling tarry by-product which binds Rh. A complex, proprietary catalyst recovery process was required to recover Rh.

If one looks at the overall process as it is integrated to cellulose acetate manufacture (equations [21] through [26]) the attraction of this process becomes apparent. By instituting a methyl acetate carbonylation, Eastman Chemical Company was able to replace the very energy intensive ketene process with a nearly thermodynamically neutral esterification as a method to remove water and subsequent exothermic carbonylation in the manufacture of cellulose acetate.

$$2\ C_{(coal)} + 2\ H_2O \longrightarrow 2\ CO + 2\ H_2 \qquad [21]$$

$$CO + 2\ H_2 \longrightarrow MeOH \qquad [22]$$

$$MeOH + AcOH \rightleftharpoons MeOAc + H_2O \qquad [23]$$

$$MeOAc + CO \longrightarrow Ac_2O \qquad [24]$$

$$Cellulose + Ac_2O \longrightarrow Cellulose\text{-}Acetate + AcOH \qquad [25]$$

Net Reaction:
$$Cellulose + 2\ C_{(coal)} + H_2O \longrightarrow Cellulose\text{-}Acetate \qquad [26]$$

While underlying chemical developments entailed in the Eastman Chemical Company Acetic Anhydride process were of significant interest, the switch from natural gas to coal as a syngas source represented a significant change and may represent the long term future of the chemical industry. While overall yields were comparable to the Monsanto Acetic Acid Process, the Eastman Chemical Company Acetic Anhydride Process was more difficult to operate. Eastman Chemical Company expanded the process in 1991. Although advantaged over the ketene process, only Eastman Chemical Company and British Petroleum currently practice the carbonylation of methyl acetate.

As of 1985, production of the acetyl chemicals had changed entirely. From that time forward, the preferred processes would contain primarily synthesis gas. Ethylene was only needed for vinyl acetate production and to address the rapidly diminishing market for acetaldehyde. (Remember that the primary outlet for acetaldehyde was to make acetic acid and acetic anhydride, so it was no longer needed for these purposes with the advent of the new technologies.) A summary of the preferred production methods as of 1985 appears in Figure 6.

382

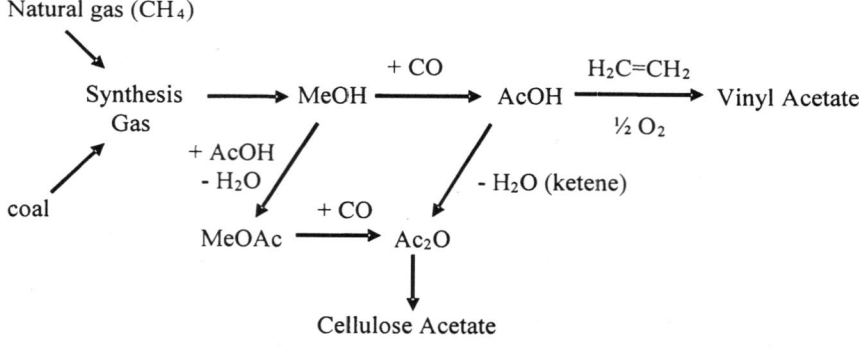

Natural gas (CH₄)

$$H_2C=CH_2 + \tfrac{1}{2} O_2 \longrightarrow AcH$$

Figure 6. Acetyl Production circa 1985.

Maturity of the Acetyl Stream. Improvements in Methanol Carbonylation. (1985-2007).

After 1985 the synthetic pathways for generating the acetyl chain would remain unchanged. The heterogeneous vinyl acetate process and liquid phase acetaldehyde process would remain virtually unchanged into the present with only minor improvements being realized (although acetaldehyde demand would become a relatively minor chemical product as a result of the successful Rh catalyzed methanol carbonylation process.) Acetic anhydride would still be made by either Rh catalyzed methyl acetate carbonylation or via acetic acid based ketene and the processes would remain unchanged after 1985. However, while the processes for acetaldehyde, vinyl acetate, and acetic anhydride remained stable and unchallenged, the methanol carbonylation would undergo some significant catalyst improvements starting in 1990.

In 1990, Celanese (then Hoechst-Celanese), , began operating a low water methanol carbonylation process that included Li as a co-catalyst and hydrogen.[1,2,11] The inclusion of Li stabilized the catalyst both within the reactor and during catalyst-product separation. It also accelerated the conversion of any methyl acetate that might be formed by esterification. This addressed the key flaw in the Monsanto process. The catalyst-production separation, while still being operated adiabatically, was more productive since there was less water being distilled prior to the onset of acetic acid distillation. It was later found that the Li was sufficiently stabilizing that a small amount of heat could be added to distill additional acetic acid, although this needs to be limited since it does not completely eliminate precipitation if the product is heated too long.

In 1996, British Petroleum announced an alternative methanol carbonylation based on an Ir-Ru-MeI catalyst. Like the Celanese Rh-Li catalyst, it was a low water process, but used about half the methyl iodide. The use of Ir as a catalyst was not new and had been disclosed by Monsanto contemporaneously with its disclosure of the Rh catalyst. However, it had very complex kinetics and was more difficult to operate. British Petroleum achieved this event by discovering a way to overcome a major shortcoming in the Ir process disclosed by Monsanto.

To understand the British Petroleum breakthrough, it is necessary to look back at the original Monsanto work on Ir. Monsanto investigated both Ir and Rh when developing their methanol carbonylation catalyst and discovered that, whereas the oxidative addition of MeI to $Ir(CO)_2I_2^-$ (equation [27]) was significantly faster than the corresponding oxidative addition of MeI to $Rh(CO)_2I_2^-$, the resultant $MeIr(CO)_2I_3^-$ was stable and, unlike $MeRh(CO)_2I_3^-$, it did not undergo spontaneous insertion of CO to form $AcIr(CO)I_2^-$ (equation [28]). Instead, $MeIr(CO)_2I_3^-$ had to exchange carbon monoxide for an iodide ligand to form $MeIr(CO)_3I_2$ (equation [29]). The $MeIr(CO)_3I_2$ species was then able to insert CO to form $AcIr(CO)_2I_2$ (equation [30]). Addition of iodide to $AcIr(CO)_2I_2$ generated $AcIr(CO)_2I_3^-$ which rapidly reductively eliminated AcI to form the necessary acetyl iodide intermediate and regenerate the $Ir(CO)_2I_2^-$ catalyst (equation [31].) The dissociation of iodide from $MeIr(CO)_2I_3^-$ was rate limiting and the reaction kinetics were very complex. The process required careful optimization of the iodine levels and was tricky to operate, especially compared to the robust Rh catalyzed analog. Therefore, the Ir based carbonylation was not chosen for commercialization by Monsanto.

$$MeI \text{ to } Ir(CO)_2I_2^- \longrightarrow MeIr(CO)_2I_3^- \qquad [27]$$

$$MeIr(CO)_2I_3^- \xrightarrow{\ \ \times\ \ } AcIr(CO)I_2^- \qquad [28]$$

$$MeIr(CO)_2I_3^- + CO \rightleftharpoons MeIr(CO)_3I_2 + I^- \qquad [29]$$

$$MeIr(CO)_3I_2 \longrightarrow AcIr(CO)_2I_2 \qquad [30]$$

$$AcIr(CO)_2I_2 + I^- \longrightarrow AcIr(CO)_2I_3^- \longrightarrow Ir(CO)_2I_2^- + AcI \qquad [31]$$

What British Petroleum discovered was that $Ru(CO)_4I_2$ (as well as some other metals) catalyzed the exchange of iodide for CO in the transformation of $MeIr(CO)_2I_3^-$ to $MeIr(CO)_3I_2$ (equation [29]) allowing the Ir catalyst to operate without being slowed by the pesky, and previously rate limiting, carbon monoxide for iodide ligand exchange. Complete mechanistic details are available for the process.[14] In the presence of Ru, Ir operated under the same conditions as the Rh based methanol carbonylation, was equally active as a catalyst, and had acceptable ease of operation. Advantages associated with the Ir process are that the Ir process used only half as much methyl iodide co-catalyst and the Ir and Ru catalyst components were (and continue to be) much less expensive than Rh.

These two derivatives of the earlier Monsanto technology are the predominant acetic acid processes today and are equally competitive in the market place. Since the advent of the Monsanto Acetic Acid process almost all new acetic acid plants are based on methanol carbonylation and acetaldehyde oxidation has been nearly phased out as a source of acetic acid. The advances in Rh and Ir based methanol carbonylation have recently been reviewed.[15]

Tomorrow - Acetyl Processes of the Future

Predicting the future of chemical technology is always dangerous since markets and resources change at unpredictable rates. However, several features of the acetyl product stream assure their future in a world with diminishing access to petroleum for chemicals. The fact that acetic acid and acetic anhydride are now derived from synthesis gas which can be generated from more widely available resources such as methane, coal, and biomass, assures access to these products in the long term. Further, one of the two large outlets, cellulose acetate, represents one of the few renewable polymer resources in the chemical industry. (Cellulose acetate consumes about 1/3 of all the acetyl units generated worldwide.) The need for cellulose acetate is also growing as it is finding new applications in electronics (flat screen displays.) Therefore, in the quest for a sustainable chemical industry, the acetyls and cellulose esters should fare well and may even serve a leadership role.

Work is already underway in various laboratories to replace the last remaining petroleum portion of the acetyl chain, namely the ethylene component of vinyl acetate, with syngas.[4,16] These processes take advantages of the old Celanese EDA approach to vinyl acetate[9] by generating EDA. Their primary drawback is that they used large volumes of acetic anhydride and acetic acid recycles. However, Eastman Chemical Company presented an interesting new scheme for the production of vinyl acetate from dimethyl ether which involves little or no recycle.[15b] The process includes a carbonylation of dimethyl ether (made by dehydration of methanol) to make acetic anhydride, a reactive distillation of acetic anhydride with acetaldehyde which reduces the Celanese EDA process to a single unit of operation, and hydrogenates acetic acid by-product (from the vinyl acetate process) to produce the acetaldehyde component. This reduces the size of the plants required by a factor of two. (Inherent in this process is a means of producing acetaldehyde from synthesis gas as well via acetic acid hydrogenation.) As ethylene becomes more expensive, these processes may be competitive.

Further, acetic acid and acetic anhydride, may serve as building blocks for additional key intermediates in a synthesis gas based chemical industry. Investigations have already been undertaken to use acetic acid as a building block for acrylic acid (by condensation with formaldehyde – a syngas molecule obtained by oxidation of methanol) and homologation of carboxylic acids to higher homologs using synthesis gas.[17]

References

1. a) Wagner, F. S., Jr., "Acetic Acid", in *Kirk-Othmer Encyclopedia of Chemical Technology*, 5th edit., John Wiley and Sons, Inc., Hoboken, NJ, Vol. 1, 2004, p. 115 and earlier editions and references cited therein. b) Cheung, H., Tanke, R. S., Torrence, G. P., "Acetic Acid", in *Ullmann's Encyclopedia of Industrial Chemistry*, Wiley-VCH Verlag GmbH and Co., KGaA, Weinhiem, Germany, 6th edit., *1*, 149 (2003) and earlier editions and references cited therein. c) McMahon, K. S. "Acetic Acid", in *Encyclopedia of Chemical Processing and Design*, McKetta, J. J., Cunningham, W. A., eds., Marcel Dekker, Inc., NY, NY, *1*, p. 216 (1976).

2. Agreda, V. H., Zoeller, J. R., eds., *Acetic Acid and Its Derivatives*, Marcel Dekker, NY, NY (1993).

3. a) Hagemeyer, H. J., "Acetaldehyde", in *Kirk-Othmer Encyclopedia of Chemical Technology*, 5th edit., John Wiley and Sons, Inc., Hoboken, NJ, *1*, p. 99 (2004) and earlier editions and references cited therein. b) Fleischman, G., Jiva, R., Bolt, H. M., Golka, K., "Acetaldehyde", in *Ullmann's Encyclopedia of Industrial Chemistry*, Wiley-VCH Verlag GmbH and Co., KGaA, Weinhiem, Germany, 6th edit., *1*, 131 (2003) and earlier editions and references cited therein. c) Aquilo, A., Penrod, J. D., in *Encyclopedia of Chemical Processing and Design*, McKetta, J. J., Cunningham, W. A., eds., Marcel Dekker, Inc., NY, NY, *1*, 114 (1976). d) Fanning, A. T., "Ethylene- and Acetylene- Based Processes", in <u>Acetic Acid and Its Derivatives</u>, Agreda, V. H., Zoeller, J. R., eds., Marcel Dekker, NY, NY , 1993, p, 15.

4. a) Wagner, F. S., Jr., "Acetic Anhydride", in *Kirk-Othmer Encyclopedia of Chemical Technology*, 5th edit., John Wiley and Sons, Inc., Hoboken, NJ, *1*, 146 (2004) and earlier editions and references cited therein. b) Held, H., Renstl, A., Mayer, D., "Acetic Anhydride and Mixed Fatty Acid Anhydrides", in *Ullmann's Encyclopedia of Industrial Chemistry*, Wiley-VCH Verlag GmbH and Co., KGaA, Weinhiem, Germany, 6th edit., *1*, 179 (2003) and earlier editions and references cited therein. c) McMahon, K. S. "Acetic Anhydride", in *Encyclopedia of Chemical Processing and Design*, McKetta, J. J., Cunningham, W. A., eds., Marcel Dekker, Inc., NY, NY, *1*, 258 (1976). d) Cook, S. L., "Acetic Anhydride", in <u>Acetic Acid and Its Derivatives</u>, Agreda, V. H., Zoeller, J. R., eds., Marcel Dekker, NY, NY , p, 145 (1993).

5. a) Cordiero, C. F., Petrocelli, F. P., "Vinyl Acetate Polymers", in *Kirk-Othmer Encyclopedia of Chemical Technology*, 5th edit., John Wiley and Sons, Inc., Hoboken, NJ, *25*, 557 (2007) and earlier editions and references

386

cited therein. b) Roscher, G., "Vinyl Esters", in *Ullmann's Encyclopedia of Industrial Chemistry*, Wiley-VCH Verlag GmbH and Co., KGaA, Weinhiem, Germany, 6th edit., *38*, 59 (2003) and earlier editions and references cited therein. c) Sumner, C. E., Zoeller, J. R., "Vinyl Acetate", in Acetic Acid and Its Derivatives, Agreda, V. H., Zoeller, J. R., eds., Marcel Dekker, NY, NY , p, 145 (1993).

6. Partin, L. R. , Heise, W. H., "Bioderived Acetic Acid" in Acetic Acid and Its Derivatives, Agreda, V. H., Zoeller, J. R., eds., Marcel Dekker, NY, NY, p. 3 (1993).

7. a) Treece, L. C., Johnson, G. I., "Cellulose Acetate" in Acetic Acid and Its Derivatives, Agreda, V. H., Zoeller, J. R., eds., Marcel Dekker, NY, NY, p. 3 (1993). b) Gedon, S., Fengi, R., "Cellulose Ester, Organic Esters", in *Kirk-Othmer Encyclopedia of Chemical Technology*, 5th edit., John Wiley and Sons, Inc., Hoboken, NJ, *5*, 412 (2007) and earlier editions and references cited therein. b) Balser, K., Hoppe, L., Eicher, T., Wandel, M., Astheimer, H.-J., Steinmeir, H., "Cellulose Esters", in *Ullmann's Encyclopedia of Industrial Chemistry*, Wiley-VCH Verlag GmbH and Co., KGaA, Weinhiem, Germany, 6th edit., *6*, 647 (2003) and earlier editions and references cited therein

8. a) British Pat. No. 283,989 (to Celanese.) b) British Pat. No. 317,867 (to Celanese.) c) British Pat. No. 343,947 (to Celanese.) d) British Pat. No. 320,457 (to Celanese.)

9. Anon., *Hydrocarbon Process., 44*, 287 (1965).

10. Kumar, D. Chen, M. S. , Goodman, D. W., *Catalysis Today*, **2007**, *123*, 77.

11. a) Howard, M. J., Jones, M. D., Roberts, M. S., Taylor, S.A., *Catalysis*, **1993**, *18*, 325. b) Zoeller, J. R., "Manufacture via Methanol Carbonylation", in Acetic Acid and Its Derivatives, Agreda, V. H., Zoeller, J. R., eds., Marcel Dekker, NY, NY, p. 35 (1993). c) Gauthier-Lafaye, J., Perron, R., *Methanol and Carbonylation*, (English Translation), Editions Technip, Paris, France (1987), chap. 6, p. 119.

12. a) Zoeller, J. R., Cloyd, J. D., Lafferty, N. L., Nicely, V. A., Polichnowski, S. W., and Cook, S. L., *Advances in Chemistry Series*, **1992**, *230*, 377 b) Agreda, V. H., Pond, D. M., and Zoeller, J. R., *Chemtech*, **1992**, *22*, 172. c) Gauthier-Lafaye, J., Perron, R., *Methanol and Carbonylation*, (English Translation), Editions Technip, Paris, France (1987), chap. 7, p. 150.

13. Agreda, V. H., Partin, L. R., Heise, W. H., *Chem. Eng. Prog.*, **1990**, *86(2)*, 40.

14. Haynes, A., Maitlis, P. M., Morris, G. E., Sunley, G. J., Adams, H., Badger, P. W., Bowers, C. M., Cook, D. B., Elliott, P. I. Ghaffer, T., Green, H., Griffin, T. R., Payne, M., Pearson, J. M., Taylor, M. J., Vickers, P. W. , Watt, R. J., *J. Amer. Chem. Soc.*, **2004**, *126*, 2847.

15. Haynes, A., *Topics in Organometallic Chemistry*, **2006**, *18*, 179.

16. a.) Gauthier-Lafaye, J., Perron, R., *Methanol and Carbonylation*, (English Translation), Editions Technip, Paris, France (1987), chap. 8, p. 201. b) Tustin, G. C., Colberg, R. D., Zoeller, J. R., *Catalysis Today*, **2000**, *58*, 281

17. Hawkins, J. A., Zoeller, J. R., "Chain Growth: Acrylics and Other Carboxylic Acids", in Acetic Acid and Its Derivatives, Agreda, V. H., Zoeller, J. R., eds., Marcel Dekker, NY, NY, p. 345 (1993).

Chapter 11

Ionic Liquids: Growth of a Field through the Eyes of the I&EC Division

Héctor Rodríguez[1], Gabriela Gurau[1], and Robin D. Rogers[1,2,*]

[1]The Quill Research Centre and School of Chemistry and Chemical Engineering, The Queen's University of Belfast, Belfast, Northern Ireland BT9 5AG, United Kingdom
[2]Center for Green Manufacturing and Department of Chemistry, The University of Alabama, Tuscaloosa, AL 35401

The last decade has witnessed an indisputable burgeoning of research within the field of ionic liquids, both in academe and industry. Part of the responsibility for this explosion of interest can be attributed to the strong commitment of the American Chemical Society's Industrial and Engineering Chemistry Division and its subdivisions. In particular, the oral and written forums sponsored by the Division have enhanced discourse between scientific and industrial partners across the world and have served as a platform for advances in the field. Here, through a series of snapshots represented by the Division's five major ionic liquids symposia, we take a look at the world of ionic liquids.

Introduction

One of us (RDR) was introduced to the I&EC Division through its Separation Science & Technology Subdivision, where service on the Program Committee began in 1992. The SS&T Subdivision and the I&EC Division had a history of allowing diverse programming and providing a forum for both fundamental and applied research. RDR co-organized his first symposium in the Division in 1994 ("Aqueous Biphasic Separations: Biomolecules to Metal Ions," with C. K. Hall) for the 207th ACS National Meeting in San Diego, CA and went on to co-organize seven other symposia through the year 2000 on a variety of topics in separations science.

In 1996, coinciding with his move back to The University of Alabama, RDR attended a NATO Advanced Study Institute meeting on crystal engineering in Digby, Nova Scotia, Canada (*1*) where he met one of the co-organizers of that meeting, Prof. Kenneth R. Seddon, from The Queen's University of Belfast. That meeting led to discussions on whether pure salts could be used as solvents for separations and green chemistry (*2*). A research program, a friendship, and a long term collaborative commitment to a field were born. Here we follow one aspect of that collaboration: commitment to facilitating communication for all practitioners in the field.

Ionic Liquids: Onto the Big Stage and Into I&EC

Ionic compounds which are liquid at room temperature have been around for quite a long time (*3*). The first room temperature ionic liquid dates back from as early as 1914, when ethylammonium nitrate was originally reported (*4*). More recently in the US, important work in the area of electrochemistry has been carried out with low melting chloroaluminate salts over at least the last thirty years or so (*5*). However, worldwide attention to the concepts of 'ionic liquids' became focused in the very last years of the 20th century, when a series of publications raised interest in ionic liquids as alternative 'green' solvents (*2,6-10*). The unique properties and promising applications of these "liquid salts" attracted an increasing number of research groups, coming from increasingly varied scientific backgrounds.

The evidence for an inflexion point in the history of ionic liquids in the late 1990s, can be clearly observed in Figure 1, where the number of publications in the literature involving this family of compounds (as retrieved by Web of Science[®] of the ISI Web of Knowledge[SM]) is plotted as a function of time. In understanding this impressive growth, it is also noteworthy to indicate that in 1999 The Queen's University Ionic Liquids Laboratory (QUILL), the first center entirely devoted to research within ionic liquids, was founded with 16 industrial members.

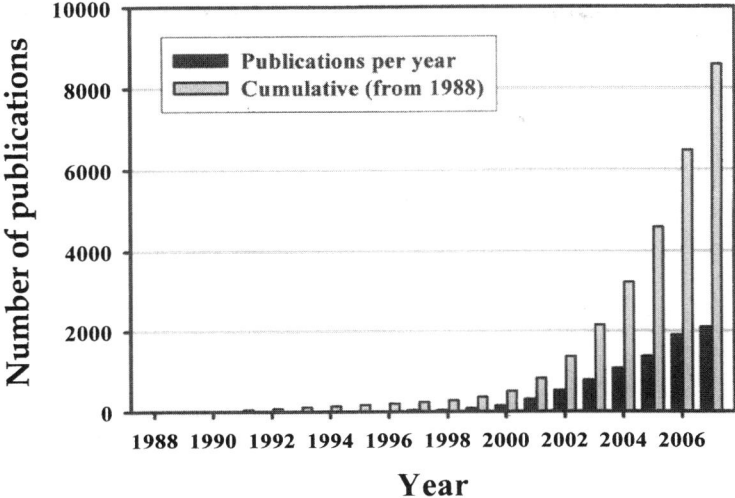

Figure 1. Number of publications per year (black bar) and cumulative number of publications (grey bar) in the period 1988-2007 containing the phrase "ionic liquid(s)" in the title, abstract, or keywords, as reported by Web of Science® of the ISI Web of Knowledge^{SM}.

In April 2000, the first international meeting devoted to ionic liquids was held in Heraklion, Crete, Greece. This NATO Advanced Research Workshop on Green Industrial Applications of Ionic Liquids (*11,12*), was a closed five-day workshop intended to increase awareness of ionic liquids, bring new research expertise into the field, and set a research agenda for its future. The organizers (Profs. R. D. Rogers, K. R. Seddon, and S. Volkov) struggled at the time to get 50 international scientists working in the field interested in attending the meeting.

Among the ten major outcomes of this meeting, the two most cited are (*12*):

- "Ionic liquids are intrinsically interesting and worthy of study for advancing science (ionic *vs.* molecular solvents) with the expectation that something useful may be derived from their study."

- "Combined with green chemistry, a new paradigm in thinking about synthesis in general, ionic liquids provide an opportunity for science/engineering/business to work together from the beginning of the field's development."

But two of the other outcomes deserve mention here:

- "There is an urgent need to increase the number, but especially the areas of expertise, of ionic liquids researchers. A model of open collaboration needs to be encouraged."

- "International collaboration, communication, and education regarding the results are needed."

These latter two outcomes pointed to the urgent need to bring people of varied backgrounds and interest together and to disseminate information. For Profs. Rogers and Seddon, this meant I&EC, and the following year they took this concept to the Program Committee.

In April 2001, one year after the NATO meeting in Crete, I&EC held its first ionic liquids symposium "Green (or Greener) Industrial Applications of Ionic Liquids" at the ACS national meeting held in San Diego, CA. Over 350 attended the opening session of this 5-day symposium; the majority from industry. In the following sections we will overview the evolution of the ionic liquids field through the analysis of the content of these meetings.

Evolution of Ionic Liquids through the Eyes of I&EC

The Meetings

The first of the symposia (Table I) on ionic liquids sponsored by I&EC was also the first open international meeting on the fundamentals and applications of ionic liquids period. "Green (or Greener) Industrial Applications of Ionic Liquids" (13) brought together the key players in this burgeoning area in a forum for the exchange of ideas between academe and industry. The high attendance, vigorous discussions, and successful proceedings publication, indicated that these meetings should be continued.

The second symposium, "Ionic Liquids as Green Solvents: Progress and Prospects" (14), was held in Boston, MA in August 2002, in the framework of the 224th ACS National Meeting. Coming as it did, only 18 months after the San Diego meeting, the varied presentations and high attendance gave notice of a rapidly evolving field. Indeed, "Progress" was rapid and the "Prospects" were good. Based on general prompting by the community, a third meeting was set for the following year to complete the trilogy!

New York, NY hosted the 226th ACS National Meeting in September 2003, and I&EC's "Ionic Liquids: Fundamentals, Progress, Challenges, and Opportunities" (15,16). Despite the short period of time elapsed from the previous symposium (12 months), the frenetic pace of the field was evident during the sessions. A broad and diverse base for activities could be identified, as well as the excitement of even newer potential opportunities for the field.

Table I. I&EC ionic liquids symposia. (See page 2 of color insert.)

I&EC Symposium	Topics Covered
Green (or Greener) Industrial Applications of Ionic Liquids *(13)* April 1-5, 2001 San Diego, CA 221st ACS National Meeting http://bama.ua.edu/~rdrogers/sandiego (last accessed March 29, 2008)	• Ionic Liquids in Context • Separations and Engineering • Green Synthesis • Nuclear and Electrochemistry • Ionic Liquid Systems • Properties of Ionic Liquids • Catalysis I • Catalysis II • Structure and Photochemistry • High-Temperature and Other Systems
Ionic Liquids as Green Solvents: Progress and Prospects *(14)* August 18-22, 2002 Boston, MA 224th ACS National Meeting http://bama.ua.edu/~rdrogers/Boston (last accessed March 29, 2008)	• Ionic Liquid Tutorial • Manufacture and Synthesis of Ionic Liquids: Industrial and Academic • Characterization and Engineering • Novel Applications of Ionic Liquids • Separations • Biotechnology • Catalytic Chemistry • Non-Catalytic Chemistry • Electrochemistry • Photochemistry and Reaction Intermediates

Continued on next page.

Table I. *Continued.* (See page 3 of color insert.)

I&EC Symposium	Topics Covered
Ionic Liquids III: Fundamentals, Progress, Challenges, and Opportunities  *(15,16)* September 7-11, 2003 New York, NY 226th ACS National Meeting http://bama.ua.edu/~rdrogers/New York (last accessed March 29, 2008)	• Ionic Liquid Tutorials • Fuels and Applications • Physical and Thermodynamic Properties • Catalysis and Synthesis • Spectroscopy • Separations • Novel Applications • Catalytic Polymers and Gels • Electrochemistry • Inorganic and Materials • General Contributions
Ionic Liquids: Not Just Solvents Anymore OR Ionic Liquids: Parallel Futures  *(17)* March 26-30, 2006 Atlanta, GA 231st ACS National Meeting http://bama.ua.edu/~rdrogers/Atlanta2006 (last accessed March 29, 2008)	• Why Are Ionic Liquids Liquid? • Structure-Activity Relationships/Modeling • Environmental Fate and Toxicity • New Industrial Applications of Ionic Liquids • Really New Ionic Liquids • Ionic Liquids and Education • Applications Based on Physical Properties • Functional Ionic Liquids/Ionic Materials • Analytical Applications of Ionic Liquids • Microengineering with Ionic Liquids

Table I. *Continued.*

I&EC Symposium	Topics Covered
Ionic Liquids: From Knowledge to Application ? August 17-21, 2008 Philadelphia, PA 236th ACS National Meeting http://bama.ua.edu/~rdrogers/Philadelphia2008 (last accessed March 29, 2008)	• Gas Separations • Liquid Separations • Interfacial Properties • Use in Sensors and Actuators • Materials Synthesis in Ionic Liquids • Thermophysical Properties • Molecular Simulations • Pharmaceutical Applications • Industrial Applications • General Session

New industrial applications acquired a more relevant presence in this third meeting.

After the trilogy of meetings in the period 2001-2003, a break in the ionic liquid symposia at the ACS meetings was taken over the next triennium. Of course, this was not due to the research in the field slowing down, but might be better understood as a timing adjustment. The first biennial International Congress on Ionic Liquids was held in Salzburg, Austria in June 2005; attracting nearly 500 delegates from around the world. In order to avoid excessive accumulation of high level meetings on the matter, the timing of I&EC symposia on the topic was changed to accommodate the newer meetings.

Thus, the fourth of the I&EC symposia was held in March 2006, at the 231st ACS National Meeting in Atlanta, GA, with the title "Ionic Liquids: Not Just Solvents Anymore OR Ionic Liquids: Parallel Futures" (*17*). With an ampler time perspective, it was observed that the intense interest in the field had led to an unprecedented increase in the knowledge of salts in general, with special focus on the manipulation of their physical and chemical properties. Parallel to this fundamental knowledge, a dramatic increase in the technological application of ionic liquids had also occurred. The variety and novelty of topics presented was truly amazing.

At the writing of this chapter, a fifth symposium has been organized for the August 2008 236th ACS National Meeting in Philadelphia, PA; "Ionic Liquids: From Knowledge to Application". The field shows no signs of slowing down and over 140 abstracts have been submitted for presentation at this symposium.

Each of the first four symposia (and perhaps the fifth) generated an ACS Symposium Series book (Table I) with key papers and authors chosen from the meetings. By simply following the titles, it is noticeable that the emphasis on the role of ionic liquids as solvents has decreased with time. Indeed, the boom in the field was prompted by the unique environment provided by ionic liquids to act as solvents in different processes; but, with the increase in scientific knowledge in academe and in industry, the impact of ionic liquids is going well beyond the borders of solvent applications. As the subtitle of the last book reads, ionic liquids are "not just solvents anymore". This conclusion can also be reached through the analysis of the topics of the sessions at each meeting.

The Topics

The featured topics for the sessions constituting each of the symposia organized within I&EC are provided in Table I. By closely reviewing these lists, insight into the evolution of the ionic liquids field over the last several years can be obtained.

A clear example of the huge speed at which the universe of ionic liquids expanded and evolved can be found in the fact that the more or less generic topics at the first symposium hardly find any correspondent session in the fourth symposium held five years later. The first three symposia in 2001, 2002, and 2003, shared a series of topics, such as separations, synthesis, electrochemistry, engineering, or catalysis. Also, the issues of characterization and physical properties were discussed in all the meetings of this symposia trilogy. There was a clear emphasis on solvent properties and solvent applications.

What happened by 2006? The generic topics were indeed discussed, but were distributed into more specialized areas. Typically, these areas were closer to final applications. From this point of view, it is interesting to note that there was no "applications" session in 2001; in 2002 a single generic topic on novel applications appeared; in 2003 this session was joined by a session on fuels and applications, by 2006, three and in 2008 five different sessions focused only on applications. Certainly, this is indicative of the amazing increase in commercial applications involving ionic liquids over the last few years (3).

This trend in the field is not only evident from these symposia. Figure 2 illustrates the growth in the number of patents over the past 10 years which utilize the term 'ionic liquid(s)'. There is little wonder that ionic liquids are no longer viewed as just 'green solvents', but as enabling concepts for truly transformational technologies!

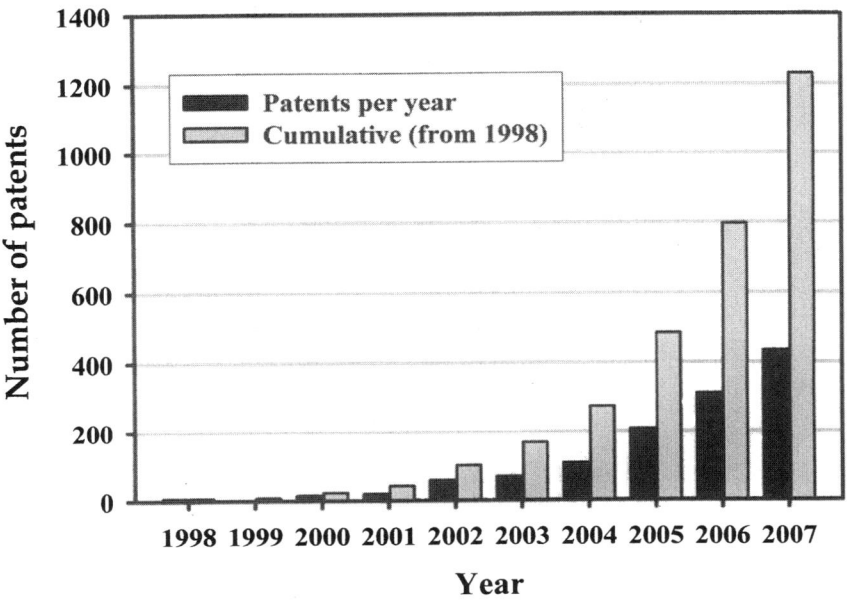

Figure 2. Number of patents per year (black bar) and cumulative number of patents (grey bar) in the period 1998-2007 containing the phrase "ionic liquid(s)", as reported by SciFinder®.

The Sponsorship

A parallel interpretation of the significance of the symposia, and subsequently of the ionic liquids field over the current decade, can be developed through an analysis of the sponsorship of these I&EC symposia. Upon review in its entirety, the sponsorship of the I&EC symposia on ionic liquids is representative of the amazing cooperation which has been key to the success of the field; including professional societies, academe, industry, and government.

The US Environmental Protection Agency's Green Chemistry Program, the US Army Research Office, and the ACS Green Chemistry Institute were among the initial sponsors of these symposia. All of the symposia were sponsored by I&EC, and in particular by its subdivisions Green Chemistry & Engineering and Separation Science & Technology (with the Novel Chemistry with Industrial Applications subdivision joining them in 2006). Also, academic research centers co-sponsored the events including the Center for Green Manufacturing at The University of Alabama, and QUILL at The Queen's University of Belfast. In 2006, the Energy Center at the University of Notre Dame added its support.

398

Perhaps the hallmark of these symposia, however, has been the industrial support and participation. Nine companies from different backgrounds sponsored the first symposium in April 2001. These initial sponsors included ACROS, BP, Covalent Associates, Chevron, Dow, Ozark Fluorine Specialties, SACHEM, Solvent Innovation, and Union Carbide. In the following years, Cytec Industries, Fluka, Merck, Strem Chemicals, BASF, and 525Solutions have also contributed.

Conclusion: A Fascinating Future Ahead

The growth of the field of ionic liquids over the last decade has truly been astonishing. A simple extrapolation to the future would indicate that ionic liquids are going to continue to be important contributors to fundamental and applied science and engineering; a perfect fit for I&EC! The new paradigms being introduced with this amazing class of compounds will become founded in a firm basis of understanding and thus lead to an ever burgeoning number of new applications, both predictable and completely unexpected at the time of writing these lines (3). While the future of ionic liquids looks bright, we must not forget that a key aspect for future success will be the continuation of open and rapid intercommunication between research groups, their industrial partners, and governmental institutions. We are confident that I&EC will continue to play a major role in this, continuing to contribute for at least the next 100 years... or more!

Acknowledgments

Profs. R. Rogers, K. Seddon, and more recently J. Brennecke, as overall symposia organizers, recognize that the true success of these symposia has been in large part to the efforts of the many individual session organizers who have tirelessly contributed their talents in putting together the many excellent sessions. These organizers include A. Robertson, J. Holbrey, R. Sheldon, T. Welton, M. Earle, W. Pitner, C. Gordon, J. Davis Jr., L. Rebelo, P. Wasserscheid, S. Pandey, B. Tumas, R. Mantz, P. Trulove, M. Watanabe, H. Ohno, D. MacFarlane, M. Deetlefs , E. Maginn , G. Lamberti, R. Bernot, B. Hembre, R. Singer, J. Wilkes, D. Armstrong, R. Allen, M. Costa Gomes, M. Shiflett, J. Anderson, P. Jessop, S. Baldelli, A. Podesta, G. Baker, X. Zeng, J. Anthony, A. Mudring, J. Magee, P. Ballone, P. Schwab, and M. Turner.

References

1. *Crystal Engineering: The Design and Application of Functional Solids*, NATO Science Series C: Mathematical and Physical Sciences, Vol. 539; Seddon, K. R.; Zaworotko, M. J., Eds.; Kluwer: Dordrecht, The Netherlands, 1999.
2. Huddleston, J. G.; Willauer, H. D.; Swatloski, R. P.; Visser, A. E.; Rogers, R. D. Room temperature ionic liquids as novel media for 'clean' liquid-liquid extraction. *Chem. Commun.* **1998**, 1765-1766.
3. Plechkova, N. V.; Seddon, K. R. Applications of ionic liquids in the chemical industry. *Chem. Soc. Rev.* **2008**, *37*, 123-150.
4. Walden, P. Ueber die Molekulargrösse und elektrische Leitfähigkeit einiger gesehmolzenen Salze. *Bull. Acad. Impér. Sci. St. Pétersbourg* **1914**, *8*, 405-422.
5. Wilkes, J. S. A short history of ionic liquids – from molten salts to neoteric solvents. *Green Chem.* **2002**, *4*, 73-80.
6. Seddon, K. R. Ionic Liquids for Clean Technology. *J. Chem. Tech. Biotechnol.* **1997**, *68*, 351-356.
7. Freemantle, M. Designer solvents – Ionic liquids may boost clean technology development. *Chem. Eng. News* **1998**, *76*[30th March], 32-37.
8. Blanchard, L. A.; Hancu, D.; Beckman, E. J.; Brennecke, J. F. Green processing using ionic liquids and CO_2. *Nature* **1999**, *399*, 28-29.
9. Holbrey, J. D.; Seddon, K. R. Ionic Liquids. *Clean Prod. Proc.* **1999**, *1*, 223-236.
10. Welton, T. Room-Temperature Ionic Liquids. Solvents for Synthesis and Catalysis. *Chem. Rev.* **1999**, *99*, 2071-2083.
11. Freemantle, M. Science/technology concentrates ionic liquids. *Chem. Eng. News* **2000**, *78*[15th May], 37.
12. *Green Industrial Applications of Ionic Liquids*, NATO Science Series II: Mathematics, Physics and Chemistry, Vol. 92; Rogers, R. D.; Seddon, K. R.; Volkov, S., Eds.; Kluwer: Dordrecht, The Netherlands, 2002.
13. *Ionic Liquids – Industrial Applications to Green Chemistry*, ACS Symposium Series, Vol. 818; Rogers, R. D.; Seddon, K. R., Eds.; American Chemical Society: Washington, DC, 2002.
14. *Ionic Liquids as Green Solvents – Progress and Prospects*, ACS Symposium Series, Vol. 856; Rogers, R. D.; Seddon, K. R., Eds.; American Chemical Society: Washington, DC, 2003.
15. *Ionic Liquids IIIA: Fundamentals, Progress, Challenges, and Opportunities – Properties and Structure*, ACS Symposium Series, Vol. 901; Rogers, R. D.; Seddon, K. R., Eds.; American Chemical Society: Washington, DC, 2005.

16. *Ionic Liquids IIIB: Fundamentals, Progress, Challenges, and Opportunities – Transformations and Processes*, ACS Symposium Series, Vol. 902; Rogers, R. D.; Seddon, K. R., Eds.; American Chemical Society: Washington, DC, 2005.
17. *Ionic Liquids IV – Not Just Solvents Anymore*, ACS Symposium Series, Vol. 975; Brennecke, J. F.; Rogers, R. D.; Seddon, K. R., Eds.; American Chemical Society: Washington, DC, 2007.

Chapter 12

The Twelve Principles of Green Chemistry

Philip G. Jessop[1], Sofia Trakhtenberg[2], and John Warner[2]

[1]Department of Chemistry, Queen's University, Kingston,
Ontario K7L 3N6, Canada
[2]Warner Babcock Institute for Green Chemistry, 66 Cummings Park,
Woburn, MA 01801

In this, the centenary year of the Industrial and Engineering Chemistry Division, we present a dozen short essays, one for each of the 12 principles of green chemistry. These principles were first articulated in 1998 as a set of tools to help the design scientist to anticipate downstream issues at the earliest stage of an R&D effort. Recognizing that most materials scientists lack the formal training necessary to deal with issues related to human health, the environment, and regulatory implications, the twelve principles serve to provide a path forward in designing products and processes that would be less environmentally damaging while maintaining or enhancing product performance and economic cost. Designing processes to minimize environmental impact has become, in recent years, essential to industrial and engineering chemistry, and is likely to shape the field for the next 100 years.

1. It is better to prevent waste than to treat or clean up waste after it is formed.

Although the Industrial and Engineering Division is having its centenary this year, the same cannot truly be said for green chemistry. The popular concept of green chemistry is quite young; the expression "green chemistry" dates from about 1990[1], receiving its most significant boost from the US Environmental Protection agency's development of it Green Chemistry program and the Presidential Green Chemistry Challenge Award. For the most part, the use of the word "green" to mean "not harmful to the environment" only dates from the early '70's.[2] However, an inspection of the literature of 1908, the year of the inception of the Industrial and Engineering Chemistry division, suggests that there were hints of the concepts of green chemistry. The U.S. Geological Survey reported that it is possible to burn coal for power production in a manner that does not produce smoke as a byproduct.[3] In the same year, Italian chemist Giacomo Ciamician's criticized organic chemists, saying that their successes in synthesis were due to a regrettable tendency to use excessively aggressive reagents.[4] However, this kind of thinking was not the norm at the time; almost every paper mentioning "pollution prevention" in the early 20th century described what we now call "pollution control" (the capture and treatment of waste), rather than methods for prevention of its formation.

The transformation of thinking from the waste-capture approach to the waste-prevention approach took most of the 20th century. Other than isolated examples such as that Geological Survey 1908 report, waste prevention only started to gain traction in the 1970's. The American Chemical Society, in a 1969 report on how chemists could help decrease pollution, gave 73 recommendations mostly on pollution control but included a call for more R&D on "nonpolluting alternative manufacturing processes".[5] In 1975, 3M Corporation instigated its "3P" program (Pollution Prevention Pays), which rewarded employees for inventing ways to reduce waste production.[6] The results included over $1 billion in first-year savings and a reduction in waste generation of 1.2 million metric tons.[7] Surprisingly, other US companies did not follow suit until 1986.[8]

Most research on green chemistry and related concepts has taken place in the past 20 years (Figure 1). A search of Chemical Abstracts for key terms related to green chemistry showed that the number of publications for each topic suddenly ramped up starting at a particular year; 1980 for "biomass conversion" and "supercritical CO_2", 1996 for "life cycle analysis" and "green chemistry", 1999 for "ionic liquids" and 2001 for "atom economy".

The green chemistry literature has jumped in only a few years from almost nothing to an enormous body of knowledge. While there are only 6,200 papers that specifically mention "green chemistry", there are many more that are in fact about green chemistry without actually mentioning those two words. For example, although a major motivation for using supercritical CO_2 ($scCO_2$) as a solvent is to reduce or eliminate organic solvent waste, only 0.9% of the 8,800 papers describing $scCO_2$ mention the words "green chemistry," while 9% of papers on ionic liquids and 51% of papers on life-cycle analysis do. This suggests that the entire body of green chemistry literature may be 10 to 100 times larger than the self-reported 8,800 papers. There is an even larger body of literature describing work that would lead to waste prevention, even if the work was not motivated by a desire to achieve waste reduction. While that kind of work doesn't really count as green chemistry because the motivation was lacking, the result is the same! Most famously, the first Nobel Prize to be awarded for green chemistry (2005, Chauvin, Grubbs, and Schrock) was awarded to three chemists who certainly did not describe their work as green chemistry, and quite likely did not think of it as such. However, thanks to researchers who describe their work as green *and* to those who don't, there are far more tools in the green chemist's toolboxtoday than there would have been in 1908.

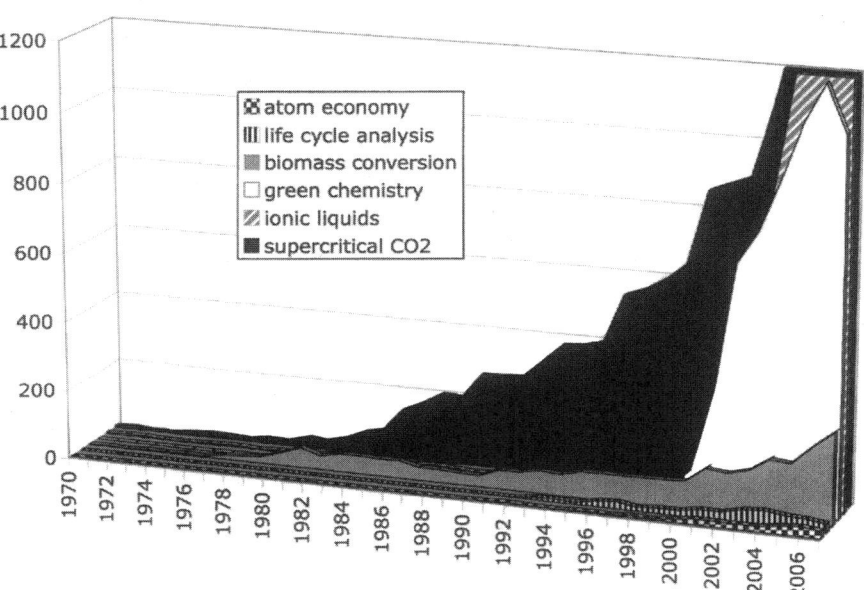

Figure 1. Number of publications per year found in Chemical Abstracts in searches for key expressions related to green chemistry.

2. Synthetic methods should be designed to maximize the incorporation of all materials used in the process into the final product.

A typical product life-cycle begins with the extraction of raw materials from the earth, followed by the functionalizing of the raw materials into useful feedstock chemicals that can be transformed into the various chemical products that we require for making products. The synthesis of the chemical products involves the use of various reagents in order to create the desired product. Many reagents are used to perform a specific function on the molecule, but are not actually incorporated into the product. The reagents that are not incorporated into the product typically result in waste in a chemical process.

Throughout the history of chemistry the quantitative success of a chemical transformation has typically focused on the concept of product yield. This number is based on the number of grams of products synthesized divided by the number of grams theoretically possible. While this measure provides some insight into the productivity of a chemical reaction, or series of reactions, much information is lost. Many synthetic transformations available in the chemist's toolbox do not merely follow the simple description of compound A is converted to compound B (Figure 2). More often than not these transformations are far more complex with a number of reactants coming together to form a number of products. While the desired product may be obtained in high yield, it is possible and in fact often the case that along with the desired product an equivalent amount of some other anticipated by-product is formed.

Idealized reaction:
A → B

Typical reaction:
A + R₁ + R₂ ...→ B + C + D...

Figure 2. An idealized versus a typical synthetic transformation (A = starting material, B = product, R = reagent, C,D = co-products or by-products)

The atom economy of a reaction is based on a calculation developed by Barry Trost:[9] the ratio of the molecular weight of the atoms used to make a product divided by the molecular weight of all the reagents and starting materials used to make the product. It is a simple measure of the amount of waste created in a chemical process based on the atoms used in a process. This principle speaks to developing molecular transformations that incorporate a maximum number of atoms into the final product thus minimizing the atoms that end up as waste.

Evaluation of different types of chemical reactions in a synthetic chemists "toolbox" (Figure 3) from the atom economy perspective can help the chemist select the most appropriate "tool". Clearly, addition and rearrangement reactions are the most atom-economical. In either case, 100% of the atoms from the starting materials and reagents end up in the product. In contrast, substitution and elimination reactions are inherently not atom-economical, since the leaving group **C** is not incorporated into the final product. Thus, while reactions of any type may produce unintended byproducts due to the possible existence of competing alternative reaction pathways, in the cases of substitution and elimination reactions byproducts must be generated in stoichiometric amounts along with the desired product. In some cases, however, these byproducts are useful chemicals in their own right, and if these byproducts do not end up in waste but instead are recovered, the apparent atom efficiency of the reaction may increase.

Rearrangement:
A → B

Addition:
A + B → C

Substitution:
A-C + B-D → A-B + C-D

Elimination:
A → B + C

Figure 3. Schematic presentation of four types of chemical reactions

Another way to increase atom efficiency of the reaction is to use efficient catalysts. Although catalysts are ideally added in minute amounts and are not consumed in the reaction, in practice many reactions require nearly stoichiometric amounts of "catalysts" or catalysts that are not recovered afterwards.

In 1997, the Presidential Green Chemistry Challenge Greener Synthetic PathwaysAward was given to BHC Company (now BASF) for a novel method of ibuprofen synthesis.[10] The new process consists of three catalytic steps, with the only byproduct being acetic acid, and has overall atom efficiency of about 80%. Since the acetic acid byproduct does not end up in the waste but is recovered, the process can be considered virtually 99% atom efficient. The older process, replaced by the award-winning one, consisted of six stoichiometric steps with an overall atom efficiency of less than 40%.

3. Wherever practicable, synthetic methodologies should be designed to use and generate substances that possess little or no toxicity to human health and the environment.

It is often overlooked that the manufacturing pathway to ultimately synthesize a product may in fact go through a series of chemical intermediates. This principle seeks to minimize the hazards associated with these intermediates. It is important to make the distinction that this principle is focused on materials that, in a perfect world, should not appear in the product. However, when considering worker exposure and the other associated costs related to handling, storage and disposal of hazardous materials, the avoidance of hazardous intermediates is extremely important. Reducing the hazards associated with the way a product is made can drastically reduce worker liabilities, contribute towards worker health and safety, and minimize the potential for chemical accidents.

In a typical manufacturing process, the final step before packaging of the product involves formulation. Formulation is the assembly of various materials acquired internally or from external vendors. Because these ingredient materials may have been synthesized elsewhere, formulators may believe that their process avoids hazardous or damaging materials while in fact in their vendors' facilities up-stream, hazardous materials are being used and generated. In a complete life-cycle analysis a "not-in-my-backyard" solution is not sufficient. Thus it is important that the entire supply chain for a product be evaluated and optimized. A typical example is plastic packaging, which is typically from an outside supplier. The plastic is selected for its ability to contain the consumer product and retain its functionality. Although most plastic materials are non-toxic, the ways of manufacturing such plastic can be hazardous to the workers who operate and handle the materials during production. Therefore, it is important to reduce the hazard during the creation of such plastic materials.

One example where a plastic material was created which accomplished the task of reducing the hazard in a manufacturing setting is that of Cargill Dow's (now NatureWorks LLC) NatureWorks[TM] PLA process. In 2002, the company won the Greener Reaction Conditions Award of the Presidential Green Chemistry Challenge Awards for developing a biobased, compostable polymer that eliminates the use of organic solvents and hazardous materials required to make many other plastic materials.[11] The NatureWorks[TM] PLA process has been used to create polylactic acid (PLA) polymers based on renewable, biobased resources. The method for creating the polymers involves a water-based natural process of fermentation.[12] Typical methods for producing other polymers include the use of organic solvents and many hazardous materials. PLA production employs low-temperature, low-toxicity processes that are much safer for the worker. PLA is currently used for many packaging applications where typical polymers once were used.

4. Chemical products should be designed to preserve efficacy of function while reducing toxicity.

How often have we found ourselves making a mistake, and then making another mistake when we rush to fix the first? "Out of the frying pan and into the fire," as they say. In our haste to correct a problem at least partly of our own making, we do not take the time to consider the consequences of our correction. A popular legend has it that the reason for the introduction and widespread adoption of the automobile was a great desire to eliminate the pollution being dropped on the roadways by legions of horses. Regardless of whether that was the original motivation, the transition solved one problem and brought us to a new and larger problem, global warming, that we might have foreseen if we had been paying close attention; Arrhenius warned us in 1896 that it would happen.[13] Can we really – if we take the time to do the research and to think things through – avoid replacing one mistake with another?

An illustrative case history is the use of solvents for de-greasing automobile parts. In automotive repair shops, gas masks are not used because of discomfort, and fume hoods are impractical, so the degreasing solvents expelled by aerosol cans are inhaled at significant concentrations by workers. Perchloroethylene, or "perc", was a principal component of such solvents, but studies have suggested a link between perc and esophageal and bladder cancer.[14] California phased out degreasing products perc, CH_2Cl_2 and trichloroethylene by 2001 and recommended nonchlorinated replacements.[15] In the rush, there was insufficient care taken in evaluating the potential replacement. The products that were sold as replacements were at first primarily hexane (from 1990) and later blends of n-hexane and acetone, often with other solvents as well.[16] Long before this replacement program was undertaken, it had been shown that long-term exposure to n-hexane causes severe polyneuropathy, to the point of victims losing the ability to walk.[17] It was also known that acetone or related ketones amplify the neurotoxicity of hexane.[16]Could a careful consideration of the literature have foreseen this debacle?

Dry cleaning is another case in point (Figure 4). The original solvents were naphtha, benzene and the like,[18] but because of flammability problems they were replaced with the slightly less flammable white spirits (Stoddard solvent, C7-C12 alkanes), a chronic nephrotoxin.[19] That was replaced in turn with carbon tetrachloride, trichloroethylene and later perchloroethylene, "perc", because they were nonflammable. Carbon tetrachloride has long been abandoned because of its carcinogenicity; trichloroethylene because of its toxicity. Perc is being phased out for the reasons cited above. Primarily as a result of pressure from California, the search is now on for a sixth-generation dry cleaning solvent. The principal candidates are CO_2, C11-C15 hydrocarbons (e.g. Chevron Phillip's EcoSolv® or Exxon-Mobil's DF-2000[TM]), and decamethylcyclopentasiloxane ("D5"). Are these replacements going to cause health or environmental problems in the future? Fortunately, relevant studies are in progress, so perhaps this time we will be more able to make the right decision.

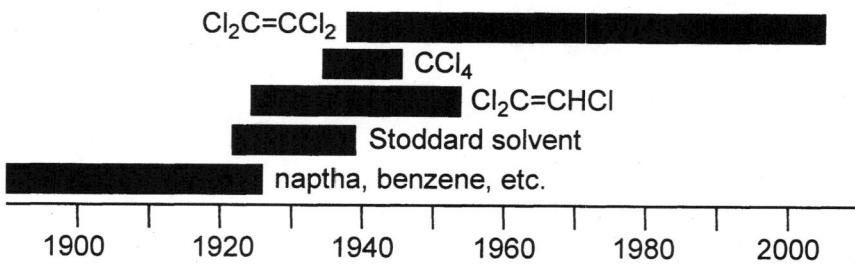

Figure 4. Approximate timeline for solvents used historically for the dry cleaning of textiles. Note that Stoddard solvent was still used many years later for specialized materials.[18, 20]

This principle is the most relevant to the consumer. Making a safe product is an obvious goal of all manufacturers. When evaluating chemical risk there are two components that are typically considered: a material's intrinsic hazard and the exposure to the hazard. While much work has been done traditionally to mitigate exposure of hazards to consumers, green chemistry seeks to address risk by dealing directly with the intrinsic hazard. Of course this is not an easy task. But the last few decades of mechanistic toxicology have provided deep insight into several of the structure-activity relationships that govern physiological, environmental and global hazards. While individual practitioners of chemistry cannot be expected to be fully versed in the vast and growing field of molecular toxicology, certain fundamental understandings and resources available must be incorporated in product design.

In some cases, the safer alternatives to the chemical products that are currently in use already exist and, as soon as the hazard is identified, the chemical of concern can be quickly replaced. Sometimes the replacement does not exist yet and first it needs to be invented. Taking into account that there exist millions of different chemicals (the chemical abstracts database currently lists more than 35,000,000 organic and inorganic substances), electronic tools for the identification of possible effective and less toxic alternatives is necessary. One such tool, Greenlist™, was developed by S.C. Johnson & Son, Inc. and won the 2006 Designing Greener Chemicals Presidential Green Chemistry Challenge Award.[21] It assisted in the reformulation of such common household products as Saran Wrap®, Windex®, and others.

With due diligence, extensive research, and the assistance of searching tools, we should be far better equipped to make wiser choices in the future.

5. The use of auxiliary substances (e.g. solvents, separation agents, etc.) should be made unnecessary whenever possible and innocuous when used.

Introduction

Why are the E-factors (mass ratios of waste todesired product)[22] of fine chemical and pharmaceutical manufacturing processes so very high? Solvents! 85-90 % of the waste[23] in fine chemical or pharmaceutical manufacturing is solvent. Such large quantities of solvents are needed because multistep chemical syntheses, where each "step" may be an extraction, a reaction, or a separation, typically operate under what could be termed *Murphy's Law of Solvents*;[24] "*The best solvent for any one process step is bad for the subsequent step.*" That is, each step only gives adequate performance (yield, rate, extraction efficiency) if one kind of solvent is used, and it is typically necessary to remove the solvent after each step in order to replace it with a different solvent that better meets the requirements for the subsequent step.

Reducing the environmental impact of solvents used in chemical manufacturing can be achieved by either reducing the volume of solvents or by selecting less environmentally-damaging solvents. Both approaches have been explored.

Volume Reduction

Volume reduction can be achieved by performing reactions and operations at higher concentrations, by finding a compromise solvent that will satisfy the requirements of two or more process steps in a row, or by defeating Murphy's Law of Solvents by creating a solvent that can be modified after a process step so that it can better meet the needs of the next step.

Compromise solvents are ideal when they can be identified, but, unfortunately, in many cases there will be no solvent that allows two consecutive steps to operate with acceptable performance. In the case of the sertraline synthesis, the last three steps required a total of four solvents (Figure 5). A compromise solvent, ethanol, was found by Pfizer to meet the needs of the last three steps, resulting in a 76% reduction in solvent volume.[25]

Solvents that can defeat Murphy's Law of Solvents have been identified and are called "switchable solvents"; these are liquids that have one set of properties (that presumably meet the needs of one process step) and that can be reversibly switched to another set of properties to meet the needs of the next step.[26] As an example of the flexibility offered by switchable solvents, consider the copolymerization of CO_2 and cyclohexene epoxide catalyzed by a Cr(salen)

410

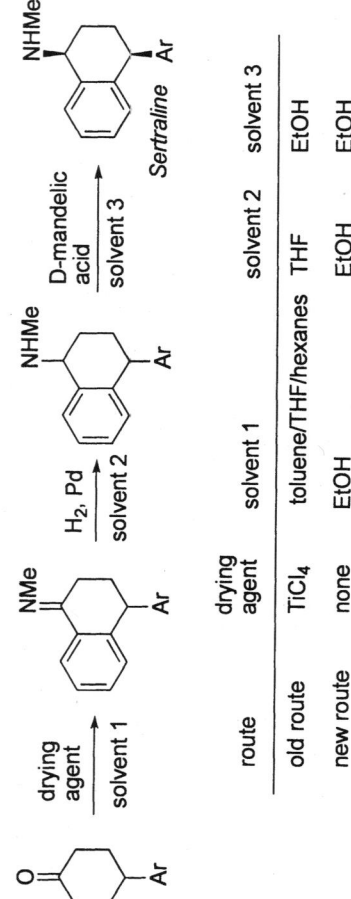

route	drying agent	solvent 1	solvent 2	solvent 3
old route	TiCl₄	toluene/THF/hexanes	THF	EtOH
new route	none	EtOH	EtOH	EtOH

Figure 5. An example of the use of a compromise solvent.

complex (Figure 6). After the solventless reaction, the product mixture is normally dissolved in a solvent of relatively low polarity (typically CH_2Cl_2), after which the Cr catalyst is cleaved from the polymer with acid (HCl) and the polymer is precipitated by the addition of a polar solvent (MeOH). Instead, one can use NHEtBu as the low polarity solvent, and then supply CO_2 at one atmosphere to generate an *in situ* acid (carbamic acid) and to switch the amine solvent into a high polarity solvent (a carbamate ionic liquid). Thus the switchable solvent replaces CH_2Cl_2, MeOH and HCl. Switchable solvents have only been known for about 4 years and are not yet in use in industry. The development of greener switchable solvents is needed.

Greener Conventional Solvents

Less environmentally-damaging solvents range from wisely-chosen conventional solvents to extremely unconventional solvents. Most of the environmental impact reduction that has been achieved in industry from solvent substitution has been by switching to greener conventional solvents.

Which of the conventional organic solvents is the greenest? Assessment of the energy required to produce and to dispose of the solvent is one method for determining the greenness of a solvent. If one assumes that the energy generated by the incineration of a solvent is used to replace energy that would have been generated by fossil fuel combustion, then the greenest solvents include hexane, pentane, diethyl ether, methanol and ethanol, all of which have calorific values that make up for more than half of the energy required for their production.[27] An alternative assessment is provided by the EHS method for combining factors such as flammability, reactivity, toxicity, irritation, persistence, air pollution and water pollution; by this measure, the greenest solvents include methyl acetate, methanol, ethanol, and ethyl acetate. In comparison to those five solvents, hexane and pentane rate poorly in the areas of irritation, chronic toxicity, persistence, and air hazard.[27]

Solvents derived from renewable resources are inherently greener than other solvents, if one neglects all other factors. Key solvents of this type include CO_2, water, methanol, ethanol, acetic acid, limonene, and fatty acid methyl esters; the first two in this list have the greatest availability.

Note that the solvents mentioned above as being greener fall into two categories; nonpolar aprotic and polar protic; there is a need for a polar aprotic green solvent.[28] Polar aprotic solvents are 6% of the solvent usage in pharmaceutical manufacture,[29] although that number might be higher if a green and inexpensive replacement were found. Dimethylformamide[30] and N-methylpyrrolidin-2-one[31] have reproductive toxicity. Most polar aprotic solvents contain nitrogen, which is unwanted in sewage treatment plants, and are difficult to separate from water. Dimethylsulfoxide (DMSO) suffers from a very

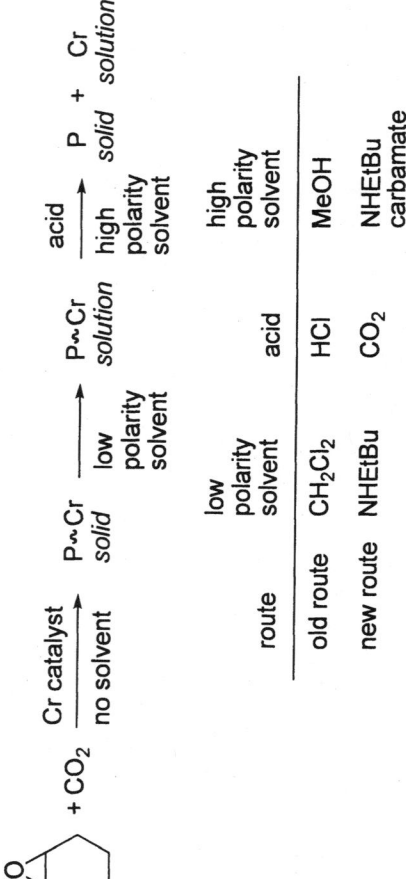

Figure 6. An example of a potential use of a switchable solvent: the copolymerization of cyclohexene oxide and CO_2. In the new route, after the polymer precipitate is removed by filtration, the ionic liquid is easily distilled from the Cr catalyst, and both the solvent and the catalyst are then recycled.

high boiling point, so that its removal is energy intensive, it is difficult to recycle, it has little calorific value for incineration. An easily removed, recyclable substitute for DMSO, piperylene sulfone, has been identified.[32]

Green Unconventional Solvents

Unconventional solvents include supercritical CO_2 ($scCO_2$),[33] CO_2-expanded liquids,[34] ionic liquids,[35] fluorous liquids,[36] and liquid polymers.[37] The advantages and disadvantages of each are summarized in Table I.

6. Energy requirements should be recognized for their environmental and economic impacts and should be minimized.

In recent years it has become impossible to escape the implications of energy utilization. With the increased cost of fuels across the board, any process that is using more energy than absolutely necessary becomes a major cost. From a global climate change perspective, this excess energy consumption is likely to be contributing to greenhouse gas production. Fundamental thermodynamics of chemical transformations requires that activation energies be overcome (Figure 7). While a discussion of spontaneity and Gibbs free energy changes is outside the realm of this chapter; simply stated, most chemical reactions need heat in order to occur. Energy is also required for many common manufacturing steps such as the drying of a material, or the cooling of an exothermic reaction. Some processes such as purification by distillation or reaction under reflux conditions require both heating and cooling to happen simultaneously. This principle seeks to develop built-in mechanisms to allow transformations to occur under ambient conditions.

Strategies that can result in increased energy efficiency include but are not limited to solventless reactions, catalysis, microwave-assisted synthesis, and synthetic photochemistry. Solventless reactions, even if they require activation (which can be either thermal or mechanical or photochemical) are relatively energy efficient because they do not require heating or agitation of a solvent. Moreover, it was demonstrated recently that when sufficiently close contact between reactants is achieved, solventless reactions are more efficient.[38] Catalysis, as will be discussed in greater detail later, increases energy efficiency by providing an alternative reaction pathway with lower activation energy. Moreover, using sufficiently selective catalysts can allow a further decrease in the energy requirements by simplifying or even foregoing purification steps. Microwave-assisted reactions typically proceed faster than the analogous thermal reactions, and it takes less time and energy to heat the reaction vessel to

Table I. Advantages and Disadvantages of Green Unconventional Solvents

Solvent	Advantages	Disadvantages
scCO$_2$	• Facile removal • Not toxic or ecotoxic • Renewable • Excellent mass transfer	• Energy for compression • Need for high pressure equipment
CO$_2$-expanded liquids	• Highly tunable • Excellent mass transfer	• Energy for compression (less than for scCO$_2$) • Need for high pressure equipment
Ionic liquids	• Nonflammable[a] • Nonvolatile[a]	• High energy and financial costs for preparation • Separation from product may be difficult • Toxicity and ecotoxicity for some examples
Fluorous liquids	• Facile removal • Nonflammable	• High energy and financial costs for preparation • Highly persistent • Bioaccumulating
Liquid polymers	• Nonflammable[a] • Nonvolatile[a] • Inexpensive • Biodegradable (for most examples)	• Separation from product may be difficult

[a] These advantages are negated if a volatile or flammable organic solvent is used to remove the product from the nonvolatile solvent. Removal of products by distillation (if volatile), product immiscibility, or extraction by water or scCO$_2$ is therefore preferred.

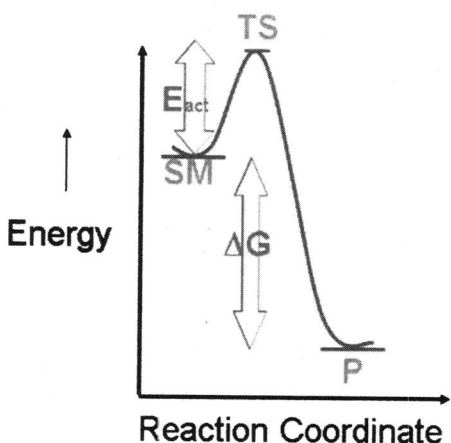

Figure 7. Reaction coordinate diagram. E_{act} = Activation Energy, TS = Transition State, SM = Starting Materials, P = Product, ΔG = Energy

a desired temperature using microwave than by conventional heating. Photochemical reactions can be highly energy efficient (especially if solar irradiation is used and thus no energy was wasted in creating the light).[39] Typically, the reaction is designed in such way that only the desired reagent can absorb the light at the required frequency; the solvent is selected to be transparent and does not interact with the light.[40] Comparing this with conventional thermal activation where most of the energy is spent on heating the solvent rather than reagents, it is clear that photoreactions (all other things being equal) are more energy efficient.

Harnessing entropically controlled phenomena, such as molecular recognition and self-assembly, can also result in energy savings. After all, it is because of intermolecular interactions that nature is able to produce such highly complicated and organized structures as biomacromolecules at ambient temperature and in aqueous media.

One example of a dramatic reduction in energy requirements is the development of a method for printing product labels directly on glass. RevTech, Inc. won the Presidential Green Chemistry Challenge Small Business Award in 2000 for developing the Envirogluv[TM] process to print product labels directly on glass, therefore replacing traditional labeling techniques.[41] The process involves the use of inks that contain no heavy metals and little to no volatile organic compounds (VOCs). The inks are biodegradable and ultraviolet light is used when the ink is adhered to the glass, replacing the energy-intensive high-temperature ovens typically used in many processes.[42]

7. A raw material or feedstock should be renewable rather than depleting, wherever technically and economically practicable.

"Renewable" is defined by the Oxford English Dictionary as "not depleted by its utilization".What feedstocks do we have that are really renewable? Those that immediately come to mind are biomass, waste CO_2 and waste plastics. While traditional discussions of renewable feedstocks exclude the wastes, they are eminently renewable (they are being renewed whether we like it or not) and are available in much larger quantities than biomass (Figure 8).

Renewable materials may be, in many cases, green and sustainable, but we should not make the mistake of assuming that this is always the case. The environmental effects of large scale use of biomass for production of materials and (especially) fuels can be quite detrimental; this is an area requiring much more research. However, renewable feedstocks have the possibility of being sustainable, whereas non-renewable feedstocks can never, regardless of the circumstances, be sustainable.

What chemical products can be made from renewables?

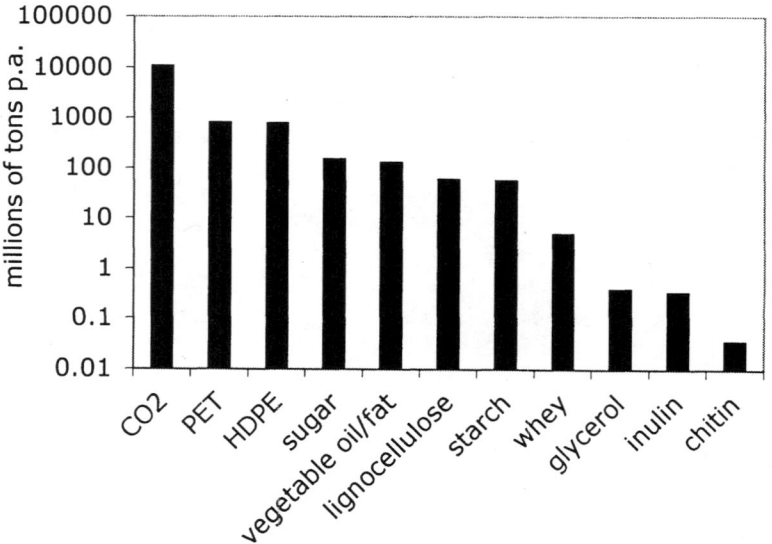

Figure 8. Annual worldwide production of renewable materials.[43] *Note that a log scale is used. Plastics data are for recycled quantities, not production quantities. CO_2 data are for emissions from electricity or heat production only,*[43b] *because most other emissions could not easily be collected for use. Whey value is the median of two literature estimates for liquid whey,*[43a] *corrected to show only whey solids (assuming 93.5% water in liquid whey). Lignocellulose data are for agriculturally produced residues not used for food or textiles.*[43e]

CO_2

There is no doubt that the amount of CO_2 at our disposal is impossibly large. There is no product that we could possibly want in such large quantities that it would consume a significant portion of the available CO_2. Why, then, should we use CO_2 as a feedstock? Because it is available, renewable, nontoxic, and free. Already, CO_2 is being industrially converted into aspirin, urea, carbamate pesticides, propylene carbonate, and methanol (Figure 9). Because methanol is used to make biodiesel, then some CO_2 is, ironically, ending up as part of a fuel!

If recent research led to industrialization, then what other products could be made from CO_2? Polycarbonates (from epoxides),[44] dimethylcarbonate (from methanol),[45] dimethylformamide (from $NHMe_2$), carboxylic acids, esters, lactones, pyrones, and many others.

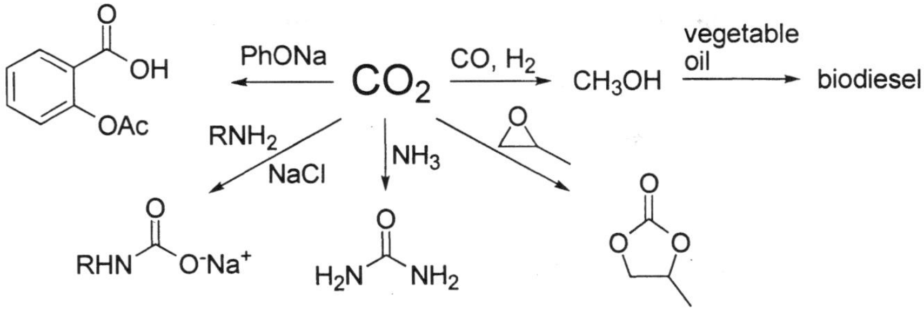

Figure 9. Chemical products currently made using CO₂.

Waste Plastics

Most polymers such as PE, PP, PS, HDPE and PET (for non-food applications) are recycled after purification while others such as PET (for food applications) are first depolymerized, then purified and repolymerized. Other polymers such as ABS and PVC are rarely recycled.[43c] Currently, most recycled polymers end up as the same polymer, but because the recycled polymer is often less pure, there can be issues with degradation of physical properties and with meeting regulatory requirements for food packaging. Therefore, there is potential for finding alternative, higher value uses for these waste materials.

Depolymerization, gasification, and thermolysis/pyrolysis are the main routes by which waste polymers could be converted into useful chemicals. The depolymerization of PE to butadiene, methane and olefins[46] and of PS to styrene[47] is known technology, but not normally practiced. Gasification of plastic wastes to syngas (CO, H_2 mix) has been reported.[48] Thermal cracking to pyrolysis oil and/or more specific organic products is also possible.[43c, 49]

Chitin

Chitin, a waste material from crab and shrimp processing, is poly(N-acetyl-D-glucosamine).[50] The de-acetylated form, chitosan, is primarily used in water purification but is also used as a chemical feedstock. For example, chitosan can be depolymerized to the monomer, which is used as a hair-setting agent in shampoo and hair conditioner.[50,51] In the future, graft polymers of chitosan may form biodegradable films. Already, carboxymethylated chitin is approved as a food wrap.

Vegetable Oils, which consist primarily of triglycerides, can be broken down into fatty acids, glycerol and, by transesterification, fatty acid methyl esters. An enormous range of products are derived from these feedstock chemicals (Figure 10).

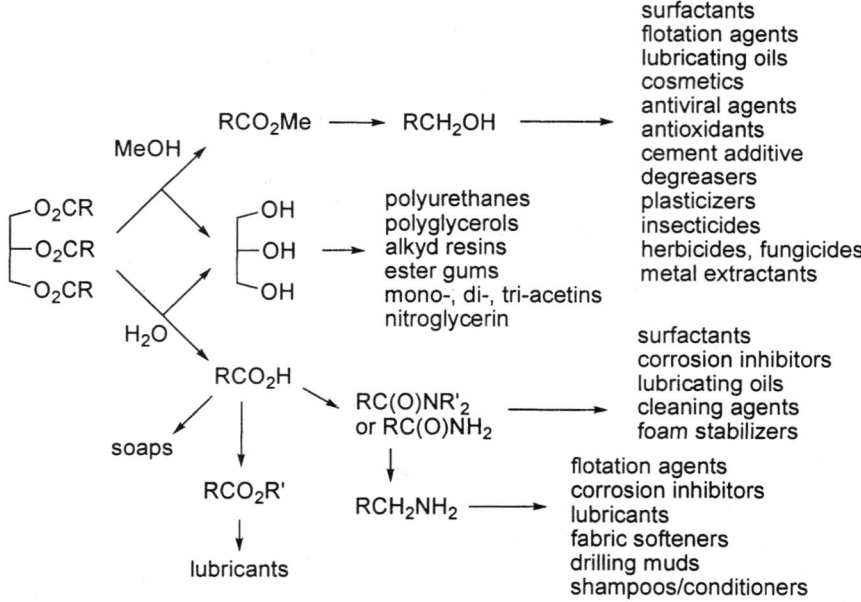

Figure 10. A selection of the products that are derived from triglycerides.[52]

Lignocellulose

This large category of biomass feedstocks includes straw, bagasse, stover, wood, vegetable/fruit food waste, grass. These materials consist of cellulose (a polymer of glucose), hemicellulose (a mixed polymer of five-carbon and six-carbon sugars), and lignin (a stable and irregular polymer containing phenol groups). After the lignin is removed, the cellulose and hemicellulose can be hydrolyzed to a mixture of sugars including glucose and xylose, or hydrolyzed to sugars that can be fermented to make acetone, butanol and/or ethanol. While the production of acetone and butanol by this route is no longer used, ethanol is made via this route in Canada by Iogen. Dehydration of xylose gives furfural and 5-hydroxymethylfurfural, opening up a route to a range of products (Figure 11). Lignin is converted by the sulfite process to lignosulfonates (concrete additives), vanillin, dimethylsulfide and dimethylsufloxide (DMSO).Pyrolysis of lignocellulosic waste to oil or gasification to syngas are alternate routes to a much wider range of products.

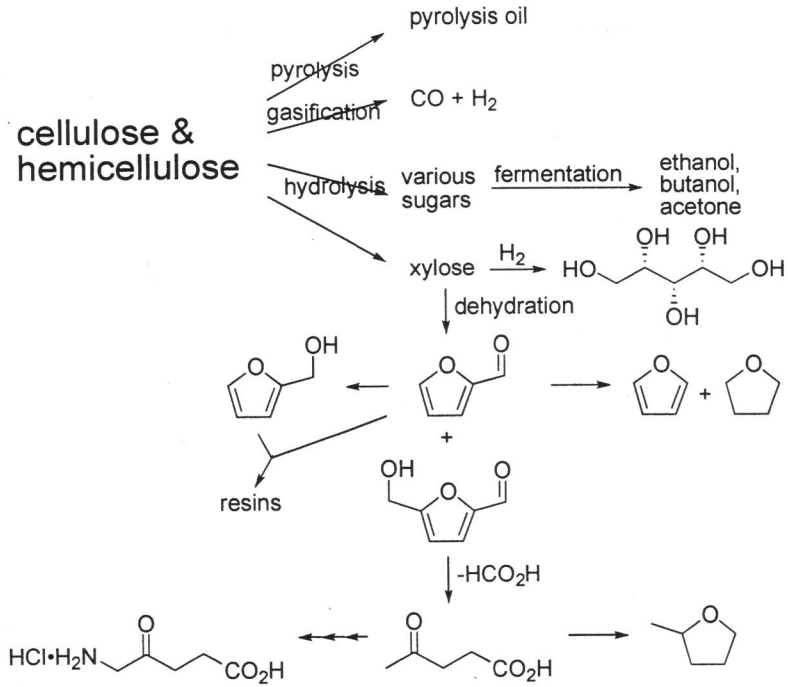

Figure 11. A selection of the chemical products currently or historically made from cellulose and hemicellulose.[52b]

Starch is a heavily used feedstock that can be converted into several sugars (Figure 12), all of which are used as sweetener. Glucose, which also comes from corn sugar, is a precursor to many other products. Ascorbic acid is made from glucose in several steps involving hydrogenations, cyclizations and biocatalysis. Glucose can also be converted by anaerobic fermentation to succinic acid and by microbial conversion to 3-hydroxypropanoic acid.[53] Lactic acid is also made from starch by fermentation. Cyclodextrins are also prepared directly from starch by enzymatic conversion.[53]

Whey, the major byproduct of cheese manufacture, is 70-80% lactose after the water has been removed. Bacteria can convert lactose to lactic acid in high yield, and yeasts can convert lactose into ethanol, although the economics of the latter process are apparently dubious.[43a]

Clearly, synthesis of chemical products from renewable feedstocks is an industrial reality, a fact that should be celebrated. Many opportunities exist for new products from these feedstocks. CO_2 fixation, after so many years of activity, is still being studied, with new example reactions being found. Waste

420

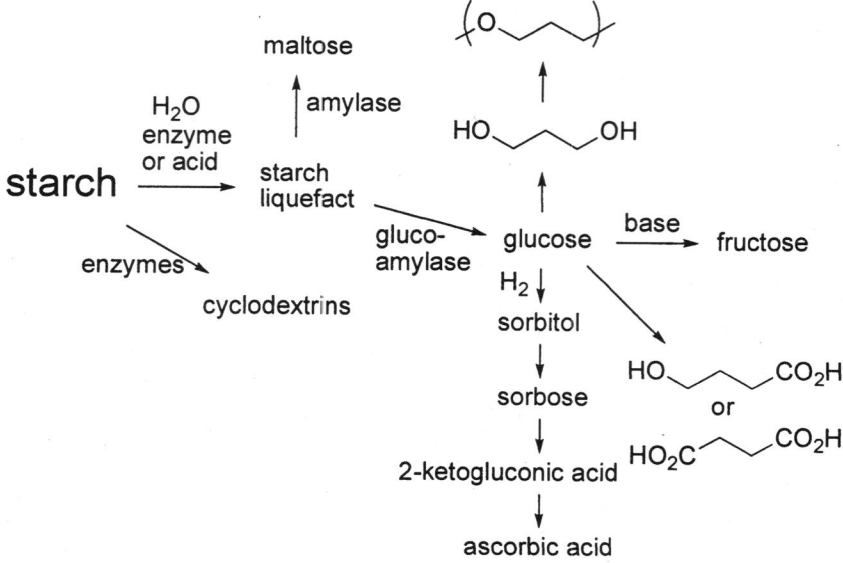

Figure 12. A selection of the chemical products currently made from starch.

polymers, especially those that are hard to recycle for their original uses, are a renewable feedstock that shouldn't be ignored any longer. Biomass conversion, while it is already doing well, could give us many more useful chemicals, and it is up to us to discover them.

8. Unnecessary derivatization (blocking group, protection/deprotection, temporary modification of physical/chemical processes) should be avoided whenever possible.

Since 1828 when urea was first synthesized in the lab from cyanic acid and ammonia[54] the development of organic synthesis has seen remarkable innovations and discoveries. Using modern state of the art techniques allows almost any molecule that can exist under normal conditions (and many of those which cannot) to be synthesized. But this near-omnipotence comes with a price. Complex organic molecules often have multiple reactive pathways available under a variety of reaction conditions. In order to develop synthetic transformations that select one specific reactive pathway, chemists have often resorted to supplemental modifications of their starting materials to "block" nondesired chemical reactions from happening. While some of these solutions are brilliant and elegant in their efficiency and selectivity, from a green chemistry perspective they must by default involve the generation of waste. After these "protecting groups" have served their purpose they must be removed via subsequent chemical reactions. From an atom economy perspective each of these protecting/deprotecting sequences decreases the overall efficiency of the transformation. Each protection and deprotection step, including product workup, is a separate chemical reaction where solvents and other auxiliaries are used (i.e. converted into waste) and energy is consumed. The costs (both material and environmental) associated with using the protection groups are therefore so high that derivatizations should be avoided whenever possible.

Material scientists are frequently turning to natural systems for inspiration.[55] In many cases this results in finding elegant and environmentally friendly solutions. In biological systems an analogue to protecting and deprotecting sequence would be interaction between an enzyme and its reversible inhibitor. While the inhibitor is bound to the enzyme, the latter is inactive, but if dissociation of the inhibitor is induced, the enzyme becomes active again. The nature of the bond between the enzyme and its reversible inhibitor is, however, very different from that which is created when a molecule is being derivatized. While the organic synthesis relies on strong covalent bonds, the reversible inhibitor binds to the enzyme via multiple weak intermolecular interactions such as hydrogen bonds or hydrophobic interactions. The low energy associated with intermolecular bonds results in the ability to form and/or break them under ambient conditions.

It has been demonstrated that manipulating the properties of a molecule using intermolecular interactions instead of altering the molecule chemically is feasible on the industrial scale.[56] The system developed involves the manipulation of a hydroquinone molecule, used for many applications including photography[57] and cosmetics[58] in order to make it less prone to oxidation. The synthetic chemistry approach would be to change its structure through traditional derivatization technique, such as blocking the hydroquinone oxygens with protecting groups. In addition, if it is desired to fine-tune its solubility and diffusivity as well, one or more of the aromatic hydrogens can be substituted by an alkyl group. This can be achieved via a multi-step organic synthesis procedure, with a number of associated purifications. The amount of hazardous materials necessary for fine-tuning the hydroquinone properties including reagents, auxiliaries and solvents was so high that alternative ways of achieving that purpose had to be found.

Harnessing bioinspired phenomena such as molecular recognition and self-assembly, a new approach termed *Non-Covalent Derivatization* was developed.[59] The properties of hydroquinone were successfully altered by incorporating the hydroquinone molecules in a matrix with compatible alkylterephthalamide molecules.[60] This was achieved via one-step solventless grinding, i.e. a reaction proceeding at room temperature, not requiring any purification steps, and not generating any waste. The auxiliary alkylterephthalamide molecules are non-toxic; they interact with hydroquinone via hydrogen and van der Waals bonds, which are easy either to make or to break. The environmental benefits of this approach make *Non-Covalent Derivatization* an example of green chemistry. [61]

The versatility of the *Non-Covalent Derivatization* approach makes it an attractive tool for the materials chemist. In the above example of the hydroquinone – alkylterephthalamide system, a variety of hydroquinone-containing matrices can be designed. Selecting an appropriate alkylterephthalamide and varying the alkylterephthalamide-to-hydroquinone ratio results in a family of structures characterized by different motifs of hydrogen bond networks and varying strengths of lipophilic interactions.[62,63] The physical properties of the auxiliary alkylterephthalamide as well as the packing of molecules in the crystal are responsible for altering the properties of hydroquinone such as its solubility and diffusivity in various media, ability to autooxidize, etc. Thus the properties of hydroquinone molecule can be fine-tuned for a specific application.

9. Catalytic reagents (as selective as possible) are superior to stoichiometric reagents.

In 1811, just under two centuries ago, Gottlieb Kirchhoff, a pharmacist in St. Petersburg, Russia, was working on a way to hydrolyze starch to glucose. Kirchhoff had identified sulfuric acid as a reagent that would hydrolyze the starch. To his, and everyone's surprise, he found that only a tiny amount (1.5 lbs) was enough to hydrolyze a large amount of starch (100 lbs)! In those days, industrialization of a new discovery could come quickly; this process was industrialized the very next year.[64] Acid-catalysts for glucose production were phased out in the 1960's in favour of enzymes. Today, glucose for U.S. consumption is made from starch in quantities of 650,000 tons per year.[65] Imagine if it were done with stoichiometric sulfuric acid rather than catalysts! That would be 10^8 liters per year, or 70 Olympic swimming pools of concentrated H_2SO_4. In principle, catalysts are greener than stoichiometric agents, all other things being equal.

Unfortunately, sometimes all other things aren't equal. Catalysts can be disastrously un-green. The villagers of Minamata Bay in Japan are famous victims of careless handling of a catalyst. An obsolete process for the synthesis of acetaldehyde from acetylene (Figure 13) used mercury(II) chloride as a catalyst. A German invention, it was used by the Chisso Corporation in Japan from 1932 to 1968. Spent catalyst was dumped in the bay, ending up in the fish and then in the villagers. Along the way, the mercury was methylated by bacteria. The first deaths were identified in 1956 and mercury was identified as the cause in 1959.[66] The number of officially-recognized victims is 2,265, with 1,784 dead. Many thousands more have some of the symptoms.[67]

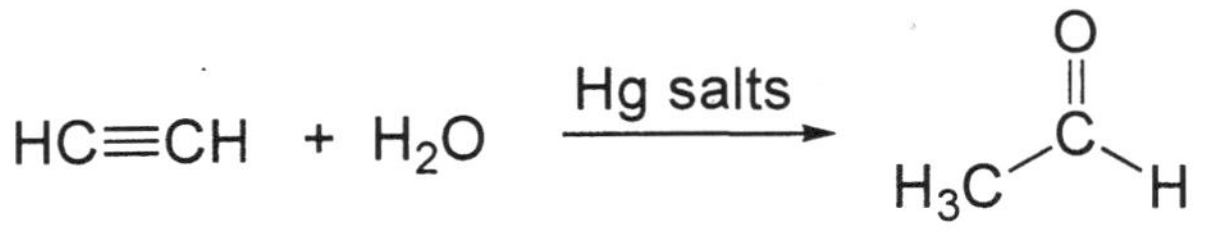

$$HC \equiv CH + H_2O \xrightarrow{\text{Hg salts}} H_3C\overset{\overset{\textstyle O}{\|}}{C}H$$

Figure 13. Synthesis of acetaldehyde from acetylene.

Fortunately, most industrializations of catalysis have been happier stories. Nevertheless, those who design catalysts must be aware that catalysts are not necessarily green; quantity is only one of many factors that determine impact on environment or health.

Catalysis has been the focus of more research than perhaps all of the other 11 principles combined. The number of papers (journals, patents, reports) published on this topic is 63,000 per year; compare this to the data in Figure 1. Likely little of this research was motivated by a conscious desire to lessen environmental impact, but many times catalysis has been selected for industrialization, over stoichiometric routes, for the simple fact that it uses less material and creates less waste. Has this brought about reductions in environmental impact?

An excellent recent example is Merck's new catalyst for the hydrogenation of unprotected enamines (Figure 14), as used in its production of sitagliptin, a diabetes treatment.[68] The process based upon this new reaction generates 220 kg less waste, per kg of product, than the prior commercial route. Thus it lowers the E-factor of the process by 220. Over the lifetime of the process, a waste reduction of 150,000 metric tons is anticipated.[69]

Many more examples of waste reduction by catalysis could be cited, but two centuries after Kirchhoff's discovery, it would require an encyclopedia-sized book!

10. Chemical products should be designed so that at the end of their function they do not persist in the environment and break down into innocuous degradation products.

There is an interesting catch-twenty-two in product design when it comes to a product's ultimate lifetime. While consumers in society demand products that

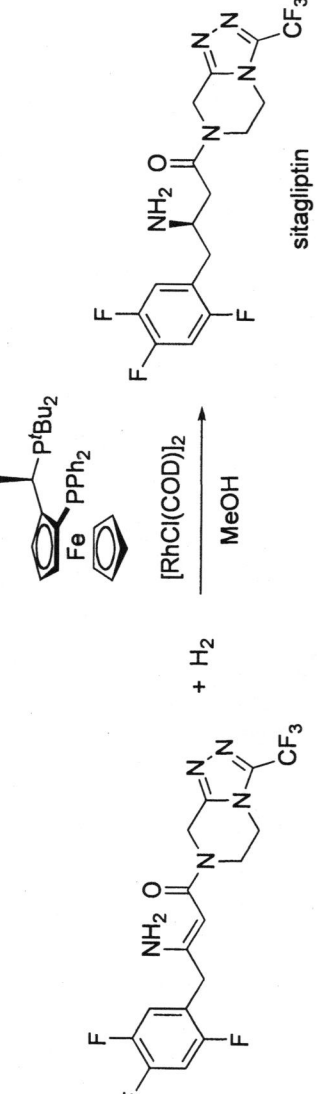

Figure 14. The asymmetric hydrogenation of an enamine producing sitagliptin.[68]

are sturdy, well-built and have appreciable life-spans we have come to find that materials with nearly infinite life-spans are a problem in the environment.

The most famous example of such controversy is dichlorodifluoromethane (Freon-12 or CFC-12), a product designed in 1928 to become a non-toxic and generally safe replacement for toxic, reactive, flammable or explosive gases, such as ammonia, sulfur dioxide and chloromethane, used as refrigerants at that time. Thomas Midgley, the inventor of the new product, demonstrated his invention at an American Chemical Society meeting in the following way: standing in front of the captivated audience he inhaled the gas and then exhaled it back to blow out a lighted candle. The new product quickly gained enormous popularity; its implementation virtually eliminated the once frequent accidents caused by leakage of refrigerant gases. Dichlorodifluoromethane was later found to be useful in other applications, for example, as an aerosol propellant. Other materials similar to dichlorodifluoromethane were designed. Termed chloroflourocarbons (CFCs), they have varying physical properties but all of them are non-toxic, non-reactive and virtually indestructible. 45 years later, trace amounts of CFCs were detected in the atmosphere.[70] In 1974, Sherwood Rowland and Mario Molina published their famous hypothesis, linking the stratospheric ozone destruction with CFCs.[71] Life-spans of a few decades allowed the CFCs to diffuse to the upper layers of the stratosphere, where under the strong UV irradiation they become a source of Cl radicals, the latter participating in chain reactions resulting in converting stratospheric ozone into oxygen. Each CFC molecule may result in destruction of 10,000 ozone molecules. Their hypothesis was later proven correct and they were awarded the Nobel Prize in 1995. Production and use of CFCs were banned under the Montreal Protocol in 1989 (with a few minor exceptions). This cautionary tale of a safe, non-toxic product becoming a major environmental threat due to its long life-span is most instructive. In most applications, CFCs have been replaced by hydrochloroflourocarbons (HCFCs) and hydrofluorocarbons (HFCs). These hydrogen–containing materials are more reactive and their life-spans are much shorter than those of CFCs. Unfortunately, both HCFCs and HFCs are greenhouse gases and their emissions contribute to global warming. Research aimed at designing truly environmentally benign alternatives to CFCs is continuing.

Another example of persistent materials becoming environmental threats is plastic debris endangering marine wildlife.[72] A number of research groups throughout the world are engaged in designing biodegradable plastic materials, such as PLA (polylactic acid), PHA (polyhydroxyalkanoate), and PHBV (poly-β-hydroxy butyrate-co-valerate). Usually they are derived from renewable biological feedstocks, which make them even more attractive alternatives to traditional plastic materials.

Compounding the problem of persistence in the environment are recent findings that many materials bioaccumulate throughout ecosystem food cycles.

Municipal water treatment systems are not equipped to remove many materials that are finding their way into the environment.[73] This principle focuses on designing materials that are incorporated in various degradative ecological cycles, especially microbial digestion and photochemical breakdown.

It is a difficult task to design materials that have "just the right" lifetime so as to provide shelf-life stability, product performance and product stability. The utilization of triggering mechanisms present in natural external systems can be employed after a product's useful lifetime.

In 2004, the Presidential Green Chemistry Challenge Small Business Award was given to Jeneil Biosurfactant Company for developing a natural, low-toxicity and biodegradable alternative to synthetic surfactants.[74] Surfactants are widely used throughout personal care products and are an essential component in many of these products. Typical surfactants have several environmental drawbacks, including toxicity and persistence in the environment. It is important for a surfactant, once released into the environment, to biodegrade in our water systems. Jeneil developed a series of rhamnolipid biosurfactant products that provide good emulsification, wetting, detergency and foaming properties. The rhamnolipid biosurfactant is a naturally occurring glycolipid that is found in soil and plants. The surfactants have reduced toxicity, are made from renewable resources, and are more biodegradable than conventional surfactants.[75]

11. Analytical methodologies need to be further developed to allow for real-time, in-process monitoring, and control prior to the formation of hazardous substances.

Why was this principleincluded in the twelve? Surely this is something to do with engineering and not chemistry? That may be, but without this principle, disastrous and unintended releases can happen. There is no more effective demonstration of this than the tragedy of Bhopal, India. Most of the students in our classes have never heard of this accident; they weren't even born yet. However, it is worth telling them about it; they will never again forget principle #11.

In December 1984, Union Carbide workers were performing maintenance work on pipes connected to a storage tank of methylisocyanate at a pesticide plant in Bhopal, India. The plant was built beside a residential area. The maintenance included washing out the pipes with water, which, due to faulty valves and unwise modifications to the plant design, made its way into the storage tank. Methyl isocyanate reacts exothermically with water, and the heat generated caused a dramatic temperature and pressure rise, which was undetected by the plant operators. The refrigeration unit, intended to keep the tank cool, had been shut down about a year earlier. By 10:45 pm, the plant supervisor knew there was a leak of some size. The staff went for tea break from midnight to 12:15 am, leaving the control room empty.[76] When the tank's rupture disk burst at 12:20 am, the hot gas flowed through a vent gas scrubber and possibly also a flare tower, neither of which were operational. The released gas killed 2,000 to 5,000 people (the exact number is disputed).[77] The alarm to warn the community was turned on briefly at 1:00 am but then switched off,[76] not to be switched back on again until 2:30 am.

Why did the accident happen? Of course, there were many contributing factors, which have been exhaustively discussed elsewhere.[76,77,78] Let us consider the accident in terms of real-time monitoring (or lack thereof). Why didn't the plant operators know what was happening? The storage tank was fitted with a temperature indicator alarm, intended to let operators know immediately if the tank temperature rose by 15 °C; however the temperature alarm had not worked properly for years. A pressure indicator in the tank was in fact working correctly at that moment, indicating a large pressure rise, but the operators distrusted it because it had a history of being faulty.[72] The level indicator had a similar history, so it was not known how much methyl isocyanate was in the tank.[79] A computerized sensing system had been installed several years earlier at Carbide's sister plant in the U.S. but not in the plant in India. The rupture disk was not equipped with any monitoring device, so its rupturing would not have raised an alarm in the operators' booth. There were no gas detectors in open spaces in the plant, so operators only found out that leaks were happening when they felt stinging in their eyes. While the complete failure of real-time monitoring equipment did not cause the accident, it prevented the control room staff from knowing what was happening. In the absence of such equipment, the only option to the staff was to go out into the plant to investigate. Given that there were no gas masks, it is not surprising that the operator declined to do so.[76] If the staff had been fully informed by real-time monitoring equipment, then the claxon for warning the neighboring community could have sounded 4 hours earlier and saved many lives.

12. Substances and the form of a substance used in a chemical process should be chosen so as to minimize the potential for chemical accidents, including releases, explosions, and fires.

It is easy to focus exclusively on the potential toxicological problems that could be encountered in a manufacturing process: carcinogenicity, endocrine disruption, mutagenicity, and other acute toxic endpoints are all too real in chemistry. This final principle acknowledges that over and above these concerns physical hazards such as explosivity and flammability must also be considered in a product design. In much the same way as structure activity relationships have been developed for physiological toxins, mechanisms of action at the molecular level have been elucidated for these physical hazards as well.

Unfortunately, there are far too many examples of cases where explosive or flammable materials were utilized and the result was a disastrous event. A recent incident involved the explosion of an ink manufacturing facility that resulted in the demolition of the facility, along with the damaging of two dozen houses and businesses in the neighborhood and leaving nearly 400 people homeless the day before Thanksgiving in 2006.[80] Fortunately this event resulted in no deaths, although a number of people were hospitalized. The explosion resulted from accidental overheating of a mixing tank containing azeotropic mixture of heptane and alcohol solvents, which was even more volatile than its individual components. Overheating resulted in release of the vapor that filled the building and, being highly flammable, subsequently ignited.

There are countless other tragedies that did result in many lives lost, including one of the most well-known disasters which happened in 1984 in Bhopal, India where a chemical plant was using methyl isocyanate as an intermediate in pesticide manufacturing. While the role of faulty real-time monitoring equipment in the tragedy was described in detail in the previous essay, it is important to look into the primary cause of the disaster. Methyl isocyanate is extremely water reactive; its reaction with water is exothermic, generating large amounts of heat (325 calories for each gram of methyl isocyanate). If the reaction mixture is not adequately cooled, its temperature is rising quickly, causing the reaction rate to accelerate, thus generating heat even faster in a positive feedback loop. Such system is referred to as a "runaway reaction". Taking into account that water is one of the most abundant substances on Earth, choosing to use large amounts of the chemical which reacts violently with water creates an opportunity for an accident to happen. In Bhopal methyl isocyanate was exposed to water, the runaway reaction resulted in a massive explosion and large amounts of toxic fumes were released to the local community, resulting in thousands of lives lost and many more injured.

Runaway reactions were an underlying cause of other industrial accidents such as that which occurred in Seveso, Italy in 1976. The nearby chemical plant was producing 2,4,5-trichlorophenol, an intermediate for manufacturing of a medical disinfectant hexachlorophene. The runaway reaction of 1,2,4,5-tetrachlorobenzene with sodium hydroxide went out of control and resulted by an explosion and release into atmosphere of an unintended byproduct of this reaction, highly toxic 2,3,7,8-tetrachlorodibenzo-p-dioxin (TCDD). The nearby communities were thus exposed to TCDD. This accident triggered industrial environmental safety regulations passed by the European Community in 1982 and termed Seveso Directive.

In order to avoid these sort of industrial tragedies, safer materials must be employed that can accomplish the same task. Therefore, flammable and highly reactive materials should be avoided whenever possible in order to ensure the safety of workers, along with the surrounding communities.

Conclusions

On this, the centenary year of the Industrial and Engineering Chemistry Division and the centenary of one of the *very* few green chemistry papers from before the 1970's, we can celebrate the fact that green chemistry has blossomed into an important field of research and development. The principles of green chemistry are increasingly being followed by industry. While there is much work to be done in the next one hundred years, let us not lose sight of the advances that have already come about.

References

1. Cathcart, C. Green chemistry in the Emerald Isle. *Chemistry and Industry* **1990**, 684.

2. Simpson, J. A. *The Oxford English Dictionary (online)*; 3rd ed.; Oxford University Press: Oxford, 2008.

3. Randall, D. T. The Burning of Coal without Smoke in Boiler Plants, *U.S. Geological Survey Bulletin* **1908**, 334.

4. (a) Ciamician, G. The Chemical Action of Light. *Bull. Soc. Chim. Fr.* **1908**, *3*, i. (b) Albini, A.; Fagnoni, M. 1908: Giacomo Ciamician and the concept of green chemistry. *ChemSusChem* **2008**, *1*, 63.

5. *Cleaning our Environment: The Chemical Basis for Action*; Cooke, L. M., Ed.; American Chemical Society: Washington, D.C., 1969.

6. Ochsner, M.; Chess, C.; Greenberg, M. Pollution prevention at the 3M corporation: Case study insights into organizational incentives, resources, and strategies *Waste Management* **1995**, *15*, 663.

7. Anonymous 2008.

8. Overcash, M. The evolution of US pollution prevention, 1976-2001: a unique chemical engineering contribution to the environment - a review. *J. Chem. Technol. Biotechnol.* **2002**, *77*, 1197.

9. Trost, B.M. The atom economy: a search for synthetic efficiency. *Science* **1991**, *254*, 1471.

10. U.S. EPA, The Presidential Green Chemistry Challenge, Summary of Award Recipients from 1996 – 2007, June 2007 [available at www.epa.gov/greenchemistry], page 110.

11. U.S. EPA, The Presidential Green Chemistry Challenge, Summary of Award Recipients from 1996 – 2007, June 2007 [available at www.epa.gov/greenchemistry], page 60.

12. NatureWorks, LLC, http://natureworksllc.com/ [accessed February 2008].

13. Arrhenius, S. On the Influence of Carbonic Acid in the Air upon the Temperature of the Ground *Philosophical Magazine* **1896**, *41*, 237.

14. Weiss, N. G. Cancer in relation to occupational exposure to perchloroethylene. *Cancer Causes and Control* **1995**, *6*, 257.

15. Anonymous; California Environmental Protection Agency, Air Resources Board: 2001.

16. Wilson, M. P.; Hammond, S. K.; Nicas, M.; Hubbard, A. E. Worker exposure to volatile organic compounds in the vehicle repair industry. *Journal of Occupational and Environmental Hygiene* **2007**, *4*, 301.

17. Takeuchi, Y. n-Hexane polyneuropathy in Japan: a review of n-hexane poisoning and its preventive measures. *Environmental Research* **1993**, *62*, 76.

18. Garfield, E. *Current Contents* **1985**, *22*, 3.

19. Daniell, W. E.; Couser, W. G.; Rosenstock, L. Occupational solvent exposure and glomerulonephritis. A case report and review of the literature. *J. Am. Med. Assoc.* **1988**, *259*, 2280.

20. Mertens, J. A. In *Kirk-Othmer Encyclopedia of Chemical Technology*; Kroschwitz, J. I., Howe-Grant, M., Eds.; Wiley: New York, NY, 1993.

21. U.S. EPA, The Presidential Green Chemistry Challenge, Summary of Award Recipients from 1996 – 2007, June 2007 [available at www.epa.gov/greenchemistry], page 20

22. Sheldon, R. A. The E Factor: fifteen years on. *Green Chem.* **2007**, *9*, 1273.

23. Constable, D. J. C.; Jimenez-Gonzalez, C.; Henderson, R. K. Perspective on Solvent Use in the Pharmaceutical Industry. *Org. Process Res. Dev.* **2007**, *11*, 133.

24. For those unfamiliar with the adage called Murphy's Law, it states that "Whatever can go wrong, will go wrong."

25. Taber, G. P.; Pfisterer, D. M.; Colberg, J. C. A New and Simplified Process for Preparing N-[4-(3,4-Dichlorophenyl)-3,4-dihydro-1(2H)-naphthalenylidene]methanamine and a Telescoped Process for the Synthesis of (1S-cis)-4-(3,4-Dichlorophenol)-1,2,3,4-tetrahydro-N-methyl-1-naphthalenamine Mandelate: Key Intermediates in the Synthesis of Sertraline Hydrochloride. *Org. Proc. Res. Dev.* **2004**, *8*, 385.

26. (a) Jessop, P. G.; Heldebrant, D. J.; Xiaowang, L.; Eckert, C. A.; Liotta, C. L. Green chemistry: Reversible nonpolar-to-polar solvent. *Nature* **2005**, *436*, 1102. (b) Yamada, T.; Lukac, P. J.; George, M.; Weiss, R. G. Reversible, Room-Temperature Ionic Liquids. Amidinium Carbamates Derived from Amidines and Aliphatic Primary Amines with Carbon Dioxide. *Chem. Mater.* **2007**, *19*, 967. (c) Phan, L.; Andreatta, J. R.; Horvey, L. K.; Edie, C. F.; Luco, A.-L.; Mirchandi, A.; Darensbourg, D. J.; Jessop, P. G. Switchable-Polarity Solvents Prepared with a Single Liquid Component. *J. Org. Chem.* **2008**, *73*, 127. (d) Phan, L.; Li, X.; Heldebrant, D. J.; Wang, R.; Chiu, D.; John, E.; Huttenhower, H.; Pollet, P.; Eckert, C. A.; Liotta, C. L.; Jessop, P. G. Switchable Solvents Consisting of Amidine/Alcohol or Guanidine/Alcohol Mixtures. *Ind. Eng. Chem. Res.* **2008**, *47*, 539.

27. Capello, C.; Fischer, U.; Hungerbühler, K. What is a green solvent? A comprehensive framework for the environmental assessment of solvents. *Green Chem.* **2007**, *9*, 927.

28. Constable, D. J. C.; Dunn, P. J.; Hayler, J. D.; Humphrey, G. R.; Johnnie L. Leazer, J.; Linderman, R. J.; Lorenz, K.; Manley, J.; Pearlman, B. A.; Wells, A.; Zaksh, A.; Zhang, T. Y. Key green chemistry research areas-a perspective from pharmaceutical manufacturers. *Green Chem.* **2007**, *9*, 411.

29. Curzons, A. D.; Constable, D. J. C.; Mortimer, D. N.; Cunningham, V. L. So you think your process is green, how do you know? - Using principles of sustainability to determine what is green - a corporate perspective. *Green Chem.* **2001**, *3*, 1.

30. Fail, P. A.; George, J. D.; Grizzle, T. B.; Heindel, J. J. Formamide and dimethylformamide: reproductive assessment by continuous breeding in mice. *Reproductive Toxicology* **1998**, *12*, 317.

31. Akesson, B. N-methyl-2-pyrrolidone. Concise International Chemical Assessment Document **2001**, 35 i-iv, 1.

32. Vinci, D.; Donaldson, M.; Hallett, J. P.; John, E. A.; Pollet, P.; Thomas, C. A.; Grilly, J. D.; Jessop, P. G.; Liotta, C. L.; Eckert, C. A. Piperylene sulfone: a labile and recyclable DMSO substitute. *Chem. Commun.* **2007**, 1427.

33. (a) *Chemical Synthesis using Supercritical Fluids*; Jessop, P. G.; Leitner, W., Eds.; VCH/Wiley: Weinheim, 1999. (b) *Green Chemistry Using Liquid and Supercritical Carbon Dioxide*; DeSimone, J. M.; Tumas, W., Eds.; New York: Oxford University Press, 2003.

34. Jessop, P. G.; Subramaniam, B. Gas-Expanded Liquids. *Chem. Rev.* **2007**, *107*, 2666.

35. Wasserscheid, P.; Welton, T. *Ionic Liquids in Synthesis*; 2nd ed.; VCH-Wiley: Weinheim, 2007.

36. *Handbook of Fluorous Chemistry*; Gladysz, J. A.; Curran, D. P.; Horváth, I. T., Eds.; VCH/Wiley: Weinheim, Germany, 2004.

37. (a) Naughton, M. J.; Drago, R. S. Supported homogeneous film catalysts. *J. Catal.* **1995**, *155*, 383 .(b) Heldebrant, D. J.; Witt, H.; Walsh, S.; Ellis, T.; Rauscher, J.; Jessop, P. G. Liquid polymers as solvents for catalytic reductions. *Green Chem.* **2006**, *8*, 807.

38. Orita, Akihiro; Okano, Junji; Uehara, Genta; Jiang, Lasheng; Otera, Junzo Importance of molecular-level contacts under solventless conditions for chemical reactions and self-assembly. *Bulletin of the Chemical Society of Japan* **2007**, *80*, 1617.

39. Oelgemoller, Michael; Jung, Christian; Mattay, Jochen Green photochemistry: production of fine chemicals with sunlight. *Pure and Applied Chemistry* **2007**, *79*, 1939.

40. Ian J. Dunkin "Photochemistry" in *Handbook of Green Chemistry and. Technology* Edited by James Clark and Duncan Macquarrie, Blackwell, Oxford, p. 416, **2002**.

41. U.S. EPA, The Presidential Green Chemistry Challenge, Summary of Award Recipients from 1996 – 2007, June 2007 [available at www.epa.gov/greenchemistry], page 76.

42. RevTech, Inc., http://www.revtechuv.com/ [accessed February 2008].

43. (a) Peters, D. Raw materials. Adv. Biochem. Engin. Biotechnol. **2007**, *105*, 1. (b) CO2 Emissions from Fuel Combustion 1971-2005; International Energy Agency (IEA): Paris, 2007. (c) Borchardt, J. K. Recycling, plastics.

Kirk-Othmer Encyclopedia of Chemical Technology **2006**, *21*, 446. (d) Nicol, S. Life after death for empty shells. *New Scientist* **1991**, *129*, 46. (e) Villas-Boas, S. G.; Esposito, E.; Mitchell, D. A. Microbial conversion of lignocellulosic residues for production of animal feeds. *Animal Feed Science and Technology* **2002**, *98*, 1.

44. Nakano, K.; Kamada, T.; Nozaki, K. Selective formation of polycarbonate over cyclic carbonate: copolymerization of epoxides with carbon dioxide catalyzed by a cobalt(III) complex with a piperidinium end-capping arm. *Angew. Chem. Int. Ed.* **2006**, *45*, 7274.

45. Aresta, M.; Dibenedetto, A.; Pastore, C.; Papai, I.; Schubert, G. Reaction mechanism of the direct carboxylation of methanol to dimethylcarbonate: experimental and theoretical studies. *Topics in Catalysis* **2006**, *40*, 71.

46. Sodero, S. F.; Berruti, F.; Behie, L.A. Ultrapyrolytic cracking of polyethylene - a high yield recycling method. *Chem. Eng. Sci.* **1996**, *51*, 2805.

47. Thomson, D. A. In *Plastics, Rubber and Paper Recycling: A Pragmatic Approach*; Rader, C. P., Baldwin, S. D., Cornell, D. D., Sadler, G. D., Stockel, R. F., Eds.; American Chemical Society: Washington, 1995, p 89-96.

48. Mackey, G. In *Plastics, Rubber and Paper Recycling: A Pragmatic Approach*; Rader, C. P., Baldwin, S. D., Cornell, D. D., Sadler, G. D., Stockel, R. F., Eds.; American Chemical Society: Washington, 1995, p 160-169.

49. (a) *Thermal Recycling of Plastics*, American Plastics Council, Energy & nvironmental Research Center, 1994. (b) Day, M.; Cooney, J. D.; Klein, C.; Fox, J. In *Polymer Durability: Degradation, Stabilization, and Lifetime Prediction*; Clough, R. L., Billingham, N. C., Gillen, K. T., Eds.; American Chemical Society: Washington, 1996, p 47-57.

50. Tharanathan, R. N.; Kittur, F. S. Chitin - the undisputed biomolecule of great potential. *Critical Reviews in Food Science and Nutrition* **2003**, *43*, 61.

51. Kripp, T. C. In *Biorefineries--Industrial Processes and Products*; Kamm, B., Gruber, P. R., Kamm, M., Eds.; Weinheim, Germany: Wiley-VCH, 2006; Vol. 2, p 409-442.

52. (a) Hill, K. *Pure Appl. Chem.* Industrial development and application of biobased oleochemicals. **2007**, *79*, 1999. (b) *Kirk-Othmer Encyclopedia of Chemical Technology*; Wiley Interscience, 2001.

53. Willke, T.; Vorlop, K.-D. Industrial bioconversion of renewable resources as an alternative to conventional chemistry. *Appl. Microbiol. Biotechnol.* **2004**, *66*, 131.

54. Hopkins, F.G. The Centenary of Wohler's Synthesis of Urea (1828-1928). *Biochem. J.* **1928**, *22*, 1341.

55. Sanchez, C.; Arribart, H.; Guille, M.M. Biomimetism and bioinspiration as tools for the design of innovative materials and systems. *Nature Mater.* **2005**, *4*, 277.

56. Guarrera, D.; Taylor, L.D.; Warner, J.C. Molecular Self-Assembly in the Solid State. The Combined Use of Solid-State NMR and Differential Scanning Calorimetry for the Determination of Phase Constitution. *Chem. Mater.* **1994**, *6*, 1293.

57. Lee, W. E.; Brown, E. R. In *The Theory of the Photographic Process*, 4th ed.; James, T. H., Ed.; Macmillan: New York, **1977**; Chapter 11.

58. Baumann, L.S.; Martin, L.K. Skin Whitening: New Hydroquinone Combination in *Handbook of Cosmetic Science and Technology, Second Edition*, Paye, M., Barel, A.O., Maibach, H.I., Eds, Taylor & Francis Group (Boca Raton, FL), 2006, p. 321.

59. "Pollution Prevention via Molecular Recognition and Self Assembly: Non-Covalent Derivatization." Warner, J. C., in *Green Chemistry: Frontiers in Benign Chemical Synthesis and Processes* Anastas, P. and Williamson, T. Eds., Oxford University Press, London. pp 336 - 346. 1998.

60. Cannon, A.S.; Warner, J.C. Noncovalent derivatization: green chemistry applications of crystal engineering. *Crystal Growth and Design* **2002**, *2*, 255.

61. Anastas, P.T.; Warner, J.C. *Green Chemistry: Theory and Practice*; Oxford University Press: London, 1988.

62. Foxman, B. M.; Guarrera, D. J.; Taylor, L. D.; Van Engen, D.; Warner, J. C. Environmentally benign synthesis using crystal engineering: steric accommodation in non-covalent derivatives of hydroquinones. *Cryst. Eng.* **1998**, *1*, 109.

63. Cannon, A. S.; Foxman, B. M.; Guarrera, D. J.; Van Engen, D.; Warner, J. C. Noncovalent Derivatives of Hydroquinone: Complexes with Trigonal Planar Tris(N,N-dialkyl)trimesamides. *Cryst. Growth Des.* **2005**, *5*, 407.

64. Völksen, W. The discovery of the saccharification of starch (acid hydrolysis) by G. S. C. Kirchhoff in the year 1811. *Die Starke* **1949**, *1*, 30.

65. Hebeda, R. E. Syrups In *Kirk-Othmer Encyclopedia of Chemical Technology*; John Wiley: 2007.

66. *Industrial Pollution in Japan*; Ui, J., Ed.; United Nations University Press: Tokyo, 1992.

67. Ministry of the Environment, Japan: Minamata City, Japan, 2008; Vol. 2008.

68. Clausen, A. M.; Dziadul, B.; Cappuccio, K. L.; Kaba, M.; Starbuck, C.; Hsiao, Y.; Dowling, T. M. Identification of Ammonium Chloride as an Effective Promoter of the Asymmetric Hydrogenation of a .beta.-Enamine Amide. *Org. Proc. Res. Devel.* **2006**, *10*, 723.

69. *The Presidential Green Chemistry Challenge Award Recipients 1996-2007*; United States Environmental Protection Agency: Washington, 2007.

70. Lovelock, J. E.; Maggs, R. J.; Wade, R. J. Halogenated hydrocarbons in and over the Atlantic. *Nature* **1973**, *241*, 194.

71. Molina, M.J.; Rowland, F. S. Stratospheric sink for chlorofluoromethanes. Chlorine atom-catalyzed destruction of ozone. *Nature* **1974**, *249*, 810.

72. Derraik, J.G.B. The pollution of the marine environment by plastic debris: a review. *Marine Pollution Bulletin* **2002**, *44*, 842.

73. Stackelberg, P.E.; Lippincott, R. L.; Furlong, E.T.; Meyer, M.T.; Zaugg, S.D. Trace-level wastewater-related compounds in New Jersey's streams and their persistence through a conventional water-treatment plant. *Proceedings - Water Quality Technology Conference* **2003**, 2138.

74. U.S. EPA, The Presidential Green Chemistry Challenge, Summary of Award Recipients from 1996 – 2007, June 2007 [available at www.epa.gov/greenchemistry], page 36.

75. Jeneil Biosurfactant Company, http://www.biosurfactant.com/ [accessed February 2008].

76. Bowonder, B.; Miyake, T. Managing hazardous facilities: lessons from the Bhopal accident. *J. Hazardous Mater.* **1988**, *19*, 237.

77. Morehouse, W.; Subramaniam, M. A. *The Bhopal Tragedy: A Preliminary Report for the Citizens Commission on Bhopal*; Council on International and Public Affairs: New York, 1986.

78. Shrivastava, P. *Bhopal: Anatomy of a Crisis*; Ballinger: Cambridge, Massachusetts, 1987.

79. de Grazia, A. *A Cloud over Bhopal*; Kalos Foundation: Bombay, 1985.

80. Cramer, M.; Daley B.; Mishra, R. A Thanksgiving Miracle. *Boston Globe*, November 23, 2006.

Indexes

Author Index

Subject Index

441